Manifolds and
Local Structures
A General Theory

Manifolds and Local Structures
A General Theory

Marco Grandis
Università di Genova, Italy

World Scientific

NEW JERSEY · LONDON · SINGAPORE · BEIJING · SHANGHAI · HONG KONG · TAIPEI · CHENNAI · TOKYO

Published by

World Scientific Publishing Co. Pte. Ltd.

5 Toh Tuck Link, Singapore 596224

USA office: 27 Warren Street, Suite 401-402, Hackensack, NJ 07601

UK office: 57 Shelton Street, Covent Garden, London WC2H 9HE

Library of Congress Control Number: 2021003543

British Library Cataloguing-in-Publication Data
A catalogue record for this book is available from the British Library.

MANIFOLDS AND LOCAL STRUCTURES
A General Theory

ISBN 978-981-123-399-9 (hardcover)
ISBN 978-981-123-400-2 (ebook for institutions)
ISBN 978-981-123-401-9 (ebook for individuals)

For any available supplementary material, please visit
https://www.worldscientific.com/worldscibooks/10.1142/12199#t=suppl

Desk Editor: Soh Jing Wen

To

Marina

Preface

Local structures, like differentiable manifolds, fibre bundles, vector bundles and foliations, can be obtained by gluing together a family (U_i) of suitable 'elementary spaces', by means of partial homeomorphisms $U_i \to U_j$ that fix the gluing conditions and form a sort of 'intrinsic atlas', instead of the more usual system of charts living in an external framework.

An 'intrinsic manifold' is defined here as such an atlas, in a suitable category of elementary spaces: open euclidean spaces, or trivial bundles, or trivial vector bundles, and so on.

This uniform approach allows us to move from one basis to another: for instance, the elementary tangent bundle of an open euclidean space is automatically extended to the tangent bundle of any differentiable manifold. The same holds for tensor calculus.

Technically, the goal of this book is to treat these structures as 'symmetric enriched categories' over a suitable basis, generally an ordered category of partial mappings. The morphisms of these 'generalised manifolds' are obtained as 'compatible profunctors' between enriched categories, which can be composed because of the existence of 'compatible joins' in the basis.

This approach to local structures is related to Ehresmann's one, based on inductive pseudogroups and inductive categories. A second source was the theory of enriched categories and Lawvere's unusual view of interesting mathematical structures as categories enriched over a suitable basis — a source of research since the 1970's.

Contents

Introduction

0.1 Classical manifolds by an external atlas

A smooth manifold is usually defined as a topological space X equipped with a C^∞-*atlas of charts*, indexed by a set I

$$u^i\colon U_i \to X_i \qquad (i \in I). \tag{0.1}$$

A *chart* u^i is a homeomorphism between an open euclidean space U_i (i.e. an open subspace of some space $\mathbb{R}^n$ with euclidean topology) and an open subspace X_i of X. We assume that these open subsets cover X and that every *transition map* (between open euclidean spaces)

$$u^i_j = u_j u^i\colon U_i \dashrightarrow U_j \qquad (i, j \in I), \tag{0.2}$$

is of class C^∞ (i.e. has continuous partial derivatives of any order).

Here $u_j\colon X_j \to U_j$ is the inverse of u^j, and the 'composite' $u_j u^i\colon U_i \dashrightarrow U_j$ (an abuse of notation) is *partially defined* on the open subset $u_i(X_i \cap X_j)$, possibly empty. We shall often distinguish partial mappings by a dot-marked arrow.

The space X is thus locally euclidean, with a locally constant dimension. It is often required to be Hausdorff paracompact, but we drop these conditions, adding them when it is the case.

If Y is also a smooth manifold, with charts $v^h\colon V_h \to Y_h$ ($h \in H$), a C^∞-*mapping* $f\colon X \to Y$ is a map such that all the *partial* mappings

$$f^i_h = v_h f u^i\colon U_i \dashrightarrow V_h \qquad (i \in I, h \in H), \tag{0.3}$$

are of class C^∞. Again, as in (0.2), there is an abuse of notation: we are 'composing' three arrows

$$u^i\colon U_i \to X_i, \qquad f\colon X \to Y, \qquad v_h\colon Y_h \to V_h,$$

which are not consecutive.

This could be fixed with Ehresmann's *pseudoproduct* (recalled below), but we want to work with ordinary composition in categories of partial mappings, along the following line.

(a) We write as $\mathcal{C}$ the category of *topological spaces and partial continuous mappings defined on open subspaces*. A morphism $f \colon X \rightarrowtail Y$ is defined on an open subspace Def f of X; for a consecutive morphism $g \colon Y \rightarrowtail Z$, the composite $gf \colon X \rightarrowtail Z$ is defined on those $x \in X$ (if any) such that $f(x) \in \text{Def}\, g$.

We write as $\mathcal{C}^\infty$ the subcategory of $\mathcal{C}$ of open euclidean spaces and partial C^∞-mappings, defined on open subspaces.

(b) We replace the homeomorphism $u^i \colon U_i \to X_i$ with the topological embedding $u^i \colon U_i \to X$; the latter has a backward morphism $u_i \colon X \rightarrowtail U_i$ in $\mathcal{C}$ (defined on X_i), characterised by the relations

$$u^i = u^i u_i u^i, \qquad u_i = u_i u^i u_i, \tag{0.4}$$

which make these morphisms *partial inverse* to each other (as in semigroup theory).

We can now replace the 'illegitimate compositions' of (0.2) and (0.3) with legitimate ones, in $\mathcal{C}$

$$u^i_j = u_j u^i \colon U_i \rightarrowtail X \rightarrowtail U_j, \qquad v_h f u^i \colon U_i \rightarrowtail X \rightarrowtail Y \rightarrowtail V_h, \tag{0.5}$$

and we can require that these composites (whose domain and codomain are open euclidean spaces) belong to the subcategory $\mathcal{C}^\infty$.

(c) More generally, for each $r \in \mathbb{N} \cup \{\infty, \omega\}$, we write as $\mathcal{C}^r$ the subcategory of $\mathcal{C}$ of *open euclidean spaces and partial* C^r-*mappings defined on open subspaces*. (C^0 means continuous and C^ω means analytic; for $0 < r < \infty$, a C^r-mapping has all continuous partial derivatives of order $\leqslant r$.)

C^r-manifolds are dealt with as above. Topological manifolds correspond to the case $r = 0$ (in which case the transition maps u^i_j automatically belong to $\mathcal{C}^0$).

(d) Categories of partial mappings will generally be denoted by calligraphic letters. The prime example is the category $\mathcal{S}$ of sets and partial mappings.

As a crucial fact, a category of partial mappings has a canonical order: for two partial mappings $f, g \colon X \rightarrowtail Y$ the relation $f \leqslant g$, means that f is a restriction of g (with Def f contained in Def g). The order is consistent with composition: $\mathcal{S}$, $\mathcal{C}$ and $\mathcal{C}^r$ are *ordered categories*.

The reader may know, or guess, that the 'categories of partial mappings' (like $\mathcal{S}$ or $\mathcal{C}$) we are using can be obtained by a general construction, starting from an 'ordinary' category (like Set or Top) and a suitable subcategory (the embeddings of the definition-sets). This point, dealt with in Section 5.1, plays here a minor role: we prefer to work directly in the relevant categories of partial mappings.

0.2 Intrinsic manifolds on ordered categories

Loosely speaking, it is possible to define a C^r-manifold in an intrinsic way, *inside the category* C^r, as a collection (U_i) of objects, equipped with a family $(u^i_j : U_i \rightarrowtail U_j)$ of transition morphisms — a system of instructions specifying how the different charts U_i should be glued together. The gluing will be realised in an external category, namely in C.

More precisely, we define an (intrinsic) *manifold* on the ordered category C^r, indexed by a set I, as a diagram

$$U = ((U_i), (u^i_j))_I$$

in C^r, consisting of objects U_i (the *charts*, for $i \in I$) and morphisms $u^i_j : U_i \rightarrowtail U_j$ (the *transition morphisms*, for $i, j \in I$), satisfying three axioms which use the canonical order of the category C^r (for $i, j, k \in I$):

(i) $u^i_i = 1_{U_i}$ (*identity law*),

(ii) $u^j_k u^i_j \leqslant u^i_k$ (*composition law*, or *triangle inequality*),

(iii) $u^i_j = u^i_j u^j_i u^i_j$ (*symmetry law*).

From a formal point of view, U is a small category enriched on the ordered category C^r, with an additional symmetry condition. The transition mapping u^i_j plays the role of $\mathrm{Hom}(i, j)$, while axiom (ii) replaces the composition mapping $\mathrm{Hom}(i, j) \times \mathrm{Hom}(j, k) \to \mathrm{Hom}(i, k)$.

All this makes sense in the theory of enriched categories on ordered categories, reviewed in Chapter 6.

Plainly, if we start from the usual charts $u^i : U_i \to X$ and define their transition morphisms u^i_j as above, in 0.5, these axioms are satisfied. Conversely, if we start from a family (u^i_j) satisfying the conditions above, we shall see (in 3.5.9) that we can reconstruct the space $X = \mathrm{gl}\, U$ as the *gluing* of the diagram U, a quotient of the disjoint union of all U_i modulo the equivalence relation produced by the transition maps. (More precisely, the pair $(X, (u^i : U_i \to X))$ will be the *lax colimit* of the diagram (u^i_j), in the ordered category C.)

The diagram U will often be written as $((U_i), (u^i_j))_I$, or as U_I. The family (u^i_j) is called the *intrinsic atlas*, or the *gluing atlas*, of the manifold.

In the ordered categories C and C^r (and in all the others used in this analysis of local structures), a prominent role will be played by the endomorphisms $e : X \rightarrowtail X$ which are restriction of identities, called *projectors*.

The projectors of X are idempotent endomorphisms and commute, forming a semilattice (i.e. an ordered set with all meets)

$$\mathsf{Prj}\,(X) = \{e \colon X \twoheadrightarrow X \mid e \leqslant \mathrm{id}\,X\}, \qquad e \wedge e' = ee' = e'e. \tag{0.6}$$

In fact, the projectors determine the order: for parallel morphisms $f, g \colon X \twoheadrightarrow Y$, the relation $f \leqslant g$, is equivalent to the existence of $e \in \mathsf{Prj}\,(X)$ such that $f = ge$. We can always take as e the *support* $\underline{e}(f)$ of f, namely the partial identity on $\mathrm{Def}\,f$, or equivalently the least $e \in \mathsf{Prj}\,(X)$ such that $fe = f$.

These projectors satisfy axioms (see 3.3.5), and supply the categories $\mathcal{C}$ and $\mathcal{C}^r$ with the structure of an *e-cohesive category*, or *e-category*, one of the main ingredients of our analysis.

More precisely, $\mathcal{C}$ and $\mathcal{C}^r$ are *totally cohesive e-categories*, which means that every family of 'compatible' morphisms $f_i \colon X \twoheadrightarrow Y$ has a join

$$\bigvee f_i \colon X \twoheadrightarrow Y,$$

and composition distributes over these joins.

Being a compatible family can be simply read as 'upper bounded', but the important fact is that this property is characterised by supports. Namely, for $f, g \colon X \twoheadrightarrow Y$, we say that f and g are *compatible*, or *linked* (written as $f \,!\, g$), if

$$f\,\underline{e}(g) = g\,\underline{e}(f), \tag{0.7}$$

which means that they coincide wherever they are both defined. If the morphisms $(f_i)_{i \in I}$ are pairwise linked, the join $f = \bigvee f_i$ is defined on $\bigcup \mathrm{Def}\,f_i$, and its graph is the union of the graphs of all f_i.

A general presentation of cohesive structures $(\leqslant, !)$ on categories can be found in Section 3.1, either determined by projectors or more general.

0.3 Morphisms of manifolds as linked profunctors

We can now define the category $\mathrm{Mf}\,\mathcal{C}^r$, of $\mathcal{C}^r$-manifolds and 'linked profunctors' between them, extending the formula (0.3), where a morphism $f \colon U \twoheadrightarrow V$ is determined by its components

$$f_h^i = v_h f u^i \colon U_i \twoheadrightarrow V_h,$$

on the charts of domain and codomain. A morphism in $\mathrm{Mf}\,\mathcal{C}^r$ will be a 'linked profunctor', that is an enriched profunctor between enriched categories, satisfying a compatibility condition.

More precisely, an (enriched) *profunctor*

$$a = (a_h^i)_{IH} \colon (U_i, u_j^i)_I \to (V_h, v_k^h)_H \tag{0.8}$$

is a family of morphisms $a_h^i \colon U_i \nrightarrow V_h$ in $\mathcal{C}^r$ such that, for all $i, j \in I$ and $h, k \in H$

(i) $a_h^j u_j^i \leqslant a_h^i, \quad v_k^h a_h^i \leqslant a_k^i$ (*profunctor laws*).

It will be said to be *linked*, or *compatible*, if it has a *resolution* $e_{ih} \in \mathsf{Prj}\, U_i$ ($i \in I$, $h \in H$), defined by the property:

(ii) $a_k^i e_{ih} = v_k^h a_h^i$ (*left linking law*),

which is meant to ensure:
- that linked profunctors can be composed,
- that the gluing of a linked profunctor gives a *single-valued* partial mapping $a \colon \mathrm{gl}\, U \nrightarrow \mathrm{gl}\, V$.

The resolution can be expressed by supports, taking $e_{ih} = \underline{e}(a_h^i)$.

The usual matrix composition of profunctors works, because $\mathcal{C}^r$ is totally cohesive: composing $a \colon U \to V$ with a consecutive linked profunctor $b \colon (V_h, v_k^h)_H \to (W_m, v_n^m)_M$, the composites

$$b_m^h a_h^i \colon U_i \nrightarrow V_h \nrightarrow W_m \qquad (h \in H),$$

form a linked family (for every $i \in I$, $m \in M$), and the component c_m^i of $c = ba \colon U \to W$ is computed as their linked join

$$c_m^i \colon U_i \nrightarrow W_m, \qquad c_m^i = \bigvee_{h \in H} b_m^h a_h^i. \tag{0.9}$$

This composition is based on resolutions (or supports): we cannot simply work in an ordered category with joins of upper bounded families of parallel morphisms — a sort of 'conditioned quantaloid'.

0.4 The interest of an intrinsic approach

This formalisation will allow us to move between different contexts.

For instance, the tangent bundle of an open n-dimensional euclidean space U is the trivial vector bundle $TU = U \times \mathbb{R}^n$. The present machinery automatically extends this obvious setting to the tangent functor of differentiable manifolds

$$T \colon \mathrm{Mf}\, \mathcal{C}^r \to \mathrm{Mf}\, \mathcal{V} \qquad (r > 0), \tag{0.10}$$

with values in the category of vector bundles, presented as intrinsic manifolds on an ordered category $\mathcal{V}$ of trivial vector bundles (in Section 4.2).

The same procedure works for tensor calculus.

In a more elementary way, the embedding of C^r in the category C of topological spaces and partial continuous mappings defined on open subspaces gives the topological realisation of C^r-manifolds

$$\text{Mf}\, C^r \to C \qquad\qquad (0.11)$$

taking into account that the second category is gluing complete (each manifold on C has a gluing space), and therefore C is equivalent to $\text{Mf}\, C$.

For a reader acquainted with the theory of enriched categories, we note that the property of Cauchy completeness of enriched categories, which is crucial in other contexts, is less important here where the morphisms are based on profunctors: replacing an intrinsic atlas by a complete one would simply give an isomorphic object (with respect to profunctors).

Furthermore, this can only be done when the basis of enrichment is a small category: it is the case for topological or differentiable manifolds, but not for fibre bundles. (Cauchy completion of enriched categories is reviewed in Chapter 6.)

0.5 An outline

Every chapter and every section has its own introduction; this is a brief synopsis, and involves topics which may be unknown to the reader.

Chapter 1 introduces the theory of ordered sets, semigroups and categories, as far as needed in this book. Some care is devoted to the classical theory of inverse semigroups. Limits and colimits in categories are only examined in their basic forms; adjoint functors are briefly presented.

Chapter 2 is devoted to ordered categories, equipped with a local order between parallel morphisms, as in Section 0.1.

The main part of the chapter deals with *inverse categories* and their canonical order, a natural extension of inverse semigroups. In fact, our categories of partial mappings, like C and C^r (see 2.1.4), have an *inverse core*, $\text{I}C$ and $\text{I}C^r$, formed of the 'partial isomorphisms' of the category (see 2.4.4, 2.4.5).

The symmetry law 0.2(iii) forces the transition maps of a manifold to belong to the inverse core; but we need the whole categories C and C^r to construct the general morphisms of manifolds, as presented above, in Section 0.3.

Some topological prerequisites are also reviewed in this chapter.

Chapter 3 introduces and studies 'cohesive categories', as ordered categories equipped with a structure which allows us to build our categories of local structures, like $\mathrm{Mf}\,\mathcal{C}^r$.

We are mainly interested in e-cohesive categories, where the order $f \leqslant g$, and the linking relation $f\,!\,g$, of morphisms are determined by their supports $\underline{e}(f)$ and $\underline{e}(g)$. But we also give a more general, unifying notion of cohesive category, because the linking relation of the inverse cores $I\mathcal{C}$ and $I\mathcal{C}^r$ is not determined in this way: it also needs cosupports, on codomains.

Theoretically, the main results are the gluing completion theorems 3.5.8 and 3.6.7, which give the universal properties of the categories of manifolds built in this chapter.

Chapter 4 shows how various concrete local structures (and their interplay) can be formalised in this way: topological and differentiable manifolds, manifolds with boundary, foliated manifolds, fibre bundles, vector bundles, G-bundles, simplicial complexes, etc.

New developments, inspired by Directed Algebraic Topology [G8], deal with 'locally cartesian ordered manifolds', which are spaces with distinguished paths, generally non-reversible (in Sections 4.3–4.5).

Chapter 5 gives further information on category theory. On this basis, Chapter 6 studies the relationship of our approach to manifolds with the general theory of enriched categories.

Finally, Chapter 7 collects the solutions of most exercises.

0.6 Manifolds by Ehresmann's pseudogroups

Loosely speaking, the approach of C. Ehresmann to differentiable manifold and other local structures relies on categories of *total mappings*, equipped with a *global order* (examined in Sections 2.8 and 6.7).

(a) First we need the groupoid $\mathrm{Iso}(\mathsf{Top}) = \mathrm{Iso}(\mathcal{C})$ of *topological spaces and homeomorphisms*, equipped with a 'global order' $f' \subset f$ on its maps, defined as follows: the homeomorphism $f'\colon X' \to Y'$ is the restriction of the homeomorphism $f\colon X \to Y$, from an open subspace X' of X to an open subspace $Y' = f(X')$ of Y.

(b) Similarly we have the groupoid $\mathrm{Iso}(\mathcal{C}^r) = \mathrm{Iso}(\mathcal{C}) \cap \mathcal{C}^r$ of *open euclidean spaces and C^r-diffeomorphisms*, with the restricted global order.

A *chart* is again a total homeomorphism in $\mathrm{Iso}(\mathcal{C})$

$$u^i\colon U_i \to X_i, \tag{0.12}$$

as in the classical definition recalled above, with inverse $u_i\colon X_i \to U_i$.

To make sense of the condition expressed in (0.2) we resort to the *extended composition*, or *pseudoproduct*

$$u_j \bullet u^i \colon u_i(X_i \cap X_j) \to u_j(X_i \cap X_j), \qquad (0.13)$$

namely the homeomorphism that takes $u_i(x)$ to $u_j(x)$, for all $x \in X_i \cap X_j$.

This extended composition turns the set of morphisms of $\text{Iso}(\mathcal{C})$ into an (inverse) semigroup; one can then require the composite to belong to $\text{Iso}(\mathcal{C}^r)$.

One constructs in this way a groupoid of manifolds and diffeomorphisms of class $\mathbf{C}^r$.

This approach makes a deep use of formal set theory, which we prefer to avoid: dealing with maps $f \colon X \to Y$, $g \colon Z \to W$ between arbitrary spaces the meaning of the relation $f \subset g$ and of the pseudoproduct $f \bullet g$ seems to be unclear, unless all these spaces are known to be subspaces of a given space.

This can be managed for Hausdorff paracompact differentiable manifolds, using the Whitney embedding theorem, according to which an n-dimensional manifold of this kind can always be embedded in $\mathbb{R}^{2n}$ (as exploited in Section 4.6). In other cases, for instance for fibre bundles, there is no opportunity of this kind.

0.7 Prerequisites, notation and conventions

This book is addressed to readers with different formation, in Topology, or Differential Geometry, or Category Theory, or Semigroup Theory, or perhaps other fields; at the cost of dealing with aspects that can be obvious to one reader or another.

We only assume as known the basic theory of topological spaces, differentiable manifolds, abelian groups, modules and vector spaces. Ordered sets, semigroups and categories are introduced and studied as far as needed here. Banach spaces occur in a marginal way. Deeper results, when used, are referred to.

The symbol $\subset$ denotes *weak* inclusion. A singleton set is often written as $\{*\}$. The equivalence class of an element x, with respect to an assigned equivalence relation, is generally written as $[x]$. A bullet in a diagram stands for an object.

We write as $|X|$ the underlying set of a structured set X, e.g. a topological space, or a semigroup. Dealing with topological spaces, the term *map* will often be used for 'continuous mapping'; neighbourhood can be abridged to 'nbd'.

A ring R is assumed to be unital, and a (left) R-module is assumed to be unitary.

The symbols $\mathbb{N}$, $\mathbb{Z}$, $\mathbb{Q}$, $\mathbb{R}$, $\mathbb{C}$ denote the sets of natural, integral, rational, real or complex numbers. The topology of the standard euclidean sphere $\mathbb{S}^n$ is reviewed in Section 2.5.

The standard compact interval $[0,1]$ is also written as $\mathbb{I}$. Open and semi-open real intervals are denoted as $]a,b[$, $[a,b[$, etc. — a notation, which distinguishes the open interval $]a,b[$ from the pair (a,b), as in Bourbaki's treatise.

Categories of partial mappings, like $\mathcal{S}$ and $\mathcal{C}$, play a central role (see Section 2.1). Our analysis of these categories will be based (as in Section 0.3) on particular idempotent endomorphisms $e\colon X \rightarrowtail X$, the 'partial identities', called *projectors*. Of course these morphisms should not be confused with the projections of a cartesian product, nor with the projection on a quotient.

A part marked with * is out of the main line of exposition. It may refer to issues dealt with in the sequel, or be addressed to readers with some knowledge of the subject, or give references to higher topics.

Most exercises have a solution or convenient hints. These can be found in Chapter 7, or — occasionally — below the exercise, if they are important for the sequel. Easy exercises may be left to the reader.

0.8 Sources and outgrowth

Our presentation of manifolds as intrinsic atlases in cohesive categories, in Chapters 3 and 4, is an expansion of matter published in two articles, in 1989–90 [G3, G4], partially based on a long work on inverse categories in Homological Algebra, in the 1970's and 1980's. (The results of the latter are summarised in [G2], and exposed in the recent book [G9].)

Presenting manifolds in an intrinsic way can be found in the literature. But the roots of our approach rely on two main domains.

(i) A first source was Ehresmann's work on local structures, in the 1960's.

(ii) Another main source is the theory of enriched categories [EiK, Kl2], and Lawvere's claim that many interesting mathematical structures (besides forming categories) *are* themselves categories, enriched on a suitable basis: a *monoidal* category as in Lawvere's article, in 1974 [Lw], or more generally a *bicategory* as in many subsequent papers [Bet, Wa1, Wa2, BetC, BetW1, BetW2].

The bases we actually use are very particular bicategories: ordered categories with 'linked' joins, preserved by composition.

Other papers of the 1980's are related to our approach, or to the structures we are using.

(a) S. Kasangian and R.F.C. Walters worked in the perspective opened by Lawvere, aiming to present differentiable manifolds as symmetric enriched categories on an involutive ordered category with all joins of parallel map, and to explore Cauchy-completeness in this context. Their research was presented in an (unpublished) talk at a Surrey meeting on Category Theory, in 1982 [KaW].

Constructing categories similar to $\mathcal{C}$ and $\mathcal{C}^r$, but having all joins (rather than the linked ones), leads to complications. Something of this kind will be presented in Section 4.6, taking advantage of the fact that n-dimensional differentiable manifolds can be embedded in $\mathbb{R}^{2n}$, where the transition maps can be treated as partial identities, and have arbitrary joins. General local structures cannot be dealt with in this way.

(b) Dominical categories and p-categories, other formalisations of categories of partial mappings, were introduced in the 1980's by R.A. Di Paola, A. Heller, G. Rosolini and E. Robinson [Di, He, DiH, Rs, RoR], making use of a monoidal structure derived from cartesian products. These categories have a natural e-structure (see 3.3.9(c)).

Finally, after the introduction of e-categories in [G3, G4], this structure has been used in computer science and category theory, under the name of 'restriction category' and equivalent axioms [CoL]: see 3.3.9(d). Later, totally cohesive e-categories have also been used, under the name of 'join-restriction categories'. A recent paper acknowledges the fact that restriction categories are the same as e-categories [CoG].

0.9 Acknowledgements

As already said, this book, is mainly indebted to Ehresmann's approach to local structures, and Lawvere's unusual view of enriched categories (partially exposed in Chapter 6).

When I presented this approach to 'manifolds', at the 1988 Prague Conference on 'Categorical Topology', Mac Lane rightly objected that the provisional term I was using, 'coherent category', might lead to confusion with categorical coherence theorems. Soon after I found a replacement: 'cohesive' instead of 'coherent', already adopted in the Proceedings of this conference [G3].

I remember with pleasure a letter by Andrée C. Ehresmann, which prompted me to sketch in [G4] a comparison between e-categories and C. Ehresmann's categories with a 'global order'. This comparison is now developed in Section 6.7.

I am pleased to acknowledge several helpful discussions with my colleague Ettore Carletti.

I would also like to thank Dr Lim Swee Cheng and Ms Soh Jing Wen, at World Scientific Publishing Co., for their kind, effective help in the publication of this book. The cover is due to Mr Ng Chin Choon.

Diagrams and figures are composed with 'xy-pic', a free package by K.H. Rose and R. Moore.

1

Order, semigroups and categories

This is an introductory chapter on ordered sets, semigroups, inverse semigroups and categories. Many points will be obvious to one reader or another, according to their interests.

The choice of arguments is aimed at the present applications, and by no means representative of Order Theory, or Semigroup Theory, or Category Theory. Complements on categories will be added in Chapters 5 and 6, related to topics that appear in the previous parts.

We recall that the symbol $\subset$ always denotes weak inclusion. The basic theory of topological spaces, groups, rings and modules is assumed to be known. Solutions of non-obvious exercises can generally be found in the last chapter. A part marked with * refers to developments which are not technically needed, or are referred to.

1.1 Preordered sets, lattices and semigroups

We begin by reviewing the basic notions of preordered sets and semigroups; we also examine the interplay of preorders and topology.

Further information on lattice theory can be found in Birkhoff [Bi] and Grätzer [Gr]; on semigroup theory in Clifford–Preston [CP], Howie [Ho] and Lawson [Ls].

1.1.1 Preordered and ordered sets

We use the following terminology.

A *preordered set* X is a set equipped with a *preorder relation* $x \prec x'$ (read as x *precedes* x'), which is assumed to be reflexive and transitive. It is an *ordered set* if the relation is *anti-symmetric*: if $x \prec x'$ and $x' \prec x$, then $x = x'$; an order relation is more often written as $x \leqslant x'$. If useful, one can write $x \prec_X x'$ and $x \leqslant_X x'$. A symmetric preorder relation is an equivalence relation, often written as $x \sim x'$.

In a *totally* ordered set any two elements are comparable: $x \leqslant x'$ or $x' \leqslant x$. An ordered set is often called a '*partially* ordered set', abbreviated to 'poset', to mean that totality is not assumed (but not excluded).

Every set X has a *discrete order* $x = x'$, which is the *finest*, or least preorder relation. It also has an *indiscrete*, or *chaotic preorder*, the relation $x, x' \in X$, which is the *coarsest*, or greatest preorder relation on X.

A preordered set X has an associated equivalence relation $x \sim x'$ defined by the conjunction: $x \prec x'$ and $x' \prec x$. The quotient $X/\!\!\sim$ has an induced order:

$$[x] \leqslant [x'] \quad \Leftrightarrow \quad x \prec x'. \tag{1.1}$$

If X is a preordered set, X^{op} is the *opposite*, or *dual* one — with reversed preorder. Every topic of the theory of preordered sets has a dual instance, which comes from the opposite preordered set, or sets.

Let X be a preordered set. The *minimum* $\min X$ is an element which precedes all the elements of X (and can exist or not, of course); the *maximum* $\max X$ is an element preceded by all the elements of X. They are determined up to the associated equivalence relation in X, and uniquely determined if X is ordered. They can also be written as $\perp$ and $\top$ (*bottom* and *top*).

Every subset of a preordered set will be equipped with the restricted preorder, by default.

A mapping $f: X \to Y$ between preordered sets is said to be *monotone*, or *preorder-preserving*, or (weakly) *increasing*, if $x \prec_X x'$ implies $f(x) \prec_Y f(x')$, for all $x, x' \in X$. It is *isotone* if it preserves and reflects the preorder: $x \prec_X x'$ if and only if $f(x) \prec_Y f(x')$. (If X is an ordered set, this implies that f is injective.)

An *isomorphism* $f: X \to Y$ of preordered sets is a bijective monotone mapping whose inverse mapping $f^{-1}: Y \to X$ is also monotone; equivalently, f is an isotone bijection. More generally, an *embedding* $f: X \to Y$ of preordered sets is an injective isotone mapping, and gives an isomorphism from X to the preordered subset $f(X) \subset Y$.

Examples (a) The set $\mathbb{R}$ of real numbers, equipped with the natural order $x \leqslant y$, is called the *ordered line*. It is a totally ordered set. Its cartesian power $\mathbb{R}^n$ has a canonical (partial) order:

$$(x_1, ..., x_n) \leqslant (y_1, ..., y_n) \quad \Leftrightarrow \quad (\text{for all } i = 1, ..., n: x_i \leqslant y_i), \tag{1.2}$$

that will be called the *cartesian order* of $\mathbb{R}^n$.

(b) More generally, any cartesian product $X = \prod_i X_i$ of preordered sets has a canonical preorder, defined componentwise as above: $(x_i) \prec_X (y_i)$ if and only if, for all indices i, $x_i \prec y_i$ in X_i. The preorder of X is the coarsest that makes all projections $X \to X_i$ monotone.

(c) For every set S, the power set $\mathcal{P}S$ is ordered by inclusion, with minimum $\emptyset$ and maximum S. This order is not total, as soon as S has more than 1 element.

(d) In the set $\mathbb{N}$ of natural numbers, the divisibility relation $n \mid m$ means that $m = nn'$ for some $n' \in \mathbb{N}$; it is an order relation, with minimum 1 and maximum 0. In the set $\mathbb{Z}$ of integers the divisibility relation is a preorder; the associated equivalence relation is $k = \pm k'$, and the associated ordered set $\mathbb{Z}/\sim$ is isomorphic to $\mathbb{N}$ (ordered by divisibility). The minimum of $\mathbb{Z}$ is 1 (or -1), the (unique) maximum is 0.

1.1.2 *Infima and suprema*

Let X be an ordered set. For $A \subset X$ and $a \in X$, the sets of their *lower bounds* and their *upper bounds* in X will be written as

$$
\begin{aligned}
L(A) &= \{x \in X \mid x \prec a, \text{for all } a \in A\}, \\
\downarrow a = L(\{a\}) &= \{x \in X \mid x \prec a\}, \\
U(A) &= \{x \in X \mid a \prec x, \text{for all } a \in A\}, \\
\uparrow a = U(\{a\}) &= \{x \in X \mid a \prec x\}.
\end{aligned}
\tag{1.3}
$$

The *infimum* $\inf_X A$ of A in X, or *greatest lower bound*, or *meet*, is defined as

$$
\inf_X A = \max(L(A)), \tag{1.4}
$$

also written as $\inf A$ or $\bigwedge A$. Dually, the *supremum* $\sup_X A$ of A in X, or *least upper bound*, or *join*, is

$$
\sup_X A = \min(U(A)) = \inf_{X^{\mathrm{op}}} A, \tag{1.5}
$$

also written as $\sup A$ or $\bigvee A$. (These outcomes can exist or not.) If A has a minimum m, the infimum also exists and $\inf_X A = \max(\downarrow m) = \min A$; if A has a maximum, $\sup_X A = \max A$.

Every element of X is (trivially) a lower bound and an upper bound of the empty subset; if X has a greatest and a least element, we have:

$$
\inf_X \emptyset = \max X = \sup_X X, \qquad \sup_X \emptyset = \min X = \inf_X X. \tag{1.6}
$$

These definitions can be extended to preordered sets, but then joins and meets are only determined up to the equivalence relation associated to our preorder.

A *meet semilattice*, or *lower semilattice*, X is an ordered set where every pair $x, y \in X$ has a *meet* $x \wedge y = \inf\{x, y\}$; a *1-semilattice* is also assumed to have a top element, written as 1 (or $\top$), which acts as a unit for the meet operation. Equivalently, a 1-semilattice is an ordered set where every finite subset A has a greatest lower bound $\bigwedge A$.

Dually, a *join semilattice*, or *upper semilattice*, has all binary joins $x \vee y$; a *0-semilattice* is also assumed to have a bottom element, written as 0 or $\bot$; in other words, a 0-semilattice is an ordered set where every finite subset A has a least upper bound $\bigvee A$.

An ordered set is said to be *filtered* (or *directed*) if every pair $x, y \in X$ has an upper bound; dually, it is *cofiltered* if every pair has a lower bound.

1.1.3 From lattices to boolean algebras

A *lattice* is an ordered set where every pair $x, y \in X$ has a *meet* $x \wedge y = \inf\{x, y\}$ and a *join* $x \vee y = \sup\{x, y\}$.

A *bounded lattice* is also required to have a top element, written as 1 (or $\top$), and a bottom element 0 (or $\bot$), that are units for the meet and join operation, respectively. (In category theory the term 'lattice' is often used in this sense; we do not follow this convention here.)

A lattice is said to be *distributive* if the meet operation distributes over the join operation

(D) $(x \vee y) \wedge z = (x \wedge z) \vee (y \wedge z)$,

or equivalently if the join distributes over the meet (see Exercise 1.1.4(a)).

A *boolean algebra* is a distributive bounded lattice where every element x has a *complement* x', defined by the following properties

$$x \wedge x' = 0, \qquad x \vee x' = 1, \tag{1.7}$$

and determined by them (as proved in Exercise 1.1.4(b)).

A *complete lattice* X is an ordered set where every subset has a join and a meet. But actually it is sufficient to assume the existence of all meets (or all joins), since one can recover the join of a subset $A \subset X$ as the meet of its upper bounds, or symmetrically (as proved in Exercise 1.1.4(d))

$$\bigvee A = \bigwedge (U(A)), \qquad \bigwedge A = \bigvee (L(A)). \tag{1.8}$$

More precisely, the first formula above means that, for any subset A of an ordered set X, $\bigvee A$ exists if and only if $\bigwedge (U(A))$ exists, and — in this case — they coincide.

1.1.4 Exercises and complements

The non-obvious solutions of these exercises can be found in Chapter 7.

(a) In a lattice X, the property 1.1.3(D) implies that the join operation distributes over the meet.

(b) In a distributive bounded lattice X, the complement of x, defined in (1.7), is unique.

(c) In an ordered set X, the lower and upper bounds of subsets $A, B \subset X$ satisfy these properties

$$A \subset B \quad \Rightarrow \quad L(A) \supset L(B) \text{ and } U(A) \supset U(B),$$

$$A \subset UL(A), \qquad A \subset LU(A), \tag{1.9}$$

$$LUL(A) = L(A), \qquad ULU(A) = U(A).$$

The mappings $L, U \colon \mathcal{P}X \to \mathcal{P}X$ form thus a contravariant Galois connection, see 1.8.8(e).

(d) Prove the formulas (1.8).

(e) A totally ordered set X is always a lattice, with $x \wedge y = \min\{x, y\}$ and $x \vee y = \max\{x, y\}$; it is a distributive lattice.

(f) The ordered line $\mathbb{R}$ and its (partially ordered) cartesian power $\mathbb{R}^n$ are distributive lattices, without minimum and maximum. They are *conditionally complete*: every non-empty upper bounded subset has a join, and every non-empty lower bounded subset has a meet. Prove that these two properties are equivalent, in every ordered set X.

(g) Prove that an ordered set X has all joins of upper bounded subsets if and only if it has all meets of non-empty subsets. The interval $[0, +\infty[$ of the ordered line has this form of 'conditioned completeness', and lacks a maximum.

(h) In a power set $\mathcal{P}S$, a subset $\mathcal{A} \subset \mathcal{P}S$ is more easily understood as an indexed family $(A_i)_{i \in I}$ of subsets of S. A lower bound of this family is any subset contained in all of them, and the greatest is $\bigcap A_i$. Symmetrically, the least upper bound of the family is $\bigcup A_i$.

$\mathcal{P}S$ is a complete boolean algebra, where joins and meets are unions and intersections, while the boolean complement of a subset A is the set-theoretical complement $X \setminus A$.

(i) The ordered set $\mathrm{Sub}(A)$ of submodules of an R-module A has also all meets, which are intersections: $\bigwedge A_i = \bigcap A_i$. Therefore $\mathrm{Sub}(A)$ is a complete lattice, and $\bigvee A_i$ is the least submodule of A containing the set-theoretical union $\bigcup A_i$.

The join is often written as ΣA_i, because it can be realised as the set of elements of A which are sums $x_1 + x_2 + ... + x_n$ of elements of $\bigcup A_i$.

(j) Show that the lattice $\mathrm{Sub}(A)$ of subgroups of an abelian group need not be distributive. (It has a weaker distributive property, dealt with in 1.1.9.) *Hints:* take $A = \mathbb{Z}^2$.

*(k) Frames and quantales will be treated in Section 6.1.

1.1.5 Semigroups

A *semigroup* S is a set equipped with an associative operation, generally written in multiplicative form, as ab or $a.b$. We do *not* exclude the *empty semigroup*, as is often the case in semigroup theory (cf. [CP], p. 1) and universal algebra.

If S is a semigroup, S^{op} is the *opposite* or *dual* one, with reversed product $a * b = ba$. As for preorders, every topic or statement of the theory of semigroups has a dual instance.

For two subsets $A, B \subset S$ and an element $x \in S$, one writes

$$AB = \{ab \mid a \in A,\, b \in B\} \subset S,$$
$$xA = \{x\}A = \{xa \mid a \in A\}, \quad Ax = A\{x\} = \{ax \mid a \in A\}. \tag{1.10}$$

In a semigroup S, a *subsemigroup* is a subset T closed under product: $TT \subset T$. The intersection $\bigcap S_i$ of a family of subsemigroups of S is a subsemigroup (also because we are not excluding the empty subsemigroup). Therefore, the set $\mathrm{Sub}(S)$ of subsemigroups of S forms a complete lattice with respect to inclusion; its minimum is the empty subsemigroup.

If A is a subset of S, the *subsemigroup $\langle A \rangle$ of S generated by A* is the least subsemigroup of S that contains A. Formally, it is the intersection of all the subsemigroups of S that contain A; concretely, it consists of all products $x_1 x_2 ... x_n$ of elements of A (for n a positive integer). The join $\bigvee S_i$ of a family of subsemigroups of S is generated by their union $\bigcup S_i$.

A *homomorphism* $h \colon S \to T$ of semigroups is a mapping that preserves the product: $h(ab) = h(a).h(b)$, for all $a, b \in S$. Its image $h(S)$ is a subsemigroup of T.

1.1.6 Unit and absorbing element

The semigroup S is called a *unital semigroup*, or a *monoid*, when it has a unit 1 (acting as an identity for the product), that is obviously unique.

The unit is also written as 1_S when useful. A *homomorphism of monoids* is assumed to preserve the unit.

Every semigroup S has an *associated monoid* S^1, obtained by adding an element $1 \notin S$ and extending the product in the obvious way: $1a = a = a1$ for all $a \in S^1$. Note that if S already has a unit e, the latter is no longer a unit in S^1.

The *universal property* of this construction says that every semigroup-homomorphism $f \colon S \to M$ with values in a monoid can be uniquely extended to a monoid-homomorphism $g \colon S^1 \to M$, letting $g(1) = 1_M$. *(Universal properties are formalised in 1.5.7.)*

(Our convention on this point differs from the usual one in semigroup theory, where one takes $S^1 = S$ when S is already unital. In this way the previous universal property is not satisfied, and the procedure $S \mapsto S^1$ cannot be extended to homomorphisms.)

An *absorbing element* for the semigroup S is an element z such that $za = z = az$ for all $a \in S$. This uniquely determined element can be written as 0 or ∞, as convenient.

Again, every semigroup S has an *associated semigroup with an absorbing element* S^∞, obtained by adding an element $\infty \notin S$ and extending the product in the obvious way: $\infty a = \infty = a\infty$ for all $a \in S^\infty$. There is again an obvious universal property.

1.1.7 Idempotents

In a semigroup S the *idempotent* elements e (such that $ee = e$) play an important role, and should be viewed as 'partial identities': see 1.2.4. *(This role will be formalised by a category, the 'idempotent completion' of S, in 2.2.5.)*

The set of idempotent elements of S, often denoted by E (or E_S) has a *canonical order* $e \leqslant f$, characterised by the following conditions, which are easily seen to be equivalent:

(i) $e = ef = fe$,

(ii) $e = fef$,

(iii) $e \in Sf \cap fS$,

(iv) $e \in fSf$.

The set E is not closed under product in S, generally. But the product of two *commuting* idempotents e, f is always idempotent

$$ef = fe \;\Rightarrow\; ef.ef = ee.ff = ef, \tag{1.11}$$

and then the element $ef = fe$ is the meet $e \wedge f$, with respect to the canonical order of E.

If S is a monoid, the unit 1 is the greatest element of E_S. An absorbing element is the least element of E_S.

Every homomorphism $h: S \to T$ of semigroups restricts to a mapping $E_S \to E_T$ that preserves the canonical order.

An *idempotent semigroup* (also called a *band* in semigroup theory) is a semigroup where each element is idempotent.

1.1.8 *Preorder and Alexandrov topologies*

The reader may have noticed a similarity of the theories of preordered sets and topological spaces. In fact, the former can be 'embedded' in the latter, so that — for instance — the finest preorder corresponds to the finest topology, the coarsest to the coarsest.

Let X be an arbitrary topological space. For $a \in X$, we write as $\bar{a}$ the closure of the singleton $\{a\}$, also called the closure of the point a. The *specialisation preorder* of the space X is defined by

$$x \prec y \iff x \in \bar{y} \iff \bar{x} \subset \bar{y}, \tag{1.12}$$

so that $\bar{a} = \{x \in X \mid x \prec a\}$.

Generally speaking, this preorder misses a large part of the information contained in the topology: for instance, in a space where all singletons are closed, this preorder is discrete: $x = y$. Thus, on the set $\mathbb{R}$, the euclidean and the discrete topology give the same order relation.

This is no longer the case if we restrict in a convenient way the topologies we are considering.

An *Alexandrov topology*, named after Pavel S. Alexandrov, is a topology where the open sets are stable under arbitrary unions *and arbitrary intersections*; equivalently, the same is true of closed sets.

In this case, for any subset $A \subset X$

$$\bar{A} = \bigcup_{a \in A} \bar{a}, \tag{1.13}$$

because this union is closed, and obviously the least closed subset containing A. Therefore, an Alexandrov topology is determined by the closure of its points, and also by the specialisation preorder. The topology is T_0 if and only if the specialisation preorder is an order relation.

The reader may now guess that preordered sets are 'the same as' spaces with Alexandrov topology, and monotone mappings amount to continuous mappings between such spaces.

More precisely, the previous argument makes the category of preordered sets isomorphic to a full subcategory of the category of topological spaces, formed of the spaces with Alexandrov topology. Similarly, ordered sets correspond to spaces with an Alexandrov T_0-topology.

All this is well known; details can be found in [G11], 3.2.7 and 6.5.4.

1.1.9 *Modular lattices*

The lattice $\mathrm{Sub}(A)$ of submodules of an R-module is not distributive, generally; but one can easily check that it always satisfies a weaker, restricted form of distributivity, called 'modularity' (as proved in Exercise (a)).

Namely, a lattice is said to be *modular* if it satisfies the following selfdual property (for all elements x, y, z)

(M) if $x \leqslant z$ then $(x \vee y) \wedge z = x \vee (y \wedge z)$.

For instance, if A is the Klein four-element group $\mathbb{Z}/2 \oplus \mathbb{Z}/2$, the lattice $\mathrm{Sub}(A)$ is isomorphic to the lattice $M = \{0, x, y, z, 1\}$ of five elements displayed in the following diagram, at the left

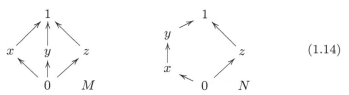

$$(1.14)$$

where the relation $a \leqslant b$, is represented by an arrow $a \to b$, and the meet (resp. join) of any two elements among x, y, z is 0 (resp. 1). This lattice is not distributive: note that x has two distinct complements, y and z. On the other hand, the lattice N displayed above is not even modular, as $x \leqslant y$ but $(x \vee y) \wedge z = z$ and $x \vee (y \wedge z) = x$.

Exercises and complements. (a) Every lattice $\mathrm{Sub}(A)$ of submodules is modular.

*(b) In figure (1.14), N is the 'basic' non-modular lattice, in the sense that: a lattice is modular if and only if it does not contain a sublattice isomorphic to N ([Bi], I.7, Theorem 12).

Similarly, M is the basic non-distributive *modular* lattice: a modular lattice is distributive if and only if it does not contain a sublattice isomorphic to M ([Bi], II.8, Theorem 13).

These two theorems can be combined, to characterise distributive lattices among all lattices.

1.2 Regular and inverse semigroups

After introducing *regular* and *inverse semigroups*, we prove a classical result: inverse semigroups are the same as regular semigroups with commuting idempotents (Vagner–Liber Theorem, in 1.2.6).

Then we briefly introduce *orthodox semigroups*, a more general notion.

The theory of orthodox, inverse and quasi-inverse semigroups was developed by B.M. Schein [Sc], N.R. Reilly and M.E. Scheiblich [ReS], T.E. Hall [Ha], M. Yamada [Ya], A.H. Clifford and G.B. Preston [CP], and others, starting in the 1960's. It can now be found in Howie's and Lawson's books [Ho, Ls].

1.2.1 Regular semigroups

One says that the elements a, b of the semigroup S are *partial inverses*, or form a *regular pair*, if

$$a = aba, \qquad b = bab. \tag{1.15}$$

The semigroup S is said to be *regular* if every element has some partial inverse (which need not be unique). This property, introduced by von Neumann [vN] in 1936 for the multiplicative semigroup of a ring, will also be called *von Neumann regularity*.

Note that each of the two relations above, in (1.15), implies that ab *and* ba *are idempotent elements of* S. On the other hand, every idempotent element e is a partial inverse of itself; therefore, an idempotent semigroup is always regular.

A homomorphism of semigroups preserves regular pairs.

We shall see, in 1.3.3, that the endo-relations of any R-module form a regular monoid.

1.2.2 Proposition (Regular semigroups)

(a) The semigroup S is regular if (and only if) for every $a \in S$ there is some $b \in S$ such that $a = aba$.

(b) Let S be a regular semigroup. If the set E_S has a maximum 1, this is a unit for S.

Note. The property in (a) is often taken as the definition of a regular semigroup.

Proof (a) If this property holds, then ab and ba are idempotents.

It follows that $b' = bab$ is a partial inverse of a:

$$ab'a = a(bab)a = (ab)(ab)a = aba = a,$$
$$b'ab' = (bab)a(bab) = b(ab)(ab)(ab) = bab = b'.$$

(b) If $1 = \max E_S$, then $e = e1 = 1e$ for every idempotent e. Any element a can be written as $a = a(ba) = (ab)a$ where ba and ab are idempotents, so that $a = a1 = 1a$.   □

1.2.3 Inverse semigroups

The semigroup S is said to be *inverse* if every element a has precisely one partial inverse, which we write as $a^\sharp$. (In semigroup theory $a^\sharp$ is usually said to be 'inverse' to a; we keep the term 'partial inverse'.)

In this case, an element a is said to be *symmetric* if $a = a^\sharp$, which is equivalent to $a = a^3$. In particular, every idempotent e is a symmetric element: $e = e^\sharp$, but there can be symmetric elements which are not idempotent (see below). Any element $a \in S$ has two associated idempotents, $a^\sharp a$ and $aa^\sharp$, which will be further analysed below (in 1.3.2); moreover, a is idempotent if and only if $a = a^\sharp a$, if and only if $a = aa^\sharp$.

By definition, an *inverse subsemigroup* of an inverse semigroup S is a subsemigroup closed under partial inverses. A homomorphism between inverse semigroups automatically preserves the partial inverses (and is thus a homomorphism of inverse semigroups).

Exercises and complements. (a) If S is a monoid, an invertible element a has a unique partial inverse: its inverse a^{-1}.

(b) Every group is an inverse semigroup; more precisely, a group is the same as an inverse monoid were the left (or the right) cancellation law holds.

(c) In a group, the symmetric elements are the involutive ones, with $a^2 = 1$, and there is only one idempotent, the unit.

(d) The monoid $S(X)$ of endomappings of any set X is a regular semigroup.

(e) The monoid $S(X)$ is not inverse, except trivial cases (to be determined).

1.2.4 Partial bijections of a set

We show now that every set X has an associated inverse semigroup $\mathcal{I}(X)$ of 'partial bijections', introduced by V.V. Vagner (also transliterated as Wagner) in 1952 [Va1], and called the *symmetric inverse semigroup* of X, by analogy with the symmetric group of permutations of X.

In analogy with Cayley theorem for groups, we shall see that every inverse semigroup S can be embedded as an inverse subsemigroup of $\mathcal{I}(S)$, in 1.3.6.

A *partial bijection* of the set X will be a bijective mapping $f\colon U \to V$ between subsets U, V of X (possibly empty). We write

$$\operatorname{Def} f = U, \qquad \operatorname{Val} f = V. \qquad (1.16)$$

(Equivalently, it is an injective mapping $U \to X$ defined on a subset of X; but we shall not use the term 'partial injection', which can mask the symmetric character of the notion.)

These mappings form the set $\mathcal{I}(X)$. Given another bijective mapping $g\colon V' \to W$ between subsets of X, the *composite* gf is again a partial bijection of X, defined where it makes sense

$$\operatorname{Def}(gf) = \{x \in X \mid x \in U, f(x) \in V'\}, \qquad (gf)(x) = g(f(x)). \qquad (1.17)$$

The operation is associative: $\mathcal{I}(X)$ is a monoid, with unit the identity mapping $\operatorname{id} X$ (everywhere defined, of course). The empty bijection 0 is the *absorbing element* of the semigroup: $0f = 0 = f0$, for every $f \in \mathcal{I}(X)$ (cf. 1.1.6).

Finally, $\mathcal{I}(X)$ is an inverse monoid: a bijective mapping $f\colon U \to V$ between subsets of X has a unique partial inverse in $\mathcal{I}(X)$, represented by the inverse mapping $f^\sharp\colon V \to U$. Let us note that the associated idempotents

$$f^\sharp f = \operatorname{id} U, \qquad f f^\sharp = \operatorname{id} V, \qquad (1.18)$$

coincide with the unit $\operatorname{id} X$ if and only if f is a *total* bijection $X \to X$. Only in this case the partial inverse $f^\sharp$ is a true inverse, and will be written as f^{-1}.

We have already seen that an element $f \in \mathcal{I}(X)$ is idempotent if and only if $f = f^\sharp f$. Therefore the idempotents of the symmetric inverse semigroup $\mathcal{I}(X)$ coincide with the partial identities $\operatorname{id}(U)$ of X, i.e. the identities of the subsets $U \subset X$.

1.2.5 Other examples

If X is a topological space we are interested in the set $\mathrm{I}\mathcal{T}(X)$ of its *partial homeomorphisms*, i.e. homeomorphisms $f\colon U \to V$ between subspaces of X, and (even more) in the subset $\mathrm{I}\mathcal{C}(X)$ of homeomorphisms $f\colon U \to V$ between *open* subspaces of X.

The notation $\mathrm{I}\mathcal{T}$, $\mathrm{I}\mathcal{C}$ will be explained in Definition 2.4.4. *(It stands for the 'inverse core' of two categories, $\mathcal{T}$ and $\mathcal{C}$, of partial continuous mappings. Similarly, $\mathcal{I}$ is the inverse core of the category $\mathcal{S}$ of sets and partial mappings.)*

Writing the underlying set of the space X as $|X|$, we have the following inclusions of inverse subsemigroups:

$$I\mathcal{C}(X) \subset I\mathcal{T}(X) \subset \mathcal{I}(|X|).$$

1.2.6 Theorem (Vagner–Liber Therem on Inverse Semigroups)

A semigroup is inverse if and only if it is regular and its idempotents commute.

In an inverse semigroup S the idempotents form a meet semilattice E_S.

Proof An implication, due to V.V. Vagner [Va2], in 1952, is easy: we assume that the semigroup S is regular and its idempotents commute, and prove the uniqueness of partial inverses.

Indeed, if b and c are partial inverses to a, then the products ab, ba, ac, ca are (commuting) idempotents, and b must coincide with c

$$b = bab = b(aca)b = b(a(cacac)a)b = (ba)(ca)c(ac)(ab)$$
$$= (ca)(ba)c(ab)(ac) = c(aba)c(aba)c = cacac = cac = c.$$

The other implication, due to A.E. Liber [Li], in 1954, is harder to prove. We suppose that S is an inverse semigroup, take two idempotents e, f and show that they commute, following the argument in Clifford–Preston [CP], Theorem 1.17.

First we prove that the product ef is idempotent. Let a be the partial inverse of ef

$$(ef)a(ef) = ef, \qquad a(ef)a = a.$$

Then $b = ae$ is also a partial inverse of ef, because

$$(ef)b(ef) = (ef)ae(ef) = (ef)a(ef) = ef,$$
$$b(ef)b = ae(ef)ae = a(ef)ae = ae.$$

It follows that $a = ae$. Similarly $fa = a$, and then a is idempotent:

$$aa = ae.fa = a(ef)a = a.$$

Therefore a is the partial inverse of itself, and coincides with ef, which is idempotent; similarly fe is idempotent. Finally, ef (being idempotent) is the partial inverse of itself; but it is also the partial inverse of fe

$$(ef)(fe)(ef) = ef.ef = ef, \qquad (fe)(ef)(fe) = fe.fe = fe,$$

and therefore $ef = fe$.

The last claim follows from (1.11): in any semigroup the product of two commuting idempotents is idempotent. □

1.2.7 Orthodox semigroups

A semigroup S is said to be *orthodox* if it is regular and its idempotents are closed under product: $E_S.E_S \subset E_S$. The name was introduced by Hall [Ha] in 1969, but the property had already been used under different names, by Schein [Sc] and Reilly–Scheiblich [ReS].

Every inverse semigroup is orthodox, by the previous theorem. On the other hand, every orthodox semigroup S has a finest 'inverse congruence' (making the quotient an inverse semigroup), as proved in [Ha].

A well-known characterisation, that will not be used here, says that a regular semigroup is orthodox if and only if every partial inverse of an idempotent element is also idempotent ([ReS], Lemma 1.3).

Orthodox categories, a natural extension of orthodox semigroups, are a main tool in the author's analysis of coherence in Homological Algebra [G9].

1.2.8 Lemma (Regularity Lemma for semigroups)

In the semigroup S we assume that:

$$a = aba, \qquad a' = a'b'a'. \tag{1.19}$$

Then the following properties are equivalent:

(i) $a'a = (a'a)(bb')(a'a)$,

(ii) the element $(b'a')(ab)$ is idempotent.

Note. A similar result can be found in [ReS], Lemma 1.1. Note also that the element $(b'a')(ab)$ is a product of idempotents; therefore *condition* (ii) *is automatically satisfied when S is an orthodox semigroup*, or more particularly an inverse semigroup.

Proof The fact that (i) implies (ii) is trivial (and independent of (1.19)):

$$(b'a'ab)(b'a'ab) = b'(a'a)(bb')(a'a)b = b'(a'a)b.$$

Conversely, if (ii) holds:

$$(a'a)(bb')(a'a) = a'(abb'a')a = (a'b'a')(abb'a')(aba)$$
$$= a'(b'a'ab)(b'a'ab)a = a'(b'a'ab)a = (a'b'a')(aba) = a'a.$$

□

1.3 The involution and order of an inverse semigroup

We introduce involutive semigroups (in 1.3.1) and ordered semigroups (in 1.3.7).

An inverse semigroup has a canonical involution, given by the partial inverses, and a canonical order, generated by the order of its idempotent elements (see 1.1.7).

1.3.1 Involutive semigroups

An *involutive semigroup* will be a semigroup S equipped with an endomapping $a \mapsto a^\sharp$ that reverses the product and is involutive

$$(ab)^\sharp = b^\sharp.a^\sharp, \qquad (a^\sharp)^\sharp = a. \qquad (1.20)$$

This mapping is an anti-automorphism of S, that is an isomorphism $S \to S^{\mathrm{op}}$. It automatically preserves the unit 1, if it exists: in fact

$$1^\sharp.a = 1^\sharp.a^{\sharp\sharp} = (a^\sharp 1)^\sharp = a^{\sharp\sharp} = a,$$

for every $a \in S$.

The involution will be said to be *regular* (in the sense of von Neumann), if for every $a \in S$

$$aa^\sharp a = a. \qquad (1.21)$$

Then we also have $a^\sharp aa^\sharp = a^\sharp$, so that a and $a^\sharp$ are partial inverses and the semigroup S is regular (cf. 1.2.1). Here we are only interested in regular involutions, that play a central role in categories of partial bijections or categories of relations of modules (and their extensions).

1.3.2 Projectors

Let S be a semigroup with regular involution. A *projector* of S will be a symmetric idempotent, i.e. an element e such that $ee = e = e^\sharp$.

They form a set

$$\mathsf{Prj}\,(S) \subset E_S. \qquad (1.22)$$

In [G3]–[G5] these elements are called 'projections', a name with too many meanings.

Every element $a \in S$ has two associated projectors, which will be written as follows:

$$\underline{e}(a) = a^\sharp a \qquad\qquad \text{(the \emph{support} of } a),$$
$$\underline{e}^*(a) = aa^\sharp = \underline{e}(a^\sharp) \qquad\qquad \text{(the \emph{cosupport} of } a). \qquad (1.23)$$

This terminology is explained by its extension to categories, where the support (resp. cosupport) of the morphism a is a projector of its domain (resp. codomain), cf. 2.2.2.

We have

$$a = a.\underline{e}(a) = \underline{e}^*(a).a = \underline{e}^*(a).a.\underline{e}(a). \qquad (1.24)$$

All the projectors of S arise as supports (and cosupports): an element $e \in S$ is a projector if and only if $e = \underline{e}(e)$, if and only if $e = \underline{e}^*(e)$.

The product of two projectors e, f is always idempotent

$$ef.ef = ef.f^\sharp e^\sharp.ef = ef.(ef)^\sharp.ef = ef, \qquad (1.25)$$

and is a projector if and only if e and f commute (because $(ef)^\sharp = f^\sharp e^\sharp = fe$).

Conversely every idempotent e is the product of two projectors:

$$e = ee^\sharp e = ee^\sharp.e^\sharp e = \underline{e}^*(e).\underline{e}(e). \qquad (1.26)$$

1.3.3 Endo-relations of modules

If A is an R-module, the endo-relations a of A (i.e. the submodules of $A \oplus A$) form a unital semigroup $\mathcal{R}(A)$. Their product is the composition of relations, defined by the following formula

$$ba = \{(x, z) \in A \oplus A \mid \exists\, y \in A \colon (x, y) \in a \text{ and } (y, z) \in b\}. \qquad (1.27)$$

This monoid has an obvious involution, given by reversing pairs:

$$a^\sharp = \{(x, y) \in A \oplus A \mid (y, x) \in a\}, \qquad (1.28)$$

and *this involution is easily proved to be regular* (see Exercise (a), below). $\mathcal{R}(A)$ is thus a regular monoid.

Exercises and complements. (a) The involution (1.28) is regular.

(b) The endo-relations of a set X, i.e. the subsets of $X \times X$, form a semigroup with a similar product and involution; the latter is not regular, provided that X has at least two elements.

*(c) As a consequence of a Coherence Theorem in Homological Algebra ([G9], Theorem 2.7.6), the following conditions on an R-module A are equivalent:

- the (regular) monoid $\mathcal{R}(A)$ of endo-relations of A is orthodox,

- the (modular) lattice $\mathrm{Sub}(A)$ of submodules of A is distributive,

- the canonical isomorphisms between the subquotients of A are closed under composition, and form a coherent system of isomorphisms.

1.3.4 Proposition and Definition

Let S be an inverse semigroup.

(a) S has a canonical involution: *the mapping $a \mapsto a^\sharp$ that takes each element to its partial inverse. This is the unique regular involution of S.*

(b) The idempotents of S are the same as its projectors for the involution. They commute and form a meet semilattice $E_S = \mathsf{Prj}\,(S)$

$$e \leqslant f \Leftrightarrow e = ef = fe, \qquad e \wedge f = ef = fe. \tag{1.29}$$

Proof (a) Plainly, if S has a regular involution $a \mapsto a^\sharp$, then $a^\sharp$ must be the partial inverse of a.

Conversely, defining $a^\sharp$ in this way, we have to verify that the partial inverse of ab is $b^\sharp a^\sharp$; this follows easily from the fact that the idempotents of S commute (by Theorem 1.2.6):

$$(ab)(b^\sharp a^\sharp)(ab) = a(bb^\sharp)(a^\sharp a)b = a(a^\sharp a)(bb^\sharp)b = ab.$$

Similarly $(b^\sharp a^\sharp)(ab)(b^\sharp a^\sharp) = b^\sharp a^\sharp$.

(One can also deduce these facts from the Regularity Lemma 1.2.8.)

(b) In an inverse semigroup every idempotent e is its own partial inverse (as already remarked in 1.2.3), and therefore a symmetric element with respect to the canonical involution: $e^\sharp = e$.

The rest follows from the commutativity of idempotents. □

1.3.5 Proposition (Characterisation of inverse semigroups, II)

For a semigroup S the following conditions are equivalent:

(i) S is inverse,

(ii) S has a regular involution and its idempotents commute,

(iii) S has a regular involution and its projectors commute,

(iv) S has a regular involution and its projectors form a subsemigroup.

Proof It is an easy consequence of Theorem 1.2.6. Indeed, the equivalence of (i) and (ii) follows from 1.2.6 and 1.3.4 (that ensures the existence of a regular involution in every inverse semigroup).

Then, condition (ii) trivially implies (iii). The converse follows from the fact that, once we have a regular involution, every idempotent is the product of two projectors (as we have seen in 1.3.2). The equivalence of (iii) and (iv) is obvious, since $(ef)^\sharp = f^\sharp e^\sharp$. □

1.3.6 *Theorem* (Vagner-Preston Representation Theorem)

Every inverse semigroup S can be embedded as an inverse subsemigroup of the symmetric inverse semigroup $\mathcal{I}(S)$ (over its underlying set, see 1.2.4), by the left translation:

$$\lambda\colon S \to \mathcal{I}(S), \qquad a \mapsto \lambda_a,$$
$$\lambda_a\colon a^\sharp S \to aS, \qquad \lambda_a(x) = ax \quad (\text{for } a \in S).$$
(1.30)

Note. In other words, every inverse semigroup is a semigroup of partial bijections, as proved by Vagner [Va2] in 1952 and independently by Preston in 1954. This result extends Cayley's theorem: every group is a group of bijections. It will not be used here.

Proof We follow the argument in Clifford–Preston [CP], Theorem 1.20, with a slightly different notation.

First, let us note that the subset $a^\sharp S = a^\sharp a S$ admits the idempotent $\underline{e}(a) = a^\sharp a$ as a left unit, while the subset $aS = aa^\sharp S$ admits the idempotent $\underline{e}^*(a) = aa^\sharp$ as a left unit.

This shows that $\lambda_a\colon a^\sharp S \to aS$ is a bijective mapping, with inverse $\lambda_{a^\sharp}\colon aS \to a^\sharp S$.

Globally, the mapping $\lambda\colon S \to \mathcal{I}(S)$ is injective. In fact, if $\lambda_a = \lambda_b$ we have $a^\sharp aS = b^\sharp bS$; but then $a^\sharp a = (a^\sharp a)(a^\sharp a) = (b^\sharp b)y$ (for some $y \in S$) and $(b^\sharp b)(a^\sharp a) = a^\sharp a$, which means that $a^\sharp a \leqslant b^\sharp b$ in the meet semilattice $E = \mathsf{Prj}\,(S)$; symmetrically, $b^\sharp b \leqslant a^\sharp a$ and these projectors coincide. Now

$$a = aa^\sharp a = \lambda_a(a^\sharp)a = \lambda_b(a^\sharp)a = ba^\sharp a = bb^\sharp b = b.$$

Finally, we have to prove that $\lambda_{ba} = \lambda_b\lambda_a$, for all a, b in S. Since $bax = b(ax)$, it suffices to show that $\mathrm{Val}\,(\lambda_{ba}) = \mathrm{Val}\,(\lambda_b\lambda_a)$ (and recall that $\mathrm{Def}\,(\lambda_a) = \mathrm{Val}\,(\lambda_{a^\sharp})$).

If $z \in \mathrm{Val}\,(\lambda_b\lambda_a)$, there is some $x \in a^\sharp S$ such that $ax \in b^\sharp S$ and $z = bax$; but then $z \in baS = \mathrm{Val}\,(\lambda_{ba})$. Conversely, let $z \in \mathrm{Val}\,(\lambda_{ba}) = baS = baa^\sharp b^\sharp S$. Then $z = ba(a^\sharp b^\sharp z')$ and the element $x = a^\sharp b^\sharp z'$ belongs to $a^\sharp S$. Its image

$$\lambda_a(x) = aa^\sharp b^\sharp z' = (aa^\sharp)(b^\sharp b)b^\sharp z' = (b^\sharp b)(aa^\sharp)b^\sharp z',$$

belongs to $b^\sharp S$; thus $\lambda_b(\lambda_a(x)) = ba(a^\sharp b^\sharp z') = z$ belongs to $\mathrm{Val}\,(\lambda_b\lambda_a)$. $\square$

1.3.7 Ordered semigroups

An *ordered semigroup* is a semigroup S equipped with an order relation $a \leqslant b$ consistent with the product:

$$\text{if } a \leqslant a' \text{ and } b \leqslant b' \text{ in } S, \text{ then } ab \leqslant a'b'. \tag{1.31}$$

Equivalently, for every a and $b \leqslant b'$ in S, we have: $ab \leqslant ab'$ and $ba \leqslant b'a$.

Examples. In a meet (or join) semilattice the order is always consistent with the operation. In an ordered ring the sum is consistent with the ordering, but only the elements $a \geqslant 0$ give a monotone multiplication:

$$\text{if } b \leqslant b' \text{ in } S, \text{ then } ab \leqslant ab' \text{ and } ba \leqslant b'a.$$

1.3.8 Theorem

Let S be an inverse semigroup and $E = \mathsf{Prj}\,(S)$ its meet semilattice of idempotents.

For $a, b \in S$ the following properties are equivalent:

(i) $a = ab^\sharp a,$

(ii) $a = b(a^\sharp a),$

(ii^*) $a = (aa^\sharp)b,$

(iii) $a = (aa^\sharp)b(a^\sharp a),$

(iv) $a \in bE,$

(iv^*) $a \in Eb,$

(v) $a \in EbE.$

Proof Trivially, (ii) implies (iv). Conversely, if $a = be$ for some $e \in E$, then

$$b(a^\sharp a) = b(eb^\sharp be) = be(b^\sharp b)e = bb^\sharp b.e = be = a.$$

Similarly (ii*) is equivalent to (iv*) and (iii) is equivalent to (v).

Now, (iv) implies (v): if $a = be$ (with $e \in E$) then $a = (bb^\sharp)be \in EbE$. Conversely, if $a = fbe$ (with $e, f \in E$) then

$$a = f(bb^\sharp)be = (bb^\sharp)fbe = b(b^\sharp fb)e$$

belongs to bE, because $b^\sharp fb$ is idempotent:

$$(b^\sharp fb)(b^\sharp fb) = b^\sharp f(bb^\sharp)fb = b^\sharp ff(bb^\sharp)b = b^\sharp fb.$$

Similarly, (iv*) is equivalent to (v), and we are left with proving that (i) is equivalent to the other properties.

The fact that (i) implies (v) is easily seen: if $a = ab^\sharp a$, then $a = a(b^\sharp bb^\sharp)a = (ab^\sharp)b(b^\sharp a)$ belongs to EbE.

Conversely, let $a = fbe$ (with e, f idempotents). Applying the Regularity Lemma 1.2.8, from $e = eee$ and $fb = (fb)(b^\sharp f)(fb)$ we deduce that $fb.e = (fbe)(eb^\sharp f)(fbe)$. Finally:

$$ab^\sharp a = (fbe)b^\sharp(fbe) = (fbe)(eb^\sharp f)(fbe) = fbe = a.$$

$\square$

1.3.9 Theorem and Definition

(a) The inverse semigroup S has a canonical order $a \leqslant b$ (consistent with the product and partial inverses), that is defined by the equivalent properties (i)–(v) of the previous theorem.

(b) This order of S extends the order in the meet semilattice $E = \mathsf{Prj}\,(S)$ of idempotent elements, defined in 1.1.7, and is generated by the latter (as an order of semigroups).

(c) If S is unital, the idempotent elements e are characterised by the condition $e \leqslant 1$. Furthermore, the semigroup-order of S is generated by the condition $e \leqslant 1$, for all idempotents e.

Proof (a) The relation $a \leqslant b$, is reflexive (use property 1.3.8(i)) and transitive (use 1.3.8(iv)); it is also antisymmetric: if $a = be$ and $b = af$ (with $e, f \in E$), then $a = afe = aef = be.ef = af = b$.

The fact that the order is consistent with the product follows from the Regularity Lemma 1.2.8. Indeed, assuming that $a \leqslant b$ and $a' \leqslant b'$, we can apply the lemma to the pairs $(a, b^\sharp)$ and $(a', b'^\sharp)$, concluding that $aa' = (aa')(b'^\sharp b^\sharp)(aa')$, which means that $aa' \leqslant bb'$.

Finally, if $a \leqslant b$, then $a^\sharp \leqslant b^\sharp$.

(b) The first claim is obvious. As to the second, let us suppose we have in S a semigroup-order $a \prec b$ that extends the order $\leqslant$ of E and prove that the canonical order is finer.

If the pair a, b satisfies the equivalent properties (i)–(v) of 1.3.8, then $ab^\sharp$ and $b^\sharp a$ are idempotent and

$$ab^\sharp = (aa^\sharp)(bb^\sharp) \leqslant bb^\sharp, \qquad b^\sharp a = (b^\sharp b)(a^\sharp a) \leqslant b^\sharp b.$$

Therefore, since the semigroup-order $\prec$ extends the order of projectors:

$$a = ab^\sharp a \prec bb^\sharp a \prec bb^\sharp b = b.$$

(c) An obvious consequence of property 1.3.8(iv). $\square$

1.4 Categories

In the second part of this chapter we review the basic notions of category theory. Examples and exercises will focus on categories of topological spaces or ordered sets, rather than algebraic structures — less present in this analysis of local structures.

The theory of categories was established by Eilenberg and Mac Lane in 1945, in a well-known paper [EiM]. The interested reader will explore with pleasure the books of S. Mac Lane [M3] and F. Borceux [Bo].

Two earlier books by B. Mitchell and P. Freyd [Mi, Fr1] are centred on abelian categories and their embedding in categories of modules. The book by J. Adámek, H. Herrlich and G.E. Strecker [AHS] gives an accurate analysis of 'concrete categories'.

At a more elementary level, the author's [G10] is a textbook for beginners, also devoted to applications in Algebra, Topology and Algebraic Topology.

Dealing with categories, one should avoid the usual paradoxes related to 'the set of all sets'. Here we make use of a particular set theory where there are *sets* and *classes*; every set is also a class, but a *proper* class is not a set: for instance the *class of all sets* and the *class of all groups* are proper (see [AHS] or the Appendix of [Ke]). This approach is followed in [Mi, Fr1, AHS].

When two levels, like sets and classes, are insufficient, one can introduce a third level of 'hyperclasses', called 'conglomerates' in [AHS].

    *Alternatively, a more flexible setting widely used for categories is ordinary Set Theory (say ZFC), with the assumption of the existence of a Grothendieck *universe*, or of a suitable hierarchy of universes (cf. [M3], Section I.6).*

1.4.1 Categories of structured sets

Loosely speaking, before giving a precise definition, a category C consists of *objects* and *morphisms*, together with a (partial) *composition law*: given three objects X, Y, Z and two 'consecutive' morphisms $f\colon X \to Y$ and $g\colon Y \to Z$, we have a composed morphism $gf\colon X \to Z$.

This partial operation is associative (whenever composition is legitimate); every object X has an identity $\operatorname{id} X$ which acts as a unit for legitimate compositions.

The prime example is the category Set *of sets* (and mappings), where:

- an object is a set,
- the morphisms $f\colon X \to Y$ between two given sets X and Y are the (set-theoretical) mappings from X to Y,

- the composition law is the usual composition of mappings: $(gf)(x) = g(f(x))$.

The following categories 'of structured sets' will often be used:

- the category Top *of topological spaces* (and continuous mappings),

- the category Ord *of ordered sets* (and monotone mappings),

- the category pOrd *of preordered sets* (and monotone mappings),

- the category Set. *of pointed sets* (and pointed mappings),

- the category Top. *of pointed topological spaces* (and pointed maps),

- the category Ab *of abelian groups* (and homomorphisms),

- the category Gp *of groups* (and homomorphisms),

- the category R Mod *of (unitary) left modules* on a given unital ring R,

- the category Ban of *Banach spaces and continuous linear mappings*,

- the category Ban_1 of *Banach spaces and linear weak contractions*.

For the category Set., let us recall that a *pointed set* is a pair (X, x_0) consisting of a set X and a *base*-element $x_0 \in X$, while a pointed mapping $f: (X, x_0) \to (Y, y_0)$ is a mapping $f: X \to Y$ such that $f(x_0) = y_0$; their composition is obvious.

Similarly, a *pointed topological space* (X, x_0) is a space with a base-point, and a pointed map $f: (X, x_0) \to (Y, y_0)$ is a continuous mapping from X to Y such that $f(x_0) = y_0$. The reader may know that the category Top. is important in Algebraic Topology: for instance, the fundamental group $\pi_1(X, x_0)$ is defined for a pointed topological space.

The categories Ban and Ban_1 will be used in a marginal, elementary way, for examples and counterexamples. It is understood that we have chosen *either* the real *or* the complex field of scalars. We also recall that a linear weak contraction is a linear mapping with norm $\leqslant 1$.

A reader interested in the categorical aspects of Banach spaces is referred to Semadeni's book [Se].

When a category is named after its objects alone (e.g. the 'category of groups'), this means that the morphisms are understood to be the obvious ones (in this case the group-homomorphisms) and the composition is understood to be the usual one.

Of course, different categories with the same objects have different names, like Ban and Ban_1.

1.4.2 Definition

A *category* C consists of the following data:

(a) a class Ob C, whose elements are called *objects* of C,

(b) for every pair X, Y of objects, a set $C(X, Y)$ (called a *hom-set*) whose elements are called *morphisms* (or *maps*, or *arrows*) of C from X to Y and written as $f: X \to Y$,

(c) for every triple X, Y, Z of objects of C, a mapping of sets, called *composition*

$$C(X, Y) \times C(Y, Z) \to C(X, Z), \qquad (f, g) \mapsto gf,$$

that gives a *partial composition law* for pairs of consecutive morphisms. The composite gf will also be written as $g.f$.

These data must satisfy the following axioms.

(i) (*Associativity*) Given three consecutive arrows, $f: X \to Y$, $g: Y \to Z$ and $h: Z \to W$, one has: $h(gf) = (hg)f$.

(ii) (*Identities*) Given an object X, there exists an endo-map $e: X \to X$ which acts as a unit whenever composition makes sense; in other words if $f: X' \to X$ and $g: X \to X''$, one has: $ef = f$ and $ge = g$.

One shows, in the usual way, that e is determined by X; it is called the *identity* of X and written as 1_X or $\mathrm{id}\, X$.

We generally assume that the following condition is also satisfied.

(iii) (*Separation*) Every map $f: X \to Y$ has a well-determined *domain* $\mathrm{Dom}\, f = X$ and a well-determined *codomain* $\mathrm{Cod}\, f = Y$.

Concretely, when constructing a category, one can forget about this condition, since one can always satisfy it *redefining* a morphism $\hat{f}: X \to Y$ as a triple $(X, Y; f)$ where f is a morphism from X to Y in the original sense (possibly not satisfying the Separation axiom).

Two morphisms f, g are said to be *parallel* when they have the same domain and the same codomain; the meaning of a pair of *consecutive* arrows has already been mentioned.

Mor C denotes the class of all the morphisms of C, i.e. the disjoint union of its hom-sets. The endomorphisms of every object X form a monoid

$$C(X) = C(X, X). \tag{1.32}$$

If C is a category, the *opposite* (or *dual*) category, written as C^{op}, has the same objects as C and 'reversed arrows', with 'reversed composition' $g * f$

$$C^{op}(X, Y) = C(Y, X), \qquad g * f = fg, \qquad \mathrm{id}^{\,op}(X) = \mathrm{id}\, X. \tag{1.33}$$

Every topic of category theory has a dual instance, which comes from the opposite category (or categories). A dual notion is often distinguished by the prefix 'co-'.

1.4.3 Small and large categories

We have assumed that a category C has a *class* $\mathrm{Ob}\,\mathsf{C}$ of objects (e.g. the class of all sets, or the class of all topological spaces) and, for every pair X, Y of objects, a *set* $\mathsf{C}(X, Y)$ of morphisms from X to Y.

The categories of structured sets that we consider, like the examples of 1.4.1, are generally *large* categories, where the objects (or equivalently the morphisms) form a *proper class* (i.e. not a set). A category C is said to be *small* if the class $\mathrm{Ob}\,\mathsf{C}$ is a set. A *finite* category is a small category whose set of morphisms is finite (then the same is true of its set of objects, since an object is determined by its identity).

A set X can be viewed as a *discrete* small category: its objects are the elements of X, and the only arrows are their (formal) identities.

A preordered set X will often be viewed as a small category, where the objects are the elements of X and the set $X(x, x')$ contains precisely one arrow if $x \prec x'$ (which can be written as $(x, x')\colon x \to x'$), and no arrow otherwise. Composition and identities are uniquely determined, as follows

$$(x', x'').(x, x') = (x, x''), \qquad \mathrm{id}\, x = (x, x), \qquad (1.34)$$

and all diagrams in such a category commute.

In this sense, categories generalise preordered sets. Their dualities agree: the category associated to the opposite preordered set X^{op} (in 1.1.1) is dual to the category associated to X. Loosely speaking, the extension from preordered sets (or classes) to categories consists in allowing different arrows between specified objects.

In particular, each finite ordinal defines a category, which is often written as $\mathbf{0}, \mathbf{1}, \mathbf{2}, \ldots$ Thus, $\mathbf{0}$ is the empty category; $\mathbf{1}$ is the *singleton category*, i.e. the discrete category on one object; $\mathbf{2}$ is the *arrow category*, with two objects (0 and 1), and one non-identity arrow, $0 \to 1$.

A monoid M can be viewed as a small category M with one object, say $*$, and set of morphisms $\mathrm{Mor}\,\mathsf{M} = \mathsf{M}(*) = M$, with composition the multiplication of M. *In this sense, categories generalise monoids.* Again, their dualities coincide (see 1.1.5). Loosely speaking, a category can be viewed as a 'many-object' extension of a monoid.

The relationship between categories and semigroups will be further explored in 2.3.1 and Section 2.8.

1.4.4 Structural remarks

The following diagrams represent a preordered set and a monoid as small categories of different kinds (as remarked above), and how a category can combine both structures

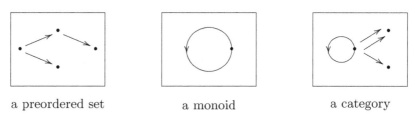

a preordered set   a monoid   a category

We only draw generators: composed arrows and identities are 'virtually present'. Thus, the second diagram represents a monoid with one generator.

We shall see how some aspects of the theory of preordered sets and the theory of semigroups can be extended to categories:

(a) (*many-arrow extensions*) infima and suprema extend to *categorical products* and *sums*, Galois connections to *adjunctions*,

(b) (*many-object extensions*) regular and inverse monoids extend to *von Neumann regular* and *inverse* categories.

In a very loose sense, the alternative between (a) and (b) is concerned with aspects of category theory where the objects play a relevant role, or not.

Of course, each of the three theories has its specific topics, whose extension or restriction can be of little interest.

1.4.5 Isomorphisms

In a category C a morphism $f\colon X \to Y$ is said to be *invertible*, or an *isomorphism*, if it has an inverse, i.e. a morphism $g\colon Y \to X$ such that $gf = 1_X$ and $fg = 1_Y$. The latter is uniquely determined by f and written as f^{-1}.

The *isomorphism relation* $X \cong Y$ between objects of C (meaning that there exists an isomorphism $X \to Y$) is obviously an equivalence relation.

For instance, the isomorphisms of Set, Top and Ab are, respectively: the bijective mappings of sets, the homeomorphisms of topological spaces and the isomorphisms of abelian groups. The isomorphisms of Ban are the linear homeomorphisms; the isomorphisms of Ban_1 are more restricted, namely the bijective linear isometries.

A *groupoid* is a category where every map is invertible. As above, a groupoid can be viewed as 'a group on many objects'.

Every category C has an *associated groupoid* Iso(C) on the same objects: its arrows are the isomorphisms of C, composed as in C.

Examples and complements. (a) If X is a preordered set, viewed as a category, two elements x, x' are isomorphic objects if and only if they correspond each other in the associated equivalence relation (in 1.1.1). The groupoid Iso(X) is this equivalence relation, viewed as a category.

(b) If M is a monoid, viewed as a category, an isomorphism amounts to an invertible element, and Iso(M) is their group.

(c) Groupoids were introduced before categories, by H. Brandt in 1927 [Bra]. Brandt dealt with 'connected' groupoids, in a single-sorted version where the only terms are the arrows, and the objects are represented by their identities. Categories can also be presented in a single-sorted version: see [M3], p. 9.

*(d) In the *fundamental groupoid* $\Pi_1(X)$ of a topological space X, an object is any point $x \in X$ and an arrow $[a] \colon x \to y$ is a class of paths in X, from x to y, up to homotopy with fixed end-points. The construction is sketched in 4.3.1(b).

1.4.6 Subcategories, quotients and products of categories

(a) Let C be a category. A *subcategory* D is defined by assigning:

- a subclass Ob D $\subset$ Ob C, whose elements are called *objects of* D,

- for every pair of objects X, Y of D, a subset D(X, Y) $\subset$ C(X, Y), whose elements are called *morphisms of* D, from X to Y,

so that the following conditions hold:

 (i) the composite in C of morphisms of D belongs to D,

 (ii) the identity in C of an object of D belongs to D.

Then D, equipped with the restricted composition law, is a category.

We say that D is a *full* subcategory of C if, for every pair of objects X, Y of D, we have D(X, Y) = C(X, Y), so that D is determined by assigning a subclass of objects. We say that D is a *wide* subcategory of C if it has the same objects.

For instance, Ab is a full subcategory of Gp, Ord is a full subcategory of pOrd, while Ban$_1$ is a wide subcategory of Ban and Iso(C) of C. Of course a full and wide subcategory must be the total one.

(b) A *congruence* $R = (R_{XY})$ in a category C consists of a family of equivalence relations R_{XY} in each set of morphisms $\mathsf{C}(X, Y)$, that is consistent with composition:

$$\text{if } f \, R_{XY} \, f' \text{ and } g \, R_{YZ} \, g', \text{ then } (gf) \, R_{XZ} \, (g'f'). \tag{1.35}$$

Then one defines the *quotient category* $\mathsf{D} = \mathsf{C}/R$: the objects are those of C, and $\mathsf{D}(X, Y) = \mathsf{C}(X, Y)/R_{XY}$; in other words, a morphism $[f] \colon X \to Y$ in D is an equivalence class of morphisms $X \to Y$ in C. The composition is induced by that of C, which is legitimate because of condition (1.35):

$$[g].[f] = [gf]. \tag{1.36}$$

For instance, in Top the homotopy relation $f \simeq f'$ is (well-known to be) a congruence of categories; the quotient category $\mathsf{hoTop} = \mathsf{Top}/\simeq$ is called the *homotopy category of topological spaces*, and is important in Algebraic Topology. Plainly, a continuous mapping $f \colon X \to Y$ is a homotopy equivalence if and only if its homotopy class $[f]$ is an isomorphism of the category hoTop.

The relation of isomorphism is wider in a quotient category (but it may coincide with the original one, also in a non-trivial quotient).

(c) If C and D are categories, one defines the *product category* $\mathsf{C} \times \mathsf{D}$. An object is a pair (X, Y) where X is in C and Y in D; a morphism is a pair of morphisms

$$(f, g) \colon (X, Y) \to (X', Y'), \qquad (f \in \mathsf{C}(X, X'), \ g \in \mathsf{D}(Y, Y')). \tag{1.37}$$

The composition of (f, g) with a consecutive morphism

$$(f', g') \colon (X', Y') \to (X'', Y'')$$

is (obviously) defined component-wise: $(f', g').(f, g) = (f'f, g'g)$.

More generally one defines the *cartesian product* $\mathsf{C} = \Pi_{i \in I} \, \mathsf{C}_i$ of a family of categories $(\mathsf{C}_i)_{i \in I}$ indexed by a set I: an object of C is a family $(A_i)_{i \in I}$ where $A_i \in \mathrm{Ob}\,(\mathsf{C}_i)$ (for every index i), and a morphism $f = (f_i) \colon (A_i) \to (B_i)$ is a family of morphisms $f_i \in \mathsf{C}_i(A_i, B_i)$; the composition is component-wise and $\mathrm{id}\,((A_i)_{i \in I}) = (\mathrm{id}\, A_i)_{i \in I}$.

1.4.7 Monomorphisms and epimorphisms

In a category, monomorphisms and epimorphisms, are defined by cancellation properties with respect to composition. In a category of structured sets, they represent an 'approximation' to the injective and surjective mappings of the category, and may coincide or not with the latter.

In a category C, a morphism $f: X \to Y$ is said to be a *monomorphism*, or *mono*, if it satisfies the following cancellation property: for every pair of maps $u, v: X' \to X$ (defined on an arbitrary object X') the relation $fu = fv$ implies $u = v$ (see the left diagram below)

$$X' \underset{v}{\overset{u}{\rightrightarrows}} X \overset{f}{\longrightarrow} Y \qquad\qquad X \overset{f}{\longrightarrow} Y \underset{v}{\overset{u}{\rightrightarrows}} Y'. \qquad (1.38)$$

The morphism $f: X \to Y$ is said to be an *epimorphism*, or *epi*, if it satisfies the dual cancellation property: for every pair of maps $u, v: Y \to Y'$ such that $uf = vf$, one has $u = v$ (see the right diagram above).

The arrows $\rightarrowtail$ and $\twoheadrightarrow$ will be used for monos and epis, respectively.

In a category of structured sets and structure-preserving mappings, an injective mapping (of the category) is obviously a monomorphism, while a surjective one is an epimorphism.

In Set, Top, pOrd, Ord and Ab, it is not difficult to prove that a monomorphism is the same as an injective mapping (of the category, of course), while an epimorphism is the same as a surjective mapping. The same is true in Gp, but here the proof for epimorphisms is not easy. All this will be examined in Section 5.1, with the associated notions of subobject and quotient.

Every isomorphism in C is mono and epi. A category is said to be *balanced* if the converse holds: every morphism which is mono and epi is invertible. This holds in Set and Ab, but not in Top where a bijective continuous mapping need not be a homeomorphism, nor in pOrd and Ord, where a bijective increasing mapping need not reflect the preorder relation.

Suppose now that we have, in a category C, two maps $m: A \to X$ and $p: X \to A$ such that $pm = \operatorname{id} A$. Then m is a monomorphism (called a *section*, or a *split monomorphism*), p is an epimorphism (called a *retraction*, or a *split epimorphism*) and one says that A is a *retract* of X. (This terminology comes in part from Topology, and in part from Algebra.)

Exercises and complements. The following exercises are left to the reader. Properties (a)–(c*) are important, and easy to prove; they deal with consecutive maps $f: X \to Y$ and $g: Y \to Z$ in an arbitrary category. Note that (a*) is dual to (a), and so on.

(a) If f and g are mono, gf is also mono; if gf is mono, f is also.

(a*) If f and g are epi, gf is also epi; if gf is epi, g is also.

(b) If f and g are split mono, gf is also; if gf is a split mono, f is also.

(b*) If f and g are split epi, gf is also; if gf is a split epi, g is also.

(c) If f is split mono and epi, then it is invertible.

(c*) If f is split epi and mono, then it is invertible.

(d) In Set a retract of a set $X \neq \emptyset$ is any non-empty subset.

(e) In $R\,$Mod, a submodule A of the module X is a retract if and only if it is a *direct summand*: there exists a submodule $B \subset X$ such that $A \cap B = 0$ and $A + B = X$.

(f) In Top there is no elementary characterisation of retracts (see 1.5.5(d), 2.6.3(d)).

1.4.8 Epi-mono factorisations

As an obvious, well-known fact, every mapping $f \colon X \to Y$ in Set has an *epi-mono factorisation* $f = mp$

$$X \xrightarrow{\;p\;} f(X) \xrightarrowtail{\;m\;} Y$$

through its image $f(X)$, letting

$$p(x) = f(x), \quad m(y) = y \qquad (x \in X,\, y \in f(X)).$$

This factorisation of f is *essentially unique*: given two epi-mono factorisations $f = mp = nq$, there is a unique mapping u that makes the diagram below commute

$$\begin{array}{ccccc} X & \xrightarrow{\;p\;} & A & \xrightarrowtail{\;m\;} & Y \\ \| & & \downarrow{\scriptstyle u} & & \| \\ X & \xrightarrow{\;q\;} & B & \xrightarrowtail{\;n\;} & Y \end{array} \qquad (1.39)$$

and it is a bijection (see Exercise (a)).

Exercises and complements. (a) Prove the existence and uniqueness of the mapping u in diagram (1.39), and its invertibility.

(b) The categories Ab, $R\,$Mod and Set$_{\bullet}$ have essentially unique epi-mono factorisations, determined up to a unique coherent isomorphism.

(c) A category C with essentially unique epi-mono factorisations is balanced.

(d) The categories Top and pOrd have epi-mono factorisations which are not essentially unique. For them, one can prefer to use a ternary factorisation, as in 1.4.9.

1.4.9 Subspaces and topological quotients

A continuous mapping $f\colon X \to Y$ of topological spaces can be uniquely factorised in three maps, as $f = mgp$, where

$$X \xrightarrow{\ f\ } Y$$
$$p\downarrow \qquad \uparrow m \qquad\qquad (1.40)$$
$$X/R_f \xrightarrow{\ g\ } f(X)$$

- p is the projection on the quotient of X modulo the equivalence relation associated to f,

- m is the inclusion of the image $f(X)$, a subspace of Y,

- $g\colon X/R_f \to f(X)$ is a continuous bijection taking the equivalence class $[x]$ to $f(x)$.

It will be useful to recall a few elementary points, about subspaces and topological quotients. (All this can be easily transferred to pOrd.)

(a) (*Subspaces and topological embeddings*) A subset A of the topological space X inherits the *subspace topology*, the coarsest that makes the inclusion $m\colon A \to X$ continuous. Concretely, the open subsets of A are the traces $U \cap A$ of the open sets of X, and the closed subsets of A are the traces of the closed sets of X.

A mapping $f\colon Z \to A$ defined on a topological space is continuous if and only if the composite $mf\colon Z \to X$ is. (This can be called 'the universal property of a subspace'.)

Generalising subspaces, up to homeomorphism, a *topological embedding* $m\colon A \to X$ is a map whose codomain-restriction $A \to m(A)$ is a homeomorphism. Equivalently, m is an injective mapping with values in a topological space, and A has the coarsest topology that makes m continuous: its open subsets are the preimages $m^{-1}(U)$ of the open sets U of X.

An open (resp. closed) injective continuous mapping is the same as a topological embedding with open (resp. closed) image.

   (Topological embeddings are the same as regular monomorphisms in Top, see 5.2.6(c).)

(b) (*Quotients and topological projections*) We recall that, in set theory, a subset $S \subset X$ is *saturated* for an equivalence relation $R \subset X \times X$ (or for the quotient projection $p\colon X \to X/R$) if

$$S = p^{-1}(p(S)).$$

Equivalently, S is a union of equivalence classes of R.

Now, the quotient X/R of a topological space X modulo an equivalence relation is the set-theoretical quotient equipped with the finest topology that makes the canonical projection $p\colon X \to X/R$ continuous.

This means that a subset V of X/R is open if and only if its preimage $p^{-1}(V)$ is open in X, if and only if V is the p-image of an open set of X, saturated for the relation R.

It is easy to see that a mapping $f\colon X/R \to Y$ with values in a topological space is continuous if and only if the composite $fp\colon X \to Y$ is. (This can be called 'the universal property of a quotient space'.)

We also note that the projection $p\colon X \to X/R$ is an open mapping if and only if the saturated $p^{-1}(p(U))$ of every open set U of X is open.

Again, topological quotients can be generalised up to homeomorphism: a *topological projection* $p\colon X \to X'$ is a surjective mapping such that X' has the finest topology that makes p continuous: a subset $V \subset X'$ is open if and only if its preimage $p^{-1}(V)$ is open in X. Equivalently, the mapping $X/R_p \to X'$ induced by p is a homeomorphism.

An open (or closed) surjective continuous mapping is always a topological projection.

(Topological projections are the same as regular epimorphisms in Top, see 5.2.6(c).)

*(c) (*The ternary factorisation*) Coming back to factorisation (1.40), we can say that in Top every morphism has a ternary factorisation $f = mgp$, where:

- p is the least regular epimorphism defined on X, through which f can be factorised,
- m is the least regular monomorphism with values in Y, through which f can be factorised.

This factorisation is determined up to a unique pair of coherent isomorphisms.

1.5 Functors and concreteness

Well-behaved mappings $F\colon \mathsf{C} \to \mathsf{D}$ between categories are called 'functors'.

Concrete categories (in 1.5.6) and universal properties (formalised in 1.5.7) will play an important role in the sequel.

Algebraic Topology is based on functors 'from topology to algebra'. Some results, related to topological and differentiable manifolds, will be recalled in 2.6.3.

1.5.1 Functors

A (covariant) *functor* $F: C \to D$ consists of the following data:

(a) a mapping $F_0: \mathrm{Ob}\,C \to \mathrm{Ob}\,D$, whose action is generally written as $X \mapsto F(X)$,

(b) for every pair of objects X, X' in C, a mapping

$$F_{XX'}: C(X, X') \to D(F(X), F(X')),$$

whose action is generally written as $f \mapsto F(f)$, so that composition and identities are preserved. In other words:

(i) if f, g are consecutive maps in C, then $F(gf) = F(g).F(f)$,

(ii) if X is in C, then $F(\mathrm{id}\,X) = \mathrm{id}\,(F(X))$.

For a second functor $G: D \to E$, one defines in the obvious way the *composed functor* $GF: C \to E$. The composition of consecutive functors is associative and has identities: the *identity functor* of each category

$$\mathrm{id}\,C: C \to C, \qquad X \mapsto X, \quad f \mapsto f.$$

A *contravariant functor* $F: C \dashrightarrow D$ can be defined as a covariant functor $C^{\mathrm{op}} \to D$.

A *functor in two variables* is an ordinary functor $F: C \times D \to E$ defined on the product of two categories. Fixing an object X_0 in C, we have a functor $F(X_0, -): D \to E$; and symmetrically.

A functor between two preordered sets, viewed as categories, is a monotone mapping. A functor between two monoids, viewed as categories, is a homomorphism of monoids.

Cat (resp. Gpd) will denote the category of small categories (resp. small groupoids) and their functors. (These categories also have a 2-dimensional structure, see Section 5.6.)

1.5.2 Isomorphic categories

An *isomorphism of categories* is a functor $F: C \to D$ which is invertible: its *inverse* is a functor $G: D \to C$ such that $GF = \mathrm{id}\,C$ and $FG = \mathrm{id}\,D$. An easy verification proves that the functor F is an isomorphism if and only if all the mappings F_0 and $F_{XX'}$ considered above are bijective.

Being isomorphic categories is an equivalence relation, written as $C \cong D$.

Categories linked by an obvious isomorphism are often perceived as 'the same thing'. For instance, everybody knows that an abelian group has a unique structure of (unitary) module on the ring $\mathbb{Z}$ of integers (where $2x = x + x$, and so on), preserved by every homomorphism of groups.

This fact readily shows that the category Ab is canonically isomorphic to the category $\mathbb{Z}$ Mod of modules on the ring of integers; one generally makes no distinction between these categories.

A topological space can be defined in many equivalent ways (based on open sets, closed sets, neighbourhoods, a closure operator, etc.); this yields many isomorphic categories which is rarely convenient to distinguish.

Yet, if $S = (X, L)$ is a set X equipped with a complete sublattice L of $\mathcal{P}X$, interpreting S as an object of the 'standard' category Top, where topologies are defined by open subsets, is (generally) different from viewing it in the isomorphic copy Top' where topologies are defined by closed subsets.

1.5.3 *Forgetful and structural functors*

(a) Forgetting the structure, or part of it, yields various examples of functors between categories of structured sets, like the following obvious instances

$$\text{Top} \to \text{Set}, \qquad\qquad R\,\text{Mod} \to \text{Ab} \to \text{Set}. \qquad (1.41)$$

The first, for instance, takes a topological space X to its underlying set $|X|$, and a continuous mapping $X \to Y$ to the underlying mapping $|X| \to |Y|$ of sets.

These *forgetful functors* are often denoted by the letter U, which refers to the *underlying* set, or *underlying* abelian group, and so on.

(b) A subcategory D of C yields an *inclusion* functor D $\to$ C, which we also write as D $\subset$ C. For instance, Ord $\subset$ pOrd and Ab $\subset$ Gp. These functors *forget properties*, of objects or morphisms, rather than structure.

(c) A congruence R in a category C yields an obvious *projection functor* $P\colon$ C $\to$ C$/R$, which is the identity on objects and sends a morphism f to its equivalence class $[f]$. For instance, the projection functor

$$\text{Top} \to \text{hoTop} = \text{Top}/\simeq$$

(see 1.4.6) is important in algebraic topology.

(d) A product category C $\times$ D has two obvious *projection functors*

$$P_1\colon \text{C} \times \text{D} \to \text{C}, \qquad P_2\colon \text{C} \times \text{D} \to \text{D}.$$

(e) Every category C has a functor of morphisms, or *hom-functor*:

$$\text{Mor}_\text{C}\colon \text{C}^\text{op} \times \text{C} \to \text{Set},$$
$$(X, Y) \mapsto \text{C}(X, Y), \qquad (h, k) \mapsto k. - .h, \qquad (1.42)$$

where, if $h: X' \to X$ and $k: Y \to Y'$ are in C, the mapping $k. - .h$

$$k. - .h: \mathsf{C}(X, Y) \to \mathsf{C}(X', Y'), \qquad f \mapsto kfh,$$

means pre-composing with h and post-composing with k.

The hom-functor (1.42) is viewed as a functor on two variables of C, 'contravariant on the first and covariant on the second'. Note that it takes values in Set because of our assumption, in 1.4.2(b), that all $\mathsf{C}(X, Y)$ be sets.

Fixing the first variable we get a covariant functor on C

$$\mathsf{C}(X_0, -): \mathsf{C} \to \mathsf{Set}, \tag{1.43}$$

that is said to be *representable*, and *represented by the object* X_0.

Dually, fixing the second variable we get a functor $\mathsf{C}(-.Y_0): \mathsf{C}^{\mathrm{op}} \to \mathsf{Set}$ that is contravariant on C, and *representable* as such. Contravariant functors are often constructed in this way (at least up to isomorphism of functors, a notion defined in 1.6.2).

1.5.4 Faithful and full functors

For a functor $F: \mathsf{C} \to \mathsf{D}$, let us consider again the mappings (of sets):

$$F_{XX'}: \mathsf{C}(X, X') \to \mathsf{D}(F(X), F(X')), \qquad f \mapsto F(f). \tag{1.44}$$

The functor F is said to be *faithful* if all these mappings are injective (for X, X' in C); F is said to be *full* if all of them are surjective. *One should not confuse these properties with the fact of being injective or surjective on all morphisms*, which is by far less important.

An isomorphism of categories is always full and faithful. The inclusion functor $\mathsf{D} \to \mathsf{C}$ of a subcategory is always faithful; it is full if and only if D is a full subcategory of C.

The forgetful functors described above, in (1.41), are faithful (and not injective on all maps, of course); but note that the obvious functor of objects Ob: Cat $\to$ Set, which can also be 'thought of' as a forgetful functor, is not faithful.

1.5.5 Exercises and complements (Preservation and reflection)

There are (obvious) properties of functors, related to preservation or reflection. The verification is left to the reader.

(a) Every functor preserves commutative diagrams, isomorphisms, retracts, split monos and split epis.

(b) A faithful functor $F: \mathsf{C} \to \mathsf{D}$ *reflects* monos, epis and commutative diagrams. The first statement means: if $f: X \to X'$ belongs to C and $F(f)$ is mono (or epi) in D, so is f in C.

(c) A full and faithful functor reflects isomorphisms, split monos and split epis.

(d) Proving that two topological spaces X, Y are not homeomorphic can be a difficult task, even when this is 'intuitively clear'. Applying point (a), it is sufficient to have at our disposition a functor $F: \mathsf{Top} \to \mathsf{Ab}$ or $F: \mathsf{Top} \to \mathsf{Gp}$ such that $F(X)$ and $F(Y)$ are not isomorphic groups.

Similarly, to prove that a topological subspace $A \subset X$ is not a retract in Top, it is sufficient to find a functor $F: \mathsf{Top} \to \mathsf{Ab}$ such that the homomorphism $F(A) \to F(X)$ associated to the inclusion is not a split mono in Ab.

Constructing 'homology functors' and 'homotopy functors' that can be used in this ways is the task of Algebraic Topology: see 2.6.3.

1.5.6 Concrete categories and lifted structures

The notion of a 'category of structured sets and structure-preserving maps' is quite useful but can be formalised in different ways, more and more complicated if we want to capture diverse structures from Algebra, Topology, Analysis, Geometry, etc.

As a clear alternative (independent of how a structure is constructed), a *concrete category* is defined as a category C *equipped* with a faithful functor $U: \mathsf{C} \to \mathsf{Set}$, called its *forgetful functor*. Then U reflects monos and epis, but need not preserve them. Concrete categories are extensively studied in [AHS]. Note, however, that non-faithful forgetful functors can also be important, see Section 5.4.

*The homotopy category hoTop *cannot* be made concrete: there exists no faithful functor from hoTop to Set, as proved by P. Freyd [Fr2].*

More generally, a category C *concrete over a category* S, or S-*concrete*, is equipped with a faithful functor $U: \mathsf{C} \to \mathsf{S}$.

The reader will note that the usual categories of structured sets are essentially constructed as Set-concrete categories. For instance, we introduce Top by defining topological spaces, with their underlying set $U(-)$, and we fix the subsets $\mathsf{Top}(X, Y) \subset \mathsf{Set}(UX, UY)$ of continuous mappings, verifying that these sets are stable under composition and identities: the ground-category Set 'takes care' of associativity and unitarity.

In general, an S-concrete category can be built extending the construction of a subcategory (in 1.4.6):

(a) we fix a class $\mathrm{Ob}\,\mathsf{C}$ of objects, and a mapping $U\colon \mathrm{Ob}\,\mathsf{C} \to \mathrm{Ob}\,\mathsf{S}$,

(b) for every pair of objects X, Y of C, we fix a subset

$$\mathsf{C}(X,Y) \subset \mathsf{S}(UX, UY),$$

whose elements are called *morphisms of* C, from X to Y, so that the following conditions are met:

(i) if $f\colon X \to Y$ and $g\colon Y \to Y$ are consecutive in C, their composite $gf\colon UX \to UZ$ (in S) belongs to $\mathsf{C}(X, Z)$,

(ii) for $X \in \mathrm{Ob}\,\mathsf{C}$, the identity of X in S belongs to $\mathsf{C}(X, X)$.

We can thus define composition and identities in C, using the composition and identities of S. C is a category and the (faithful) functor $U\colon \mathsf{C} \to \mathsf{S}$ is defined on the morphisms letting

$$U(f\colon X \to Y) = f\colon UX \to UY. \tag{1.45}$$

We say that this procedure *lifts to* C *the categorical structure of* S. We shall also lift to C additional structures of S, like order (in Chapter 2), or 'cohesion' (in Chapter 3).

A careful reader will note that this lifting takes no care of the Separation axiom 1.4.2(iii), even when it holds in S. As always, one *can* fix this point by redefining a morphism $X \to Y$ of C as a triple $(X, Y; f)$, where $f\colon UX \to UY$ belongs to the distinguished subset of $\mathsf{S}(UX, UY)$. In practice, we leave all this as understood.

1.5.7 Universal arrows

There is a general way of formalising universal properties, based on a functor $U\colon \mathsf{A} \to \mathsf{C}$ and an object X of C. This will be particularly interesting for a reader with a taste for abstraction and a drive towards the 'root of things'. On the other hand, a reader more oriented to concrete aspects may prefer to view the many universal properties appearing in this book as individual things, whose familiarity need not be inserted in a formal frame.

A *universal arrow from the object* X *to the functor* U is a pair

$$(A, \eta\colon X \to UA)$$

consisting of an object A of A and an arrow η of C which is universal, in the sense that every similar pair $(B, f\colon X \to UB)$ factorises uniquely through (A, η): namely, there is a unique map $g \colon A \to B$ in A such that the following triangle commutes in C

$$X \xrightarrow{\ \eta\ } UA$$
$$f \searrow \quad \downarrow Ug \qquad Ug.\eta = f. \qquad (1.46)$$
$$UB$$

Dually, a *universal arrow from the functor U to the object X* is a pair

$$(A, \varepsilon \colon UA \to X)$$

consisting of an object A of A and an arrow ε of C such that every similar pair $(B, f \colon UB \to X)$ factorises uniquely through (A, ε): there is a unique $g \colon B \to A$ in A such that the following triangle commutes in C

$$UA \xrightarrow{\ \varepsilon\ } X$$
$$Ug \uparrow \quad \nearrow f \qquad \varepsilon.Ug = f. \qquad (1.47)$$
$$UB$$

Universal arrows for 2-dimensional categories are considered in 5.6.3.

Exercises and complements. (a) Universal arrows are determined, up to a unique coherent isomorphism.

(b) Construct the universal arrow from a set X to the forgetful functor $U \colon R\,\mathsf{Mod} \to \mathsf{Set}$. *Hints:* use the free R-module generated by the set; examine what happens with the trivial ring R.

(c) Construct the universal arrow from a group G to the inclusion functor $\mathsf{Ab} \to \mathsf{Gp}$. *Hints:* use the subgroup of commutators of G.

1.6 Natural transformations and equivalences

Given two *parallel* functors $F, G \colon \mathsf{C} \to \mathsf{D}$ (between the same categories) there can be 'second-order arrows' $\varphi \colon F \to G$, called 'natural transformations', and natural isomorphisms when they are invertible. The latter give rise to the crucial notion of equivalence of categories (in 1.6.3).

'Category', 'functor' and 'natural transformation' are the three basic items of category theory, since the very beginning of the theory in [EiM].

It is interesting to note that only the last name is taken from the common language: Eilenberg and Mac Lane introduced categories and functors because they wanted to formalise the natural transformations that they were encountering in algebra and algebraic topology (as remarked in [M3], at the end of Section I.6). Much in the same way as a general theory of *continuity* (a familiar name for a familiar notion) is based on *topological spaces* (a theoretical term).

A reader acquainted with basic homotopy theory can take advantage of a formal parallelism, where *spaces* correspond to *categories, continuous mappings* to *functors, homotopies* of mappings to *invertible natural transformations* of functors, and *homotopy equivalence* of spaces to *equivalence* of categories. This analogy is even deeper in the domain of Directed Algebraic Topology, where *directed homotopies* correspond to *natural transformations*, see [G8].

1.6.1 Natural transformations

Given two functors $F, G \colon \mathsf{C} \to \mathsf{D}$ between the same categories, a natural transformation $\varphi \colon F \to G$ (also written as $\varphi \colon F \to G \colon \mathsf{C} \to \mathsf{D}$) consists of the following data:

- for each object X of C, a morphism $\varphi X \colon FX \to GX$ in D (also written as φ_X, or φ), called the *component* of φ on X,

so that, for every arrow $f \colon X \to X'$ in C, we have a commutative square in D (the *naturality condition* of φ on the morphism f)

$$
\begin{array}{ccc}
FX & \xrightarrow{\varphi X} & GX \\
{\scriptstyle Ff}\big\downarrow & & \big\downarrow{\scriptstyle Gf} \\
FX' & \xrightarrow[\varphi X']{} & GX'
\end{array}
\qquad \varphi X'.F(f) = G(f).\varphi X \ \ (= \varphi(f)), \qquad (1.48)
$$

whose diagonal will be written as $\varphi(f)$, when useful.

For instance, the definition of homotopic maps is based on the two embeddings of a space X as the lower or upper basis of its standard cylinder $I(X) = X \times [0, 1]$

$$
\partial^\alpha \colon X \to X \times [0, 1], \qquad x \mapsto (x, \alpha) \quad (\alpha = 0, 1), \qquad (1.49)
$$

which are the components of two natural transformations $\partial^\alpha \colon \mathrm{id} \to I \colon \mathsf{Top} \to \mathsf{Top}$. (The standard interval $[0, 1]$ has the euclidean topology, by default.)

Backwards, we have an obvious natural transformation $e \colon I \to \mathrm{id}$, with $e(x, t) = x$.

The identity of a functor $F \colon \mathsf{C} \to \mathsf{D}$ is the natural transformation $\mathrm{id}\, F \colon F \to F$, with components $(\mathrm{id}\, F)X = \mathrm{id}\,(FX)$.

Exercises and complements. (a) One can easily prove that every scalar λ of the *centre* of a ring R (this means that λ commutes with any scalar) determines a natural endo-transformation

$$\hat{\lambda} \colon \mathrm{id} \to \mathrm{id} \colon R\,\mathsf{Mod} \to R\,\mathsf{Mod} \tag{1.50}$$

of the identity functor of left R-modules.

(b) Less easily, if more interestingly, one can prove that the correspondence $\lambda \mapsto \hat{\lambda}$ between these scalars and these transformations is bijective. *Hints:* use the component of a transformation on the object R, as a module on itself.

*(c) As a consequence, *the category $R\,\mathsf{Mod}$ determines the centre C of the ring R in a structural way* (up to isomorphism of categories and rings). The ring itself is determined, if it is commutative.

1.6.2 Two operations

Two natural transformations $\varphi \colon F \to G$ and $\psi \colon G \to H$ between the same categories have a *vertical composition* $\psi\varphi \colon F \to H$

$$\mathsf{C} \xrightarrow[\begin{subarray}{c} \downarrow\psi \end{subarray}]{\begin{subarray}{c} F \\ \downarrow\varphi \\ \\ H \end{subarray}} \mathsf{D} \qquad\qquad (\psi\varphi)(X) = \psi X . \varphi X \colon FX \to HX. \tag{1.51}$$

which is associative; the identities of functors act as units.

There is also a *whisker composition* of natural transformations with functors, or *reduced horizontal composition*, in the following situation

$$\mathsf{C}' \xrightarrow{\;H\;} \mathsf{C} \xrightarrow[\begin{subarray}{c} \downarrow\varphi \\ G \end{subarray}]{\begin{subarray}{c} F \end{subarray}} \mathsf{D} \xrightarrow{\;K\;} \mathsf{D}'$$

$$\begin{aligned} K\varphi H &\colon KFH \to KGH \colon \mathsf{C}' \to \mathsf{D}', \\ (K\varphi H)(X) &= K(\varphi(HX)) \qquad (X \text{ in } \mathsf{C}'). \end{aligned} \tag{1.52}$$

This '2-dimensional structure' of categories, where natural transformations play the role of 2-dimensional arrows between functors, will be analysed in Section 5.6. In particular, the properties of the previous operations can be found in 5.6.1 and 5.6.2.

An *isomorphism of functors*, or *functorial isomorphism*, or *natural isomorphism*, is a natural transformation $\varphi \colon F \to G$ which is invertible for the vertical composition (see Exercise (a)). Then we write $F \cong G$; it is obviously an equivalence relation between parallel functors.

(The original term of 'natural equivalence', in [EiM], is now abandoned, as it can be confused with an equivalence of categories.)

Exercises and complements. (a) A natural transformation $\varphi\colon F \to G\colon$ C → D is invertible if and only if all the components φX are invertible in the category D.

(b) For the natural transformations $\hat{\lambda}$ of 1.6.1(a), the vertical composition corresponds to the product of scalars; $\hat{\lambda}$ is invertible if and only if the scalar λ is invertible.

1.6.3 Equivalence of categories

Isomorphism of categories was defined in 1.5.2. The relation of equivalence of categories is more general, and more important.

An *equivalence of categories* is a functor $F\colon$ C → D which is invertible *up to isomorphism of functors* (in 1.6.2), i.e. there exists a functor $G\colon$ D → C such that $GF \cong$ id C and $FG \cong$ id D. The latter can be said to be *pseudo inverse* to F; it is obviously determined up to functorial isomorphism.

An *adjoint equivalence of categories* is a 'coherent version' of this notion, namely a four-tuple $(F, G, \eta, \varepsilon)$ where:

- $F\colon$ C → D and $G\colon$ D → C are functors,

- $\eta\colon$ id C → GF and $\varepsilon\colon FG \to$ id D are isomorphisms of functors,

- $F\eta = (\varepsilon F)^{-1}\colon F \to FGF, \qquad \eta G = (G\varepsilon)^{-1}\colon G \to GFG$

(the *coherence conditions*).

The following conditions on a functor $F\colon$ C → D are equivalent, forming a very useful *characterisation of the equivalence of categories*:

(i) F is an equivalence of categories,

(ii) F can be completed to an adjoint equivalence of categories $(F, G, \eta, \varepsilon)$,

(iii) F is faithful, full and *essentially surjective on objects*.

The last property means that: for every object Y of D there exists some object X in C such that $F(X)$ is isomorphic to Y in D. The proof of the equivalence of these three conditions is rather complex and requires the axiom of choice: see [M3], Section V.4, or [G10], Theorem 1.5.8. Hints at the main part of the proof can be found below, in 1.6.5(a).

We say that two categories C, D are *equivalent*, written as C $\simeq$ D (not to be confused with the notation C $\cong$ D, for isomorphic categories) if there is an equivalence of categories C → D, if and only if there is an *adjoint* equivalence of categories (because of the previous result).

This is indeed an equivalence relation, as follows easily from condition (iii) above (or can be proved directly, without resorting to the axiom of choice).

1.6.4 Examples and complements

The following examples of equivalence of categories are a straightforward application of the previous characterisation. Other cases related to our analysis of manifolds, will be presented in 2.1.5.

(a) The category of finite sets (and mappings between them) is equivalent to its full subcategory of finite cardinals, which is small (and therefore cannot be isomorphic to the former).

(b) A category is said to be *skeletal* if it has no pair of distinct isomorphic objects. Every category C has a *skeleton*, i.e. an equivalent skeletal category; it can be obtained by *choosing* one object in every class of isomorphic objects, and taking their full subcategory. A category is *essentially small* if its skeleton is small.

We have described in (a) a skeleton of the category of finite sets that can be formed without any choice, even though we do need the axiom of choice to prove that the inclusion of this skeleton has a pseudo inverse.

(c) In a different way, a preordered set X has a natural skeleton *formed by a quotient*, the associated ordered set $X/\sim$ modulo the equivalence relation associated to the preorder. Now the axiom of choice proves that the projection $X \to X/\sim$ has a pseudo inverse.

*(d) Two rings are said to be *Morita equivalent* [Mo] if their categories of left modules are equivalent. This is an important notion in ring theory, that becomes trivial in the domain of commutative rings: if two rings are Morita equivalent one can prove that their centres are isomorphic (using Exercise 1.6.1(c)); but commutative rings can be Morita equivalent to non-commutative ones, like their rings of square matrices.

Thus, studying left modules on any matrix ring $M_n(\mathbb{R})$ is equivalent to studying real vector spaces.

*(e) Two categories are equivalent if and only if they have isomorphic skeletons. Loosely speaking, this says that an equivalence of categories amounts to multiplying or deleting isomorphic copies of objects, even though there may be no canonical way of doing this.

(For an interested reader, the proof is easy: one only has to prove that two equivalent skeletons are isomorphic.)

1.6.5 Equivalence of categories, continued

(a) It may be useful to give some hints at the proof of (iii) $\Rightarrow$ (ii), in 1.6.3. (Note that (ii) $\Rightarrow$ (i) is obvious, while (i) $\Rightarrow$ (iii) is an easy exercise.)

Assuming that the functor $F\colon \mathsf{C} \to \mathsf{D}$ is faithful, full and essentially surjective on objects, we can 'construct' with the axiom of choice a pseudo inverse G. For every object Y of D we choose an object $G(Y)$ in C *and* an isomorphism $\varepsilon Y\colon F(G(Y)) \to Y$ in D. The rest of the procedure is determined by this choice.

For every morphism $g\colon Y \to Y'$ in D, we define $G(g)\colon GY \to GY'$ as the unique morphism of C such that

$$F(G(g)) = (\varepsilon Y')^{-1}.g.\varepsilon Y\colon F(GY) \to F(GY'),$$

in order to make the following square commute in D

$$\begin{array}{ccc} FGY & \xrightarrow{\ \varepsilon Y\ } & Y \\ {\scriptstyle FGg}\downarrow & & \downarrow{\scriptstyle g} \\ FGY' & \xrightarrow[\ \varepsilon Y'\ ]{} & Y' \end{array} \qquad g.\varepsilon Y = \varepsilon Y'.FG(g). \tag{1.53}$$

It is easy to verify that G is a functor. The family $\varepsilon = (\varepsilon Y)_Y$ of isomorphisms is a natural transformation $\varepsilon\colon FG \to \mathrm{id}\,\mathsf{D}$, by (1.53), and therefore a functorial isomorphism.

For every X in C there is precisely one morphism $\eta X\colon X \to GFX$ such that $F(\eta X) = (\varepsilon FX)^{-1}\colon FX \to FG(FX)$; moreover ηX is invertible, by 1.5.5(c).

Using again the faithfulness of F, one shows that the family $\eta = (\eta X)_X$ is a natural transformation $\mathrm{id}\,\mathsf{C} \to GF$, and therefore a functorial isomorphism. We already have one coherence condition: $F\eta = (\varepsilon F)^{-1}$; verifying the other is more delicate — an interesting challenge for a motivated reader.

(b) Suppose now that C is a full subcategory of D, and $J\colon \mathsf{C} \to \mathsf{D}$ its (full) embedding. Then J is an equivalence of categories if and only if every object of D is isomorphic (in D) to some object of C.

If this is the case, we can always choose a pseudo inverse $G\colon \mathsf{D} \to \mathsf{C}$ such that

$$GJ = \mathrm{id}\,\mathsf{C} \qquad \eta = 1_{\mathrm{id}\,\mathsf{C}}\colon \mathrm{id}\,\mathsf{C} \to GJ \qquad (JG \cong \mathrm{id}\,\mathsf{D}). \tag{1.54}$$

In fact, in the argument above, we can adopt this constraint on the choice of G and ε:

- if the object Y of D belongs to the subcategory C, we take $G(Y) = Y$ *and* $\varepsilon Y = \mathrm{id}\,Y$.

Then, for every X in C, $G(JX) = X$ and $\eta X\colon X \to G(JX)$ is the identity, because

$$J(\eta X) = (\varepsilon JX)^{-1} = (\mathrm{id}\, JX)^{-1} = \mathrm{id}\, JX = J(\mathrm{id}\, X).$$

1.6.6 Categories of functors and presheaves

Let S be a small category and $S = \mathrm{Ob}\, \mathsf{S}$ its set of objects. For any category C, one writes as C^{S} the category whose objects are the functors $F\colon \mathsf{S} \to \mathsf{C}$ and whose morphisms are the natural transformations $\varphi\colon F \to G\colon \mathsf{S} \to \mathsf{C}$, with vertical composition (see 1.6.2).

In particular, the arrow category **2** (with two objects and one non-identity arrow $0 \to 1$, see 1.4.3) gives the *category of morphisms* $\mathsf{C}^{\mathbf{2}}$ of C, where a map $f = (f', f'')\colon x \to y$ is a commutative square of C, as in the left diagram below

$$
\begin{array}{ccc}
X' & \xrightarrow{\;f'\;} & Y' \\
\scriptstyle x\big\downarrow & & \big\downarrow\scriptstyle y \\
X'' & \xrightarrow[\;g'\;]{} & Y''
\end{array}
\qquad\qquad
\begin{array}{ccccc}
X' & \xrightarrow{\;f'\;} & Y' & \xrightarrow{\;f''\;} & Z' \\
\scriptstyle x\big\downarrow & & \big\downarrow\scriptstyle y & & \big\downarrow\scriptstyle z \\
X'' & \xrightarrow[\;g'\;]{} & Y'' & \xrightarrow[\;g''\;]{} & Z''
\end{array}
\qquad (1.55)
$$

These maps are composed *by pasting* commutative squares, as in the right diagram above. The identity of x is: $\mathrm{id}\,(x) = (\mathrm{id}\, X', \mathrm{id}\, X'')\colon x \to x$.

A functor $\mathsf{S}^{\mathrm{op}} \to \mathsf{C}$, defined on the opposite category S^{op}, is also called a *presheaf* of C on the (small) category S. They form the presheaf category $\mathrm{Psh}(\mathsf{S}, \mathsf{C}) = \mathsf{C}^{\mathsf{S}^{\mathrm{op}}}$.

S is canonically embedded in the latter, by the *Yoneda embedding*

$$y\colon \mathsf{S} \to \mathsf{C}^{\mathsf{S}^{\mathrm{op}}}, \qquad y(i) = \mathsf{S}(-, i)\colon \mathsf{S}^{\mathrm{op}} \to \mathsf{Set}, \qquad (1.56)$$

which sends every object i to the corresponding *representable* presheaf $y(i)$ (see 1.5.3).

Exercises and complements. (a) $\mathsf{C}^{\mathbf{0}}$ is always the singleton category **1** (also for $\mathsf{C} = \mathbf{0}$). $\mathsf{C}^{\mathbf{1}}$ is isomorphic to C.

(b) A natural transformation $\varphi\colon F \to G\colon \mathsf{C} \to \mathsf{D}$ can be viewed as a functor $\mathsf{C} \times \mathbf{2} \to \mathsf{D}$ or equivalently as a functor $\mathsf{C} \to \mathsf{D}^{\mathbf{2}}$.

*(c) Taking as S the category Δ of finite positive ordinals (and monotone maps), one gets the category $\mathrm{Smp}\mathsf{C} = \mathsf{C}^{\Delta^{\mathrm{op}}}$ of simplicial objects in C, and – in particular – the well-known category of simplicial sets $\mathrm{Smp} = \mathrm{SmpSet}$ (see [May]). Here, the Yoneda embedding sends the ordinal n to the simplicial set Δ^n, freely generated by one simplex of dimension n.

1.7 Basic limits and colimits

We shall often use cartesian products, disjoint unions and quotient objects in categories of sets, or ordered sets, or topological spaces. Each of these constructions can be formalised in a general category C, by a definition based on a universal property.

Here we define *products* and *equalisers* in a category C; then we deal with the dual notions, *sums* and *coequalisers*. We end this section with a brief introduction to biproducts and additive categories.

Various examples are presented, as exercises. A reader unfamiliar with these notions will find it useful to complete the missing details, and work out similar results for other categories of structured sets.

The general definition of categorical limits (including products and equalisers) and colimits (including sums and coequalisers) is deferred to Section 5.2, where we shall see that all of them can be constructed with the basic ones, dealt with here.

1.7.1 Products in a category

In a category C, the *product* (or cartesian product) of a family $(X_i)_{i \in I}$ of objects, indexed by a set I, is defined as an object X equipped with a family of morphisms $p_i \colon X \to X_i$ $(i \in I)$ called (cartesian) *projections*, which satisfy the following universal property:

$$\begin{array}{ccc} X & \xrightarrow{\ p_i\ } & X_i \\ {\scriptstyle f} \big\uparrow & \nearrow & \\ Y & {\scriptstyle f_i} & \end{array} \tag{1.57}$$

(i) for every object Y and every family of morphisms $f_i \colon Y \to X_i$ there is a unique morphism $f \colon Y \to X$ such that, for all $i \in I$, $p_i f = f$.

The map f will often be written as (f_i), by its *components* (although it is not the same as the family of the components). The formal aspect of the universal property is made clear in Exercise 1.7.2(f). Let us note that a cartesian projection need not be an epimorphism, even though it is 'often' the case: see 1.7.2(a).

The product of a family need not exist, but is determined up to a unique *coherent* isomorphism, in the sense that if also Y is a product of the family $(X_i)_{i \in I}$ with projections $q_i \colon Y \to X_i$, then the unique morphism $f \colon X \to Y$ which commutes with all projections (i.e. $q_i f = p_i$, for all indices i) is an isomorphism.

In fact, there is also a unique morphism $g\colon Y \to X$ such that $p_i g = q_i$, for all indices i. Moreover $gf = \operatorname{id} X$ (because $p_i(gf) = q_i f = p_i(\operatorname{id} X)$, for all i) and similarly $fg = \operatorname{id} Y$.

We generally speak of *the* product of the family (X_i), denoted as $\prod_i X_i$. We say that a category C *has products* (resp. *finite products*) if every family of objects indexed by a set (resp. by a finite set) has a product in C.

In particular, the product of the empty family of objects $\emptyset \to \operatorname{Ob} \mathsf{C}$ means an object X (equipped with no projections) such that for every object Y (equipped with no maps) there is a unique morphism $f\colon Y \to X$ (satisfying no conditions). The solution is called the *terminal* object of C; if it exists, it is determined up to a unique isomorphism, and can be written as $\top$.

1.7.2 Exercises and complements (Products)

Verifying the following statements is easy, and left to the reader. Point (g) requires some work, which may be pleasant according to one's interests.

(a) In Set and many categories of structured sets (like Top, Ord and Ab) all products exist and are realised as the usual cartesian ones; the terminal object is the singleton (with the unique, appropriate structure).

In Set, Top and Ord a cartesian projection $p_i\colon X \to X_i$ is 'nearly always' surjective (and epi), but this fails when some factors are empty and others are not. In Ab and Gp all cartesian projections are epi (cf. 1.7.9(a)).

We recall that, in Top, a cartesian product $X = \prod_i X_i$ has the coarsest topology making the projections continuous. A *basic open set* is a product $U = \prod_i U_i$ where each U_i is open in X_i, and all of them are total subsets (i.e. $U_i = X_i$) except for a finite subset of indices $i \in I$. The basic open subsets form a basis of the product topology: every open subset of X is a union of basic ones.

(b) Products in the categories $\mathsf{Set}_\bullet$ and $\mathsf{Top}_\bullet$ are also obvious

$$\Pi_i(X_i, \overline{x}_i) = (\Pi_i X_i, (\overline{x}_i)). \qquad (1.58)$$

(c) In the category X associated to a preordered set (see 1.4.3), the categorical product of a family of points $x_i \in X$ amounts to their greatest lower bound $\inf(x_i)$, and the terminal object amounts to the greatest element of X (when such elements exist). We already noted that they are determined up to the equivalence relation associated to our preorder, and uniquely determined in an ordered set.

(d) Products in Cat have been considered in 1.4.6; the terminal object is the singleton category $\mathbf{1}$.

(e) In 'less usual' categories, *categorical products can look quite different.* The reader might now have a look at what happens in the category $\mathcal{S}$ of sets and partial mappings (in 2.1.5(b)), and in the category Rel Set, of sets and relations (in 2.2.3(c)).

(f) (*Products as universal arrows*) Formally, a family $(X_i)_{i \in I}$ is an object of the cartesian power $\mathsf{C}^I = \prod_{i \in I} \mathsf{C}$. A product $(X, (p_i \colon X \to X_i))$ in the category C is a universal arrow $(X, p \colon \Delta X \to (X_i))$ from the *diagonal functor* $\Delta \colon \mathsf{C} \to \mathsf{C}^I$ to the object $(X_i)_{i \in I}$. Here ΔX is the constant family $(X)_{i \in I}$ at the object X; similarly, $\Delta(f) = (f)_{i \in I}$.

In particular, the power C^0 with empty exponent is the singleton category $\mathbf{1}$, and the terminal object $\top$ of C is a universal arrow $(\top, \Delta\top \to 0)$ from the (unique) functor $\Delta \colon \mathsf{C} \to \mathbf{1}$ to the (unique) object 0 of $\mathbf{1}$.

*(g) A category has finite products if and only if it has binary products $X_1 \times X_2$ and a terminal object.

Hints. Unary products always exist, trivially. A ternary product $X_1 \times X_2 \times X_3$ can be obtained as $(X_1 \times X_2) \times X_3$, with adequate projections. One can be satisfied at this point, or prove the general result by induction.

1.7.3 Equalisers

Another basic limit is the *equaliser* of a pair $f, g \colon X \to Y$ of parallel maps of C. This is an object E with a map $m \colon E \to X$ such that $fm = gm$ and the following universal property holds:

$$
\begin{array}{ccc}
E & \xrightarrow{\ m\ } & X \xrightarrow[g]{f} Y \\
& \underset{w}{\diagdown} & \uparrow h \\
& & Z
\end{array}
\tag{1.59}
$$

(ii) every map $h \colon Z \to X$ such that $fh = gh$ factors uniquely through m (i.e. there is a unique map $w \colon Z \to E$ such that $mw = h$).

The equaliser morphism m is necessarily a monomorphism (as proved in Exercise 1.7.4(f)).

Also here the universal property of the pair (E, m) can be interpreted as in 1.5.7: a universal arrow for the diagonal functor $\Delta \colon \mathsf{C} \to \mathsf{C}'$, with values in a suitable category of diagrams over C. Writing out the details is left to an interested reader, who can later see the general formulation of limits and colimits as universal arrows, in 5.2.4.

1.7.4 Exercises and complements (Equalisers)

(a) In Set the natural solution for the equaliser of a pair of mappings $f, g \colon X \to Y$ is the subset

$$E = \{x \in X \mid f(x) = g(x)\}, \qquad (1.60)$$

with the inclusion $m \colon E \subset X$.

(b) In Top, Ord, Ab, Set. and Top. (and many other concrete categories) we construct the equaliser in the same way, putting on the subset $E \subset X$ the structure induced by X as a subspace, or an ordered subset, or a subgroup, etc.

(c) In Top, the equaliser of two maps $f, g \colon X \to Y$ with values in a Hausdorff space is a closed subspace of X.

As a familiar consequence, two continuous functions $f, g \colon \mathbb{R} \to \mathbb{R}$ which coincide on a dense subset (for instance the rational numbers) coincide everywhere.

(d) In Cat, the equaliser of two functors $F, G \colon \mathsf{C} \to \mathsf{D}$ is the subcategory of the objects and arrows of C on which F and G coincide.

(e) In the category associated to a preordered set X, two parallel maps $x \to x'$ always coincide and their equaliser is the identity 1_x.

(f) In any category, an equaliser morphism m is necessarily a monomorphism.

(g) In any category, every split monomorphism is an equaliser.

1.7.5 Sums

The notion dual to a product is called a *sum*, or *coproduct*. It is important to see how this topic can have far different realisations, in different concrete categories (cf. 1.7.6).

The *sum* of a family $(X_i)_{i \in I}$ of objects of the category C is an object X equipped with a family of morphisms $u_i \colon X_i \to X$ $(i \in I)$, called *injections*, which satisfy the following universal property:

$$
\begin{array}{ccc}
X_i & \xrightarrow{\ u_i\ } & X \\
 & \searrow & \big\downarrow f \\
 & {\scriptstyle f_i} & Y
\end{array}
\qquad (1.61)
$$

(i*) for every object Y and every family of morphisms $f_i \colon X_i \to Y$, there is a unique morphism $f \colon X \to Y$ such that, for all $i \in I$, $f u_i = f_i$.

The map f can be written as $[f_i]$, by its *co-components*.

If the sum of the family (X_i) exists, it is determined up to a unique coherent isomorphism, and denoted as $\Sigma_i X_i$, or $X_1 + ... + X_n$ in a finite case. The sum of the empty family is the *initial* object $\perp$: this means that every object X has a unique map $\perp \to X$.

Dualising 1.7.2(f), a sum $(X, (u_i \colon X_i \to X))$ in the category C is a universal arrow $(X, u \colon (X_i) \to \Delta X)$ from the object $(X_i)_{i \in I}$ to the diagonal functor $\Delta \colon \mathsf{C} \to \mathsf{C}^I$.

1.7.6 Exercises and complements (Sums)

These points are easy (except the *-marked ones), and left to the reader.

(a) In Set the initial object is the empty set. A sum $X = \Sigma_i X_i$ is realised as a disjoint union of isomorphic copies of all summands.

A simple way of doing this is letting

$$X = \bigcup_{i \in I} X_i \times \{i\}, \qquad u_i(x) = (x, i) \quad \text{(for } x \in X_i). \tag{1.62}$$

If all X_i are equal to a set Y, this gives $\Sigma_{i \in I} X_i = Y \times I$. On the other hand, if all X_i are disjoint, we can simply take $X = \bigcup_i X_i$; of course this does not work in the previous case.

(b) In Top, a sum $\Sigma_i X_i$ is the set-theoretical sum equipped with the finest topology that makes the injections continuous: the open sets are the unions $\bigcup_i U_i$ of open sets of the summands (an abuse of notation, for $\bigcup_i u_i(U_i)$). The initial object is the empty space.

(c) In Ord, $\Sigma_i X_i$ is the set-theoretical sum equipped with the finest order consistent with the injections: $(x, i) \leqslant (y, j)$ if and only if $i = j$ and $x \leqslant y$ in X_i.

(d) In categories of 'algebras', categorical sums are *not* realised on the disjoint union of the underlying sets. In Ab and R Mod, they are realised as 'direct sums'

$$\bigoplus_{i \in I} A_i \\ = \{(a_i) \in \Pi A_i \mid a_i = 0 \text{ except for finitely many } i \in I\}, \tag{1.63}$$

and the initial object is the null group or module. Note that, for a finite set of indices, $\bigoplus A_i = \Pi A_i$ — a point dealt with in 1.7.9.

(e) In Set. and Top. a sum $\Sigma_i (X_i, \overline{x}_i)$ can be constructed as a quotient of the unpointed sum $\Sigma_i X_i$, by identifying all the base points $\overline{x}_i$; their class gives the new base point.

In Set$_{\bullet}$ the sum can also be realised as a subset of the cartesian product

$$\sum_{i \in I} (X_i, \overline{x}_i)$$
$$= \{(x_i) \in \Pi_i X_i \mid x_i = \overline{x}_i \text{ for all } i \text{ except one at most}\}, \qquad (1.64)$$

namely the union of all the 'cartesian axes' of the product, pointed again at the 'origin' $\overline{x} = (\overline{x}_i)$.

In Top$_{\bullet}$ a finite sum can be similarly realised as a subspace of a cartesian product. For instance, $(\mathbb{R}, 0) + (\mathbb{R}, 0)$ can be realised as the subspace of the cartesian plane formed of the two cartesian axes.

(f) In a preordered set X (viewed as a category), categorical sums $\sum_i x_i$ amount to joins. A bounded lattice (as defined in 1.1.3) is the same as an ordered set with finite (categorical) products and sums.

(g) The injections of a sum are monomorphisms in all the cases considered above. In Set^{op} this need not be the case, as follows from 1.7.2(a). In the category $0\,\mathsf{Mod} \cong \mathbf{1}$ of modules on the trivial ring, all the injections of a sum are the identity of the only object.

*(h) Categorical sums of commutative rings are realised as tensor products over $\mathbb{Z}$: see 5.4.5(d).

1.7.7 Coequalisers

The *coequaliser* of a pair $f, g \colon X \to Y$ of parallel maps of C is a map $p \colon Y \to C$ such that $pf = pg$ and:

$$X \underset{g}{\overset{f}{\rightrightarrows}} Y \overset{p}{\longrightarrow} C$$

$$\qquad\qquad\qquad h \downarrow \quad \swarrow w \qquad\qquad (1.65)$$

$$\qquad\qquad Z$$

(ii*) every map $h \colon Y \to Z$ such that $hf = hg$ factorises uniquely through p (i.e. there exists a unique map $w \colon C \to Z$ such that $wp = h$).

A coequaliser morphism is always an epimorphism.

1.7.8 Exercises and complements (Coequalisers)

(a) In Set the natural solution of the coequaliser of a pair $f, g \colon X \to Y$ of parallel mappings is the projection

$$p \colon Y \to Y/R, \qquad (1.66)$$

on the quotient modulo the equivalence relation spanned by the pairs $(f(x), g(x)) \in Y^2$, for $x \in X$.

Note that every quotient Y/R can be obtained as a coequaliser in Set

$$R \underset{f_2}{\overset{f_1}{\rightrightarrows}} Y \overset{p}{\longrightarrow} Y/R \qquad (1.67)$$

viewing the relation R as a subset of $Y \times Y$ (satisfying the usual conditions) and letting f_i be the restriction of the cartesian projection $p_i \colon Y \times Y \to Y$ (for $i = 1, 2$).

(b) In Top the coequaliser of a pair $f, g \colon X \to Y$ is constructed in the same way, putting on Y/R the quotient topology, namely the finest that makes the mapping p continuous (see 1.4.9).

Again, a topological quotient Y/R can be obtained as the coequaliser of diagram (1.67), viewing R as a subspace of $Y \times Y$.

(c) In pOrd, coequalisers are computed in the same way, putting on Y/R the induced preorder, namely the finest that makes p monotone. In Ord we first compute the coequaliser Y/R in pOrd, and then take the associated ordered set. (This construction will be extended to all colimits.)

(d) Compute coequalisers in Ab and Gp.

(e) Compute coequalisers in Set. and Top..

(f) It is interesting to examine the coequaliser of the following pair of mappings, with values in the standard interval $\mathbb{I} = [0, 1]$

$$\partial^0, \partial^1 \colon \{*\} \rightrightarrows \mathbb{I}, \qquad \partial^0(*) = 0, \quad \partial^1(*) = 1, \qquad (1.68)$$

in the categories Set, Top, pOrd and Ord.

Here the interval $\mathbb{I}$ has no structure, or the euclidean topology, or the natural order, according to the case. *This coequaliser will also play a role for 'directed spaces', in 4.3.5 and 4.5.7.*

*(g) The coequaliser in Cat of two functors $F, G \colon \mathsf{C} \to \mathsf{D}$ is the quotient of D modulo the *generalised congruence* generated by this pair; the latter also involves equivalent objects.

One can avoid giving a 'construction' of the coequaliser category (necessarily complicated) and just prove its existence by the Adjoint Functor Theorem, mentioned in 1.8.4.

1.7.9 Biproducts and additive categories

A *zero object* of a category, often written as 0, is both initial and terminal.

This exists in Ab, R Mod, Gp, Set., Top., Ban and Ban_1; it does not in Set, Top and Cat, where the initial and terminal objects are distinct.

A category with zero object is said to be *pointed*; then each pair of objects A, B has a *zero-morphism* $0_{AB}: A \to B$ (also written as 0), which is given by the composite $A \to 0 \to B$.

A zero object is thus, at the same time, an empty product and an empty sum (cf. 1.7.1, 1.7.5). But we already noted that in Ab and R Mod, this coincidence holds for every *finite* family of objects (A_i): the product $\prod A_i$ and the sum $\sum A_i$ are realised as the same object $A = \bigoplus A_i$, which satisfies:

- the property of the product, by a family of projections $p_i: A \to A_i$,

- the property of the sum, by a family of injections $u_i: A_i \to A$,

- the equations $p_i u_i = \mathrm{id}\, A_i$ and $p_j u_i = 0: A_i \to A_j$ (for $i \neq j$).

Such an object, in a pointed category, is called a *biproduct*. The empty biproduct is the zero object.

As a related notion, a *preadditive* (or $\mathbb{Z}$-*linear*) category is a category C where every hom-set $\mathsf{C}(A, B)$ is equipped with a structure of abelian group, generally written as $f + g$, so that composition is additive in each variable (or bilinear over $\mathbb{Z}$)

$$(f + g)h = fh + gh, \qquad k(f + g) = kf + kg, \tag{1.69}$$

where $h: A' \to A$ and $k: B \to B'$. A functor $F: \mathsf{C} \to \mathsf{D}$ between preadditive categories is said to be *additive*, or $\mathbb{Z}$-*linear*, if it preserves the sum of parallel morphisms.

A preadditive category on a single object 'is' a unital ring R. An additive functor $R \to S$ between such categories is a homomorphism of unital rings. An additive functor $R \to$ Ab amounts to an R-module.

*An *additive category* C is a preadditive category with finite products, or equivalently with finite sums, which are then biproducts; *in this case*, the sum of parallel maps is determined by the categorical structure. These results are proved in [M3], Section VIII.2.

Typical examples are Ab and R Mod, with the pointwise sum of homomorphisms

$$(f + g)(x) = f(x) + g(x) \qquad (f, g: A \to B;\ x \in A). \tag{1.70}$$

This is also true of Ban, and not of Ban_1, as shown in the exercises below.*

Exercises and complements. (a) In a pointed category, every cartesian projection is a split epi. Dually, every injection in a sum is a split mono.

(b) In a pointed category with binary sums and products, there is a canonical morphism $f: X_1 + X_2 \to X_1 \times X_2$. (This has been used in 1.7.6(e).)

*(c) Compute binary products and sums, in Ban$_1$ and Ban. *Hints:* use the direct sum of vector spaces, with suitable norms.

*(d) Ban has a natural additive structure.

*(e) Ban$_1$ has none.

1.8 Adjoint functors and Galois connections

A functor $F\colon \mathsf{C} \to \mathsf{D}$ can have a left adjoint $\mathsf{D} \to \mathsf{C}$, or a right one; possibly both, or none.

Adjunctions, a global way of presenting universal properties and a crucial step in category theory, were introduced in 1958 by D. Kan [Ka]. They appear everywhere in Mathematics, from simple instances as the free R-module on a set to more complex ones, as the completion of a metric space or the universal compactification of a space. Adjunctions between ordered sets, viewed as categories, are called (covariant) Galois connections.

We give here a brief presentation of this topic, mostly without proofs (referred to [M3, G10]).

Adjoint functors have a limited role in this book, and will be made explicit in a marginal way, not to make the categorical framework heavier than necessary. A reader interested in category theory will easily recognise their presence, and find further information in any book on category theory. Otherwise, one can skip this section and come back when referred to.

1.8.1 Introducing adjunctions

We begin with some well known examples.

(a) The forgetful functor $U\colon \mathsf{Top} \to \mathsf{Set}$ has two obvious 'best approxima-tions' to an inverse which does not exist, the functors $D, C\colon \mathsf{Set} \to \mathsf{Top}$, where DX (resp. CX) is the set X with the discrete (resp. indiscrete, or chaotic) topology: they provide the set X with the finest (resp. the coars-est) topology.

For every set X and every space Y we can 'identify' the hom-sets

$$\mathsf{Top}(DX, Y) = \mathsf{Set}(X, UY), \qquad \mathsf{Set}(UY, X) = \mathsf{Top}(Y, CX), \qquad (1.71)$$

since every mapping $X \to Y$ becomes continuous if we put on X the discrete topology, and every mapping $Y \to X$ becomes continuous if we put on X the indiscrete one.

These two facts tell us that D is *left adjoint* to U (written as $D \dashv U$) and C is *right adjoint* to U (i.e. $U \dashv C$), respectively.

(The name comes from the relations (1.71) and their similarity to adjoint operators between Hilbert spaces.)

All this can be easily transferred to the forgetful functor U: pOrd $\to$ Set. Now the left adjoint D: Set $\to$ pOrd gives to a set X the discrete order, while the right adjoint C: Set $\to$ pOrd gives to X the indiscrete, or chaotic, preorder.

There is no coarsest order on a set with at least two elements. Accordingly, the forgetful functor Ord $\to$ Set has no right adjoint: see 1.8.4(e). Similarly, the forgetful functor Hsd $\to$ Set of Hausdorff spaces has no right adjoint.

(b) In many cases it may be convenient to follow another approach, starting from one of the functors (the simpler) and constructing the other by universal arrows of the former.

To wit, let us start from the forgetful functor U: R Mod $\to$ Set. (This functor has a left adjoint, and no right adjoint.)

We have already seen, in 1.5.7(b), that, for every set X, there is a universal arrow from X to the functor U

$$F(X) = \bigoplus_{x \in X} R, \qquad \eta\colon X \to UF(X), \qquad (1.72)$$

consisting of the free R-module $F(X)$ on the set X, with the insertion of the basis η. The latter is rewritten as ηX, or η_X, when convenient.

The fact that this universal arrow exists *for every X in* Set allows us to construct a 'backward' functor F: Set $\to$ R Mod, *left adjoint* to U.

F is already defined on the objects. For a mapping $f\colon X \to X'$ in Set

$$
\begin{array}{ccc}
X & \xrightarrow{\eta X} & UF(X) \\
{\scriptstyle f}\downarrow & & \downarrow{\scriptstyle U(Ff)} \\
X' & \xrightarrow[\eta X']{} & UF(X')
\end{array}
\qquad (1.73)
$$

the universal property of ηX implies that there is a unique homomorphism $F(f)\colon F(X) \to F(X')$ such that $U(F(f))$ makes the previous square commute. ($F(f)$ is just the linear extension of f; but the general pattern is better perceived if we proceed in a formal way.)

Further applications of the universal property show that F preserves composition and identities (which is obvious, here). Let us note that we have extended the given object-function $F(X)$ in the only way that makes the family (η_X) into a natural transformation

$$\eta\colon 1 \to UF\colon \text{Set} \to \text{Set},$$

called the *unit* of the adjunction.

(c) The relationship between these functors U and F (assuming now that we have defined both) can be equivalently expressed in a form similar to that used in (a), by a family of bijective mappings

$$\varphi_{XA} \colon R\,\mathsf{Mod}(FX, A) \to \mathsf{Set}(X, UA), \qquad (1.74)$$

which is natural in X (varying in Set) and A (in $R\,\mathsf{Mod}$), as will be made precise in 1.8.2.

Concretely, $\varphi_{XA}(g \colon FX \to A)$ is the restriction of the homomorphism g to the basis X of FX; its bijectivity means that every mapping $f \colon X \to UA$ has a unique extension to a homomorphism $g \colon FX \to A$, as in the universal property of $\eta \colon X \to UFX$.

Formally, given the family (ηX), we define

$$\varphi_{XA}(g \colon FX \to A) = Ug.\eta X \colon X \to UA.$$

The other way round, given the family (φ_{XA}), we define

$$\eta X = \varphi_{X,FX}(\mathrm{id}\, FX) \colon X \to UFX.$$

(d) Coming back to the forgetful functor $U \colon \mathsf{Top} \to \mathsf{Set}$, the adjunction $D \dashv U$ in (a) can also be presented as a 'collage' of universal properties of the functor U (even too obvious, in this case).

In fact, for every set X we have a universal arrow

$$(DX, \eta \colon X \to UDX)$$

from X to the functor U, where DX is the set X with the discrete topology and $\eta = \mathrm{id}\, X$. The universal property says that every mapping $f \colon X \to UY$ with values in the underlying set of a space 'lies under' a (unique) continuous mapping $g \colon DX \to Y$ (i.e. $f = Ug = Ug.\eta$). The left adjoint D is then completed, defining its action on morphisms by the universal property.

Dually, the other adjunction $U \dashv C$ can also be presented as a family of universal properties of the functor U: for every set X we have a universal arrow $(CX, \varepsilon \colon UCX \to X)$ from the functor U to the object X, with $\varepsilon = \mathrm{id}\, X$.

1.8.2 Definition (Adjunction)

Extending the previous cases, an *adjunction* $F \dashv G$ between a functor $F \colon \mathsf{C} \to \mathsf{D}$ (the *left adjoint*) and a functor $G \colon \mathsf{D} \to \mathsf{C}$ (the *right adjoint*), can be equivalently presented in four main forms.

(a) We assign two functors $F\colon \mathsf{C} \to \mathsf{D}$ and $G\colon \mathsf{D} \to \mathsf{C}$, together with a family of bijections

$$\varphi_{XY}\colon \mathsf{D}(FX,Y) \to \mathsf{C}(X,GY) \qquad (X \text{ in } \mathsf{C}, Y \text{ in } \mathsf{D}),$$

which is natural in X, Y. More formally, the family (φ_{XY}) is an invertible natural transformation

$$\varphi\colon \mathsf{D}(F(-),=) \to \mathsf{C}(-,G(=))\colon \mathsf{C}^{\mathrm{op}} \times \mathsf{D} \to \mathsf{Set}.$$

(b) We assign a functor $G\colon \mathsf{D} \to \mathsf{C}$ and, for every X in C, a universal arrow

$$(F_0X, \eta X\colon X \to GF_0X) \qquad \text{from the object } X \text{ to the functor } G.$$

(b*) We assign a functor $F\colon \mathsf{C} \to \mathsf{D}$ and, for every Y in D, a universal arrow

$$(G_0Y, \varepsilon Y\colon FG_0Y \to Y) \qquad \text{from the functor } F \text{ to the object } Y.$$

(c) We assign two functors $F\colon \mathsf{C} \to \mathsf{D}$ and $G\colon \mathsf{D} \to \mathsf{C}$, together with two natural transformations

$$\eta\colon \mathrm{id}\,\mathsf{C} \to GF \quad (\text{the } \textit{unit}), \qquad \varepsilon\colon FG \to \mathrm{id}\,\mathsf{D} \quad (\text{the } \textit{counit}),$$

which satisfy the *triangular identities*: $\varepsilon F.F\eta = \mathrm{id}\,F$ and $G\varepsilon.\eta G = \mathrm{id}\,G$

$$F \xrightarrow{F\eta} FGF \xrightarrow{\varepsilon F} F \qquad G \xrightarrow{\eta G} GFG \xrightarrow{G\varepsilon} G \qquad (1.75)$$
$$\underbrace{\phantom{F \xrightarrow{F\eta} FGF \xrightarrow{\varepsilon F} F}}_{\mathrm{id}\,F} \qquad \underbrace{\phantom{G \xrightarrow{\eta G} GFG \xrightarrow{G\varepsilon} G}}_{\mathrm{id}\,G}$$

A proof of the equivalence — based on the axiom of choice — can be found in [M3], Section IV.1, Theorem 2, or [G10], Theorem 3.1.5. Essentially:

- given (a), one defines $\eta X = \varphi_{X,FX}(\mathrm{id}\,FX)\colon X \to GFX$ and $\varepsilon Y = (\varphi_{GY,Y})^{-1}(\mathrm{id}\,GY)\colon FGY \to Y$,

- given (b), one defines $F(X) = F_0X$, the morphism $F(f\colon X \to X')$ by the universal property of ηX, and $\varphi_{XY}(g\colon FX \to Y) = Gg.\eta X$,

- given (c), one defines the mapping $\varphi_{XY}\colon \mathsf{D}(F(X),Y) \to \mathsf{C}(X,G(Y))$ as above, a backward mapping $\psi_{XY}(f\colon X \to GY) = \varepsilon Y.Ff$, and proves that they are inverse to each other by the triangular identities.

1.8.3 Remarks

The previous forms have different features.

Form (a) is the classical definition of an adjunction, and is at the origin of the name, by analogy with adjoint maps of Hilbert spaces. Form (b) is used when one starts from a given functor and wants to construct its left adjoint (possibly less easy to define). Form (b*) is dual to the previous one, and used in a dual way. The 'algebraic' form (c) is adequate to the formal theory of adjunctions (and makes sense in an abstract 2-category, cf. 5.6.2).

Duality of categories interchanges left and right adjoint.

An adjoint equivalence (defined in 1.6.3) is the same as an adjunction where the unit and counit are invertible.

1.8.4 Main properties of adjunctions

(a) *Uniqueness and existence.* Given a functor, its left adjoint (if it exists) is uniquely determined up to isomorphism ([G10], Theorem 3.1.6).

A crucial theorem for proving the existence (under suitable hypothesis) is the Adjoint Functor Theorem of P. Freyd: see [M3], Section V.6, Theorem 2, or [G10], Theorem 3.5.2.

(b) *Composing adjoint functors.* Given two consecutive adjunctions

$$F: \mathsf{C} \rightleftarrows \mathsf{D} : G, \qquad \eta: 1 \to GF, \quad \varepsilon: FG \to 1,$$
$$H: \mathsf{D} \rightleftarrows \mathsf{E} : K, \qquad \rho: 1 \to KH, \quad \sigma: HK \to 1, \tag{1.76}$$

there is a composed adjunction from the first to the third category:

$$HF: \mathsf{C} \rightleftarrows \mathsf{E} : GK,$$
$$G\rho F.\eta: 1 \to GF \to GK.HF, \tag{1.77}$$
$$\sigma.H\varepsilon K: HF.GK \to HK \to 1,$$

whose unit and counit are computed with the operations of natural transformations (in 1.6.2).

This can be verified using the algebraic form (c) of Definition 1.8.2.

There is thus a category AdjCat of small categories and adjunctions, with morphisms $(F, G; \eta, \varepsilon): \mathsf{C} \nrightarrow \mathsf{D}$.

(c) *Faithful and full adjoints.* Suppose we have an adjunction $F \dashv G$, with counit $\varepsilon: FG \to 1$. As proved in [M3], Section IV.3, Theorem 1, or [G10], Theorem 3.2.5:

  (i) G is faithful if and only if all the components εY of the counit are epi,

  (ii) G is full if and only if all the components εY are split mono,

  (iii) G is full and faithful if and only if the counit ε is invertible.

(d) *Adjoints and limits.* Adjoints have a strong relationship with categorical limits and colimits (whose general form is deferred to Chapter 5).

In fact:

- a left adjoint preserves (the existing) colimits, in particular sums and coequalisers,

- a right adjoint preserves (the existing) limits, in particular products and equalisers.

The (easy) proof can be found in the solution of Exercise 5.2.3(a). In particular, an equivalence of categories preserves all limits and colimits. The case of Galois connections is in 1.8.9.

(e) The forgetful functor Ord → Set does not preserve coequalisers (by an example in 1.7.8(f)), and cannot have a right adjoint.

1.8.5 Reflective and coreflective subcategories

A subcategory $C' \subset C$ is said to be *reflective* if the inclusion functor U: $C' \to C$ has a left adjoint, and *coreflective* if U has a right adjoint.

Exercises and complements. (a) Ab is reflective in Gp. *Hints*: use Exercise 1.5.7(c).

(b) The full subcategory of Ab formed by the torsion abelian groups is coreflective in Ab.

(c) The full subcategory of Ab formed by the torsion-free abelian groups is reflective in Ab.

1.8.6 Idempotent adjunctions

This part of the theory of adjoint functors is not always dealt with in books on category theory. An interested reader can see [G10], Section 3.8.

An adjunction $(F, G; \eta, \varepsilon)$: $C \rightarrow D$ is said to be *idempotent* if the four natural transformations which appear in the triangular identities

$$F\eta \colon F \to FGF, \qquad\qquad \varepsilon F \colon FGF \to F,$$
$$\eta G \colon G \to GFG, \qquad\qquad G\varepsilon \colon GFG \to G, \tag{1.78}$$

are functorial isomorphisms, so that $F\eta$ is inverse to εF, and ηG to $G\varepsilon$.

*This property has many interesting characterisations (see [G10], Theorem 3.8.2): for instance, it is sufficient to know that *one* of the natural transformations above is invertible. It is also sufficient to know that $G\varepsilon F \colon GF \to GFGF$ is invertible (so that the functor GF: $C \to C$ is idempotent, up to isomorphism: an 'idempotent monad' on C).*

For an idempotent adjunction $F \dashv G$, we single out one point of this topic, that will be used in Chapter 4.

We let Alg(C) denote the full subcategory of C consisting of its *algebraic objects*: namely, the objects X of C such that $\eta X \colon X \to GFX$ is invertible. Dually, Coalg(D) will be the full subcategory of D of *coalgebraic objects* Y, for which $\varepsilon Y \colon FGY \to Y$ is invertible.

Then one can form three adjunctions (as in [G10], Theorem 3.8.7)

$$\mathsf{C} \;\underset{G'}{\overset{F'}{\rightleftarrows}}\; \mathrm{Alg}(\mathsf{C}) \;\underset{G^\sharp}{\overset{F^\sharp}{\rightleftarrows}}\; \mathrm{Coalg}(\mathsf{D}) \;\underset{G''}{\overset{F''}{\rightleftarrows}}\; \mathsf{D} \tag{1.79}$$

whose composition gives back the original adjunction, up to isomorphism. The reader can write this out, as an exercise, directly or with the help of the following outline.

(a) The functor $G' \colon \mathrm{Alg}(\mathsf{C}) \to \mathsf{C}$ is the full embedding. For X in C, $F'X = GFX$ is an algebraic object (as any GY), and the universal arrow

$$(F'X, \eta X \colon X \to G'F'X)$$

gives the reflector $F' \colon \mathsf{C} \to \mathrm{Alg}(\mathsf{C})$, a codomain-restriction of $GF \colon \mathsf{C} \to \mathsf{C}$. Note that Alg(C) is a reflective subcategory of C.

(b) Dually, $F'' \colon \mathrm{Coalg}(\mathsf{D}) \to \mathsf{D}$ is the full embedding. For Y in D, $G''Y = FGY$ is a coalgebraic object (as any FX), and the universal arrow

$$(G''Y, \varepsilon Y \colon F''G''Y \to Y)$$

gives the coreflector $G'' \colon \mathsf{D} \to \mathrm{Coalg}(\mathsf{D})$, a codomain-restriction of $FG \colon \mathsf{D} \to \mathsf{D}$. The subcategory Coalg(D) is coreflective in D.

(c) Now $F^\sharp \colon \mathrm{Alg}(\mathsf{C}) \to \mathrm{Coalg}(\mathsf{D})$ is a restriction of $F \colon \mathsf{C} \to \mathsf{D}$, and $G^\sharp$ a restriction of G. These functors form an adjoint equivalence, because the components

$$\eta^\sharp X = \eta X \colon X \to GFX, \qquad \varepsilon^\sharp Y = \varepsilon Y \colon FGY \to Y$$

are invertible, for every algebraic object X, and every coalgebraic object Y, respectively.

(This equivalence is even an isomorphism if GF is the identity over all algebraic objects and FG is the identity over all the coalgebraic ones, as is often the case).

(d) Composing the three adjunctions above we get

$$F''F^\sharp F' = FGF \colon \mathsf{C} \to \mathsf{D},$$

isomorphic to F (by $F\eta = (\varepsilon F)^{-1}$), and

$$G'G^\sharp G'' = GFG \colon \mathsf{D} \to \mathsf{C},$$

isomorphic to G (by $\eta G = (G\varepsilon)^{-1}$).

1.8.7 Galois connections

Finally we consider adjunctions between ordered sets, viewed as categories; the previous points take now a simpler form, and all results can be easily verified, directly.

Given a pair X, Y of ordered sets, a (covariant) *Galois connection* between them can be equivalently presented in the following ways.

(i) We assign two monotone mappings $f\colon X \to Y$ and $g\colon Y \to X$ such that, for every $x \in X$ and $y \in Y$:

$$f(x) \leqslant y \text{ in } Y \quad \Leftrightarrow \quad x \leqslant g(y) \text{ in } X.$$

(ii) We assign a monotone mapping $g\colon Y \to X$ such that, for every $x \in X$, there exists in Y:

$$f(x) = \min \{y \in Y \mid x \leqslant g(y)\}.$$

(ii*) We assign a monotone mapping $f\colon X \to Y$ such that, for every $y \in Y$, there exists in X:

$$g(y) = \max \{x \in X \mid f(x) \leqslant y\}.$$

(iii) We assign two monotone mappings $f\colon X \to Y$ and $g\colon Y \to X$, such that $\mathrm{id}\, X \leqslant gf$ and $fg \leqslant \mathrm{id}\, Y$.

By these formulas g determines f (called its *left adjoint*) and f determines g (its *right adjoint*). We write $f \dashv g$, as in the general notation of adjoints in category theory.

Of course an isomorphism of ordered sets is, at the same time, left and right adjoint to its inverse. More generally, a monotone mapping *may* have one or both adjoints, which can be viewed as 'best approximations' to an inverse, of two different kinds.

One can also consider Galois connections between *preordered* sets; the results are similar, if less sharp — up to the equivalence relation associated to the relevant preorder.

1.8.8 Exercises and complements

(a) Prove the equivalence of conditions (i)–(iii), above.

(b) The embedding of ordered sets $i\colon \mathbb{Z} \to \mathbb{R}$ has both adjoints. In other words, the subcategory $\mathbb{Z} \subset \mathbb{R}$ is reflective and coreflective, see 1.8.5.

(c) The embedding $\mathbb{Q} \to \mathbb{R}$ of the rational numbers has neither a left nor a right adjoint.

(d) A mapping $f \colon A \to B$ of sets gives two monotone mappings, of *direct* and *inverse image*, between the power sets of A and B

$$f_* \colon \mathcal{P}A \rightleftarrows \mathcal{P}B \colon f^*,$$

$$f_*(X) = f(X), \qquad f^*(Y) = f^{-1}(Y) \qquad (X \subset A, Y \subset B).$$

(1.80)

These mappings form a Galois connection $f_* \dashv f^*$.

(e) A *contravariant* (or antitone) *Galois connection* between the ordered sets X, Y is the same as a Galois connection between X and Y^{op}. Then f and g are antitone mappings (between X and Y) and the adjunction is expressed in a symmetric way: $\mathrm{id}\, X \leqslant gf$ and $\mathrm{id}\, Y \leqslant fg$.

We have already seen such a case, in 1.1.4(c): if X is an ordered set, the antitone mappings $L, U \colon \mathcal{P}X \dashrightarrow \mathcal{P}X$ of lower and upper bounds satisfy the conditions $A \subset UL(A)$ and $A \subset LU(A)$.

Important examples, in Galois theory and Algebraic Geometry, come out naturally in contravariant form.

1.8.9 Properties

Let us come back to a general Galois connection $f \dashv g$ between ordered sets X, Y.

The mapping f *preserves all the existing joins* (also infinite), while g *preserves all the existing meets*. It is a particular case of 1.8.4(d), which again can be easily verified directly.

In fact, if $x = \bigvee x_i$ in X then $f(x_i) \leqslant f(x)$ (for all indices i). Supposing that $f(x_i) \leqslant y$ in Y (for all i), it follows that $x_i \leqslant g(y)$ (for all i); but then $x \leqslant g(y)$ and $f(x) \leqslant y$.

Furthermore, the relations $\mathrm{id}\, X \leqslant gf$ and $fg \leqslant \mathrm{id}\, Y$ imply that $f = fgf$ and $g = gfg$.

A Galois connection is always an idempotent adjunction, in a strict sense: the natural transformations of (1.78) are identities.

As a consequence, the connection restricts to an isomorphism (of ordered sets) between the sets of *fixed elements* of X and of Y, defined as follows

$$\mathrm{Fix}(X) = \{x \in X \mid x = gf(x)\} = g(Y),$$

$$\mathrm{Fix}(Y) = \{y \in Y \mid y = fg(y)\} = f(X).$$

(1.81)

For $x \in X$, $\bar{x} = gf(x)$ is the least fixed element of X greater than x. For $y \in Y$, $y^{\circ} = fg(y)$ is the greatest fixed element of Y smaller than y.

(This isomorphism corresponds to the adjoint equivalence of 1.8.6(c), between algebraic elements in X and coalgebraic elements in Y.)

An adjunction $f \dashv g$ can be written as a dot-marked arrow $(f, g) \colon X \to Y$, conventionally directed as the left adjoint $f \colon X \to Y$. These arrows have an obvious composition

$$(f', g').(f, g) = (f'f, gg') \qquad (\text{for } (f', g') \colon Y \to Z), \qquad (1.82)$$

and form the category AdjOrd *of ordered sets and Galois connections.* It is a selfdual category, by the anti-isomorphism

$$X \mapsto X^{\mathrm{op}}, \qquad ((f, g) \colon X \to Y) \mapsto ((g, f) \colon Y^{\mathrm{op}} \to X^{\mathrm{op}}). \qquad (1.83)$$

2

Inverse categories and topological background

Our analysis of manifolds will rely on ordered categories of partial mappings, provided with an 'inverse subcategory' of partial bijections, the *inverse core*. The first aspect is examined in Section 2.1, the second in Sections 2.2–2.4 and 2.7.

Various topics of semigroup theory related to partial inverses have a natural, less known extension to categories. We review here part of this extension, which was studied in a series of papers of the present author in the 1970's and 1980's, then summarised in [G2] and reorganised in the book [G9]. These works use regular and orthodox categories of relations, to solve problems of coherence in Homological Algebra and provide universal models of spectral sequences.

The second part of this chapter reviews several points of Topology, as another background for the study of local structures. The chapter ends with a brief review of Ehresmann's pseudogroups.

2.1 Ordered categories, partial mappings and topology

We introduce a first list of categories of partial mappings, or partial continuous mappings; all of them are ordered 'by restriction'. These ordered categories, and others, will be used in the next chapters for the intrinsic atlases of local structures, and their morphisms.

2.1.1 Ordered categories

An *ordered category* C is a category equipped with a (local) *order of categories*: this means that every set of morphisms $\mathsf{C}(X, Y)$ has an order relation $f \leqslant f'$, consistent with the composition:

$$\text{if } f \leqslant f' \text{ in } \mathsf{C}(X, Y) \text{ and } g \leqslant g' \text{ in } \mathsf{C}(Y, Z), \text{ then } gf \leqslant g'f'. \qquad (2.1)$$

In other words:

- all hom-sets $C(X, Y)$ are ordered sets,

- all the composition-mappings

$$C(X, Y) \times C(Y, Z) \to C(X, Z), \qquad (f, g) \mapsto gf, \qquad (2.2)$$

are morphisms of the category Ord (already considered in 1.4.1). Of course, the product $C(X, Y) \times C(Y, Z)$ must be taken in the category Ord, i.e. equipped with the product order: $(f, g) \leqslant (f', g')$ whenever $f \leqslant f'$ and $g \leqslant g'$.

In an ordered category C all semigroups of endomorphisms $C(X)$ are ordered semigroups.

A functor $F \colon C \to D$ between ordered categories is said to be *monotone*, or *order-preserving*, or *increasing*, if it preserves their orderings: $f \leqslant f'$ in C implies $F(f) \leqslant F(f')$ in D.

More generally, a *lax functor* $F \colon C \to D$ between ordered categories satisfies the following conditions (assuming that all compositions make sense):

$$F(g).F(f) \leqslant F(gf), \qquad \mathrm{id}\, FX \leqslant F(\mathrm{id}\, X),$$
$$f \leqslant f' \Rightarrow F(f) \leqslant F(f'). \qquad (2.3)$$

A *colax functor* satisfies the order-dual conditions: $F(g).F(f) \geqslant F(gf)$, $\mathrm{id}\, FX \geqslant F(\mathrm{id}\, X)$ and the same third condition (invariant by order-duality).

2.1.2 Remarks and examples

(a) An order relation $f \leqslant f'$ in a category is intended as a *local notion*, dealing with *parallel* morphisms. This is the prevalent use in category theory.

    *Formally, an ordered category is a category *enriched on the category* Ord (see 5.5.4(b)); it is also the same as a locally ordered 2-category (see 5.6.2).*

(b) There is also a *global notion*, used in Ehresmann's work on local structures and recalled in 2.8.1. It will play a limited role here, in Sections 2.8 and 6.7.

(c) The category Ord of ordered sets and increasing mappings is itself an ordered category, in the obvious, pointwise way: for $f, g \colon X \to Y$ we let $f \leqslant g$ if $f(x) \leqslant g(x)$ for all $x \in X$.

(d) Categories of relations have a natural order, as we shall see in 2.2.3 and 2.2.4.

(e) We shall also see, in Section 2.4, that every inverse category has a canonical order.

2.1.3 Categories of partial mappings

Partial mappings, between sets or structured sets, will often be written as dot-marked arrows $X \dashrightarrow Y$, to distinguish them from everywhere-defined mappings. Their categories, naturally ordered 'by restriction', will usually be denoted by calligraphic letters and will play an important role in this book.

The prime example is now the category $\mathcal{S}$ *of sets and partial mappings,* where a morphism $f \colon X \dashrightarrow Y$ is a mapping $f_0 \colon \operatorname{Def} f \to Y$ defined on a subset $\operatorname{Def} f \subset X$, possibly empty or total.

Composition works as in the well-known case of real functions $\mathbb{R} \dashrightarrow \mathbb{R}$: given a partial mapping $g \colon Y \dashrightarrow Z$, the composite $gf \colon X \dashrightarrow Z$ is 'defined where it makes sense'

$$(gf)(x) = g(f(x)),$$
$$\operatorname{Def}(gf) = \{x \in X \mid x \in \operatorname{Def} f,\ f(x) \in \operatorname{Def} g\}. \tag{2.4}$$

The composition is associative, and the identity mapping $\operatorname{id} X$ of the set X (everywhere defined) acts as a unit: for every partial mapping $f \colon X' \dashrightarrow X$ or $g \colon X \dashrightarrow X''$ we have $(\operatorname{id} X).f = f$ and $g.(\operatorname{id} X) = g$. Set is a wide subcategory of $\mathcal{S}$; moreover these two categories have the same isomorphisms: the bijective (everywhere defined) mappings.

The category $\mathcal{S}$ has a natural order: for $f, g \colon X \dashrightarrow Y$ we define $f \leqslant g$, when f is a restriction of g; this means that $\operatorname{Def} f \subset \operatorname{Def} g$ and $f(x) = g(x)$ whenever $x \in \operatorname{Def} f$.

An $\mathcal{S}$-*concrete category* is a category C equipped with a faithful functor $U \colon \mathsf{C} \to \mathcal{S}$ (see 1.5.6); many examples below are of this kind. Obviously, this perspective is of interest when the image of U is not contained in Set.

An $\mathcal{S}$-concrete category C is canonically ordered, *lifting* the order of $\mathcal{S}$. Namely, we let $f \leqslant g$, in C when f, g are parallel morphisms and $Uf \leqslant Ug$ in $\mathcal{S}$.

The category $\mathcal{S}$ *should not be viewed as essentially different* from the categories of structured sets and 'total' mappings that we have examined above: we shall see (in 2.1.5) that $\mathcal{S}$ is equivalent to the category Set. of pointed sets, introduced in 1.4.1, and essentially behaves as the latter: for instance categorical products and sums in $\mathcal{S}$ (more generally all categorical limits and colimits) are easily computed in Set.

The subcategory $\mathcal{I}$ of sets and partial bijections will be examined in Section 2.3.

Exercises and complements (Proper morphisms). Let C be an $\mathcal{S}$-concrete category; the following points are obvious.

(a) A morphism $f \colon X \to Y$ is said to be a *proper morphism* of C if Uf is an everywhere defined mapping, i.e. belongs to Set. Proper morphisms form a wide subcategory Prp C.

(b) Any proper morphism h is maximal in C: $h \leqslant f$ implies $h = f$ (for every f in C); in particular, $f \geqslant \operatorname{id} X$ implies $f = \operatorname{id} X$.

(c) Prp C inherits a discrete order from C.

2.1.4 Partial continuous mappings

We also use various $\mathcal{S}$-concrete categories of partial continuous mappings, with the canonical order lifted from $\mathcal{S}$.

(a) We write as $\mathcal{T}$ the category *of topological spaces and partial continuous mappings*, where a morphism $f \colon X \nrightarrow Y$ is a continuous mapping Def $f \to Y$ defined on a subspace Def f of X.

(b) If we restrict the morphisms of $\mathcal{T}$ requiring that the subspace Def f be open in X (a condition consistent with composition) we obtain the wide subcategory $\mathcal{C}$ of *topological spaces and partial continuous mappings defined on an open subspace*.

(c) We write as $\mathcal{D}$ another wide subcategory of $\mathcal{T}$, formed of the partial continuous mappings defined on a closed subspace.

(d) For a fixed $r \in \mathbb{N} \cup \{\infty, \omega\}$, $\mathcal{C}^r$ will denote the subcategory of $\mathcal{C}$ of *open euclidean spaces* (i.e. open subspaces of some euclidean space $\mathbb{R}^n$) and *partial C^r-mappings defined on open subspaces*.

$\mathcal{C}^0$ is thus a full subcategory of $\mathcal{C}$; all $\mathcal{C}^r$ form a nested family of wide subcategories of $\mathcal{C}^0$, beginning with partial analytic mappings

$$\mathcal{C}^\omega \subset \mathcal{C}^\infty \subset \ldots \mathcal{C}^{r+1} \subset \mathcal{C}^r \subset \ldots \mathcal{C}^0 \subset \mathcal{C}.$$

An open euclidean space has a topological dimension: see 2.6.4.

(e) The categories Top, $\mathcal{C}$, $\mathcal{D}$ and $\mathcal{T}$ have the same isomorphisms, namely the (total) homeomorphisms of topological spaces. We see below (in 2.1.5) that $\mathcal{C}$ and $\mathcal{D}$ are 'equivalent' to suitable full subcategories of the category Top. of pointed topological spaces; this is not the case of $\mathcal{T}$.

The inclusion $\mathcal{C}^r \subset \mathcal{C}$ makes $\mathcal{C}^r$ concrete over $\mathcal{C}$. Our analysis of manifolds, fibre bundles, vector bundles, etc. will be based on $\mathcal{C}$-concrete categories.

(f) We shall see in Section 5.1 that these 'categories of partial morphisms' can be obtained by a general construction $\mathsf{P_M C}$ based on an 'ordinary' category C equipped with a wide subcategory M of monomorphisms satisfying some conditions: for instance, we get $\mathcal{C}$ as $\mathsf{P_M Top}$, letting M be the wide subcategory of Top of open topological embeddings.

Yet, we prefer to work *directly* in our categories of partial morphisms, which are the natural framework of intrinsic atlases of manifolds.

2.1.5 Exercises and complements (Some equivalences)

The following exercises should help to better understand the crucial notion of equivalence of categories, *and* to better understand the categories of partial mappings. Solutions can be found below.

(a) The category $\mathcal{S}$ of sets and partial mappings is equivalent to the category $\mathsf{Set_{\bullet}}$ of pointed sets (introduced in 1.4.1), by a canonical functor

$$F\colon \mathsf{Set_{\bullet}} \to \mathcal{S}, \qquad F(X, x_0) = X \setminus \{x_0\}, \tag{2.5}$$

that sends the pointed mapping $f\colon (X, x_0) \to (Y, y_0)$ to its (partially defined) restriction

$$\begin{aligned} F(f)\colon X \setminus \{x_0\} &\twoheadrightarrow Y \setminus \{y_0\}, \\ \mathrm{Def}\,(Ff) = f^{-1}(Y \setminus \{y_0\}), \qquad (Ff)(x) &= f(x). \end{aligned} \tag{2.6}$$

A pseudo inverse $G\colon \mathcal{S} \to \mathsf{Set_{\bullet}}$ is obtained by choosing, for every set X, a pointed set $G(X) = (X \cup \{x_0\}, x_0)$, where $x_0 \notin X$. Then we define G on a partial mapping $f\colon X \twoheadrightarrow Y$ as

$$\begin{aligned} G(f)\colon G(X) &\to G(Y), \\ (Gf)(x) = f(x), \text{ if } x \in \mathrm{Def}\,f, \qquad (Gf)(x) &= y_0, \text{ otherwise,} \end{aligned} \tag{2.7}$$

where $G(Y) = (Y \cup \{y_0\}, y_0)$. This gives $FG = 1$ and $GF \cong 1$.

(b) The category $\mathsf{Set_{\bullet}}$ has all products (see 1.7.2(b)), which can be used to construct products in $\mathcal{S}$, working back-and-forth with the equivalence between these two categories. This gives 'unusual' results on finite sets!

(c) On the other hand, the order of $\mathcal{S}$ gives an order of $\mathsf{Set_{\bullet}}$, where, for $f, g\colon (X, x_0) \to (Y, y_0)$

$$f \leqslant g \;\Leftrightarrow\; (\text{for every } x \in X,\ f(x) = y_0 \text{ or } f(x) = g(x)). \tag{2.8}$$

This will be used in 3.1.6.

(d) The category $\mathcal{C}$ of topological spaces and partial continuous mappings defined on an open subspace is equivalent to a full subcategory $\mathsf{Top}'_{\bullet}$ of the category $\mathsf{Top_{\bullet}}$ of pointed topological spaces (cf. 1.4.1), formed of those objects (X, x_0) where the base point x_0 is closed and adherent to every other point (i.e. the only open set containing x_0 is X).

(e) Similarly the category $\mathcal{D}$ of topological spaces and partial continuous mappings defined on a closed subspace is equivalent to another full subcategory $\mathsf{Top}''_\bullet$ of $\mathsf{Top}_\bullet$, consisting of those objects (X, x_0) where the base point x_0 is open and dense (i.e. the only closed set containing x_0 is X, or equivalently every non-empty open set contains x_0).

Solutions. (a) The only remaining point, i.e. the isomorphism $GF \cong \mathrm{id}$, is easy and left to the reader.

There is a canonical choice for $G(X)$: one can take $x_0 = X$, since Set Theory says that $X \notin X$. This may be more confusing than useful. *Yet, the reader may know that all finite ordinals, after $0 = \emptyset$, are inductively constructed letting $n + 1 = n \cup \{n\}$.*

(b) Products in $\mathsf{Set}_\bullet$ are described in 1.7.2

$$\Pi_i\,(X_i, \overline{x}_i) = (\Pi_i\,X_i, \overline{x}), \qquad \overline{x} = (\overline{x}_i). \tag{2.9}$$

As to products in $\mathcal{S}$, for a family (S_i) of sets, we let $G(S_i) = (X_i, \overline{x}_i)$ and $F(X_i, \overline{x}_i) = S_i$; then

$$\Pi_i\,S_i = (\Pi_i\,X_i) \setminus \{\overline{x}\}. \tag{2.10}$$

In particular, the product in $\mathcal{S}$ of two finite sets of m and n elements has

$$(m + 1)(n + 1) - 1 = mn + m + n$$

elements; the product of two singletons has three elements.

A reader acquainted with categorical limits (cf. Section 5.2) will now gather that $\mathcal{S}$ has all limits and colimits, that can be similarly constructed using back-and-forth the categorical equivalence with $\mathsf{Set}_\bullet$. In fact, the colimits are easily constructed directly: see 2.1.6.

(d) Again, the equivalence comes from a canonical functor

$$F \colon \mathsf{Top}'_\bullet \to \mathcal{C}, \qquad F(X, x_0) = X \setminus \{x_0\}, \tag{2.11}$$

whose action on maps is defined as above, in (2.6).

Of course $X \setminus \{x_0\}$ has now the subspace topology. A pointed map $f \colon (X, x_0) \to (Y, y_0)$ is sent to a partial map $F(f)$ defined on the *open* subspace $\mathrm{Def}\,(Ff) = f^{-1}(Y \setminus \{y_0\})$, because $\{y_0\}$ is closed in Y.

We now construct a pseudo inverse $G \colon \mathcal{C} \to \mathsf{Top}'_\bullet$.

For a space S we form a disjoint union $X = S \cup \{x_0\}$ with the topology whose open sets are those of S and X itself: then $\{x_0\}$ is closed and X is the only open set containing the base point.

Given a partial map $g \colon S \rightharpoonup T$ defined on an open subspace U, we extend it to a total map $f \colon S \cup \{x_0\} \to T \cup \{y_0\}$ letting $f(x) = y_0$ for $x \notin U$.

In fact, if V is open in $T \cup \{y_0\}$, then:

- either $y_0 \notin V$ and $f^{-1}(V) = g^{-1}(V)$ is open in $X \setminus \{x_0\}$, and in X,

- or $y_0 \in V$, in which case V must be total and $f^{-1}(V) = S \cup \{x_0\}$.

(e) The equivalence comes again from a canonical functor $F \colon \mathsf{Top}''_\bullet \to \mathcal{D}$, where $F(X, x_0) = X \setminus \{x_0\}$.

Here the pseudo inverse functor $G \colon \mathcal{D} \to \mathsf{Top}''_\bullet$ takes a space S to a disjoint union $X = S \cup \{x_0\}$ with the topology whose *closed* sets are those of S and X itself; now $\{x_0\}$ is open and X is the only closed subset containing the base point.

The action on partial maps is as above, and the argument on continuity is proved in a similar way, working with closed subsets instead of the open ones.

2.1.6 Exercises and complements

The following colimits in $\mathcal{S}$, $\mathcal{C}$ and $\mathcal{D}$ can be easily constructed, and are related to the gluing of manifolds, in Chapter 4. *Hints:* the Cover Lemmas of the next subsection can simplify some points.

(a) All sums and coequalisers exist in $\mathcal{S}$.

(b) All sums and coequalisers exist in $\mathcal{C}$.

(c) Finite sums and all coequalisers exist in $\mathcal{D}$.

*(d) We shall not use products and equalisers in $\mathcal{C}$, but it is not difficult to prove that they exist, working in the equivalent category $\mathsf{Top}'_\bullet$ (in 2.1.5(d)), where products and equalisers are computed as in $\mathsf{Top}_\bullet$.

Taking out topologies, these results hold in $\mathcal{S}$.

*(e) The category $\mathcal{D}$ has finite products and equalisers. Again, it is easier to prove this in the equivalent category $\mathsf{Top}''_\bullet$, where these limits are computed as in $\mathsf{Top}_\bullet$.

2.1.7 Exercises and complements (Cover Lemmas)

Let $(X_i)_{i \in I}$ be a *cover* of the space X: this simply means that X is the union of its subsets X_i. The continuity of a mapping $f \colon X \to Y$ with values in a topological space Y can be reduced to the continuity of its restrictions $f_i \colon X_i \to Y$ in two main cases.

(a) (*The Open Cover Lemma*) Let $(X_i)_{i \in I}$ be an *open cover* of the space X (i.e. all X_i are open subsets of X). Then $f \colon X \to Y$ is continuous (on X) if and only if all its restrictions $f_i \colon X_i \to Y$ are.

(b) (*The Finite Closed Cover Lemma*) Let $(X_i)_{i \in I}$ be a *finite closed cover* of the space X (i.e. all X_i are closed subsets of X, and I is finite). Then the mapping $f \colon X \to Y$ is continuous if and only if all its restrictions $f_i \colon X_i \to Y$ are.

*(c) The general situation is a space X and a cover $(X_i)_{i \in I}$ such that X has the finest topology that makes all inclusions $u_i \colon X_i \to X$ continuous. Again, $f \colon X \to Y$ is continuous if and only if all its restrictions $f_i \colon X_i \to Y$ are.

Categorically, this means that X is the colimit of the diagram formed by the inclusions $X_i \to X$.

Solutions. (a) Obvious: if V is open in Y, $f^{-1}(V) = \bigcup_i f_i^{-1}(V)$.

(b) Use the preimage $f^{-1}(C) = \bigcup_i f_i^{-1}(C)$ of a closed subset of Y.

*(c) By hypothesis, U is open in X if and only if every trace $u_i^{-1}(U)$ is open in X_i. If all the restrictions $f_i = f u_i \colon X_i \to Y$ are continuous and V is open in Y, then every $u_i^{-1} f^{-1}(V) = f_i^{-1}(V)$ is open in X_i, and $f^{-1}(V)$ is open in X.

The converse is trivial: all restrictions of continuous mappings are continuous.

2.2 Involutive categories and categories of relations

We now examine involutive categories. The categories of relations are an outstanding example, of which we only consider elementary instances, based on sets, or abelian groups, or R-modules.

General constructions of categories of relations, for regular categories, or abelian categories and their extensions, can be found in [Bo, G10] and their references.

2.2.1 Involutive categories

An *involutive category* will be a category C equipped with an *involution*, i.e. a contravariant endo-functor that is the identity on objects and involutive on arrows.

In other words, the involution sends every morphism $a \colon X \to Y$ of C to a morphism $a^\sharp \colon Y \to X$ so that:

$$(ba)^\sharp = a^\sharp b^\sharp, \qquad (\mathrm{id}\, X)^\sharp = \mathrm{id}\, X, \qquad (a^\sharp)^\sharp = a. \tag{2.12}$$

This involution is an anti-automorphism of C, i.e. an isomorphism between C and C^{op}; therefore, C is *selfdual*.

Here we are mainly interested in *regular* involutions (in the sense of von Neumann), which means that for every arrow a

$$aa^\sharp a = a. \tag{2.13}$$

Then we also have $a^\sharp aa^\sharp = a^\sharp$. (This will mean that a and $a^\sharp$ are partial inverses in C, and the category C is regular in the sense of von Neumann, as defined in 2.3.2.)

2.2.2 Projectors

Let C be a category with a regular involution $(-)^\sharp$. For every object X, the monoid $C(X)$ inherits a regular involution.

As in 1.3.2, a *projector* of X in C is a symmetric idempotent endomorphism $e\colon X \to X$, satisfying the conditions $e = ee = e^\sharp$ (or equivalently $e = e^\sharp e$, or $e = ee^\sharp$). They form a subset $\mathsf{Prj}\,(X) \subset C(X)$.

We recall that the product of two projectors is always idempotent (by (1.25)), and every idempotent endomorphism e is the product of two projectors: $e = ee^\sharp . e^\sharp e$.

Every morphism a has two associated projectors (extending 1.3.2):

$$\underline{e}(a) = a^\sharp a \colon X \to X \qquad\qquad \text{(the \emph{support} of } a\text{)},$$
$$\underline{e}^*(a) = aa^\sharp = \underline{e}(a^\sharp) \colon Y \to Y \qquad \text{(the \emph{cosupport} of } a\text{)}. \tag{2.14}$$

The support (resp. cosupport) of a is a projector of the domain (resp. codomain) of a, which behaves as a right (resp. left) unit for a

$$a = a.\underline{e}(a) = \underline{e}^*(a).a = \underline{e}^*(a).a.\underline{e}(a). \tag{2.15}$$

All the projectors of C arise in this way: indeed an arrow e is a projector if and only if $e = \underline{e}(e)$, if and only if $e = \underline{e}^*(e)$.

Because of the regularity of the involution, for every morphism a:

(i) a is mono $\;\Leftrightarrow\; a^\sharp a = 1 \;\Leftrightarrow\;$ a is a split mono $\;\Leftrightarrow\; a^\sharp$ is epi,

(ii) a is epi $\;\Leftrightarrow\; aa^\sharp = 1 \;\Leftrightarrow\;$ a is a split epi $\;\Leftrightarrow\; a^\sharp$ is mono,

(iii) a is iso $\;\Leftrightarrow\; (aa^\sharp = 1$ and $aa^\sharp = 1) \;\Leftrightarrow\;$ a is mono and epi.

(Split monos and epis are defined in 1.4.7.)

Exercises and complements. Let C be a category with a regular involution.

(a) C is balanced.

(b) If C has epi-mono factorisations (see 1.4.8), these are essentially unique.

2.2.3 Relations of sets

Categories of relations are important involutive categories, that will be used here in a marginal way, often as a source of counterexamples. In fact, they have properties in part similar, in part different with respect to the $\mathcal{S}$-concrete categories, which are the focal point of the present analysis.

The category Rel Set of *sets and relations* (or *correspondences*) is well known.

A relation $a\colon X \nrightarrow Y$ (often denoted in the literature by a dot-marked or dash-marked arrow) is a subset of the cartesian product $X \times Y$. It can be viewed as a 'partially defined, many-valued mapping', that sends an element $x \in X$ to the subset $\{y \in Y \mid (x, y) \in a\}$ of Y.

The identity relations and the composite of a with $b\colon Y \nrightarrow Z$ are

$$\begin{aligned}
\operatorname{id} X &= \{(x, x) \mid x \in X\}, \\
ba &= \{(x, z) \in X \times Z \mid \exists y \in Y\colon (x, y) \in a \text{ and } (y, z) \in b\}.
\end{aligned} \tag{2.16}$$

We have thus a category, that contains $\mathcal{S}$ and Set (as wide subcategories), by identifying a partial (or total) mapping $f\colon X \nrightarrow Y$ with its graph, the single-valued relation $\{(x, y) \in X \times Y \mid y = f(x)\}$.

Rel Set is an *involutive ordered category*, in the following sense. First it is an involutive category as defined in 2.2.1: the *opposite relation* $a^\sharp\colon Y \nrightarrow X$ is obtained by reversing pairs. Second, it is an ordered category, in the sense of 2.1.1: given two parallel relations $a, a'\colon X \nrightarrow Y$ we say that $a \leqslant a'$ if $a \subset a'$ as subsets of $X \times Y$. Third, these structures are consistent: if $a \leqslant b$, then $a^\sharp \leqslant b^\sharp$ (and conversely, as a consequence).

The involution makes Rel Set into a selfdual category. The involution is not regular (in the sense of 2.2.1), as we have already seen in 1.3.3(b).

Exercises and complements. (a) Verify that Rel Set is an ordered category.

(b) A set-theoretical sum $X = \Sigma\, X_i$ is still a categorical sum in Rel Set, with the same injections $u_i\colon X_i \to X$.

(c) As Rel Set is selfdual, $\Sigma\, X_i$ is also a categorical product in Rel Set, with projections $u_i{}^\sharp\colon X \nrightarrow X_i$ (that are not mappings, generally).

2.2.4 Relations of abelian groups

The category Rel Ab *of abelian groups and (additive) relations* is used in Homological Algebra, since Mac Lane's book on Homology [M1] and Hilton's paper [Hi]. We have already considered its (involutive) monoids of endomorphisms $\mathcal{R}(A) = \operatorname{Rel} \operatorname{Ab}(A, A)$, in 1.3.3.

A relation $a\colon A \rightarrowtail B$ is a subgroup of the direct sum $A \oplus B$. Again, it can be viewed as a 'partially defined, many-valued homomorphism', that sends an element $x \in A$ to the subset $\{y \in B \mid (x, y) \in a\}$ of B, in a way consistent with the algebraic structure. (This means that, if (x, y) and (x', y') are a-related, also $(x + x', y + y')$ is.)

Given a consecutive $b\colon B \rightarrowtail C$, the composed relation $ba\colon A \rightarrowtail C$ is the subgroup of $A \oplus C$ defined as in (2.16).

As in 2.2.3, we have an ordered category, where $a \leqslant a'$ means that $a \subset a'$ as subgroups of $A \oplus B$.

The category Rel Ab has a *regular* involution, by reversing pairs:

$$a^{\sharp}\colon B \rightarrowtail A, \qquad a^{\sharp} = \{(y, x) \in B \oplus A \mid (x, y) \in a\}. \tag{2.17}$$

The regularity is proved as in 1.3.3(a). The ordering is consistent with the involution, in the sense that $a \leqslant b$, is equivalent to $a^{\sharp} \leqslant b^{\sharp}$.

The category Ab is a wide subcategory of Rel Ab, identifying a homomorphism $f\colon A \to B$ with its graph — an everywhere defined, single-valued relation.

Similarly we have the category $\mathrm{Rel}\,(R\,\mathsf{Mod})$ *of R-modules and (linear) relations*, where a relation $a\colon A \rightarrowtail B$ is a submodule of the direct sum $A \oplus B$.

*Every abelian category A can be similarly embedded in an ordered category Rel A, with the same objects and a regular involution (see [G10], Section 6.6).

2.2.5 The idempotent completion

As a well-known construction, that will appear below in many contexts and variations, a category C has an *idempotent completion* C^E, also called the *Karoubi envelope* or *Cauchy completion* of C (the last name will be explained in 6.5.4).

Formally, C^E can be defined as a category of functors (see Exercise (a)). Concretely, the *objects* of C^E are the idempotent endomorphisms $e, f \ldots$ of C. A *morphism* $a\colon e \to f$ between two idempotents $e\colon X \to X$ and $f\colon Y \to Y$ is an arrow $a \in \mathsf{C}(X, Y)$ such that $a = ae = fa$, or equivalently $a = fae$. (More formally, we can write this morphism as $(e, f; a)\colon e \to f$.)

The *composite* of $a\colon e \to f$ with $b\colon f \to g$ is defined as $ba\colon e \to g$. The *identity* of the projector e is $e\colon e \to e$, which indeed acts as a unit for legitimate compositions in C^E.

The original category C has an obvious full embedding in C^E

$$X \mapsto 1_X, \qquad (a\colon X \to Y) \mapsto (a\colon 1_X \to 1_Y). \tag{2.18}$$

The crucial property of C^E is that *its idempotents split*.

In fact an idempotent endomorphism $\varepsilon\colon e \to e$ of C^E is determined by two idempotents $\varepsilon \leqslant e$ in a semigroup $\mathsf{C}(X)$, and factorises as a composite of a split epi ε_1 and a split mono ε_2

$$\varepsilon_1 = (\varepsilon\colon e \to \varepsilon), \qquad \varepsilon_2 = (\varepsilon\colon \varepsilon \to e),$$

$$\varepsilon_2\varepsilon_1 = \varepsilon, \qquad \varepsilon_1\varepsilon_2 = (\varepsilon\colon \varepsilon \to \varepsilon) = 1_\varepsilon. \tag{2.19}$$

(The splitting of idempotents will be examined in 5.2.5.)

In particular, this procedure can be applied to any monoid S (viewed as a category on one formal object, as in 2.3.1(a)), producing a small *category* S^E where S is embedded as the monoid of endomorphisms of the object 1_S.

Actually, every *semigroup* S has an associated category S^E constructed as above, with $\mathrm{Ob}\,(S^E) = E_S$; but now we should not expect of embedding S in S^E: the latter category can even be empty, if S has no idempotents. However, if S is an inverse semigroup, the idempotent completion S^E is an important construction that will be used later (see 2.8.5).

Exercises and complements. (a) Define C^E as a category of functors, as in 1.6.6.

*(b) An interested reader can verify the *universal property of the idempotent completion*: every functor $F\colon \mathsf{C} \to \mathsf{D}$ with values in a category where all idempotents split factorises through the canonical embedding $\mathsf{C} \to \mathsf{C}^E$, by a functor $G\colon \mathsf{C}^E \to \mathsf{D}$ *determined up to functorial isomorphism*. This property will not be used here.

*(c) More precisely, this should be called a *biuniversal* property (see 5.6.3), because of the *weak* uniqueness condition on the functor G, which only determines C^E *up to equivalence* of categories.

2.2.6 The projector completion

Similarly, a category C equipped with a regular involution has a *projector completion* C^P, namely the full subcategory of C^E whose objects are the projectors of C (its symmetric idempotents).

The category C^P is equivalent to C^E (see Exercise (b)) but has the advantage of inheriting a regular involution, with $(a\colon e \to f)^\sharp = (a^\sharp\colon f \to e)$; the canonical embedding $\mathsf{C} \to \mathsf{C}^P$ preserves the involution.

The term 'projector' (for a 'special' idempotent endomorphism) will also be used for other kind of categories, not equipped with a regular involution: for pre-inverse categories (in 2.4.4) and 'prj-cohesive categories' (a particular case, in Section 3.3). The 'projector completion' of C will still be the full subcategory of C^E whose objects are the 'projectors' of C, in the specified sense.

Again, this procedure can be applied to any regular involutive semigroup S (possibly without a unit) producing a small involutive category S^P with $\text{Ob}(S^P) = \text{Prj}(S)$ and a regular involution.

Because of the regular involution, a morphism $a: e \to f$ is an isomorphism of the category S^P if and only if its composites with $a^\sharp: f \to e$ are the identities of e and f, if and only if we have in S

$$a^\sharp a = e, \qquad aa^\sharp = f. \tag{2.20}$$

Two projectors e, f have in S^P a *canonical morphism* from e to f

$$fe: e \to f. \tag{2.21}$$

It is an isomorphism of S^P if and only if $ef: f \to e$ is its inverse, which is equivalent to saying that (e, f) *is a regular pair* of projectors, in S

$$efe = e, \qquad fef = f. \tag{2.22}$$

In other words, two projectors of S are partial inverses if and only if they are related by a canonical isomorphism in S^P. However, saying that they are 'canonically isomorphic objects' of S^P *would be misleading and a source of errors, unless the relation expressed in* (2.22) *is transitive*

This is the case if and only if the semigroup S is orthodox: see [G9], Theorem 2.7.5.

Exercises and complements. Let C be a category with a regular involution.

(a) Every object of C^E is isomorphic to an object of C^P.

(b) The category C^E is equivalent to its subcategory C^P.

(c) C^P can be defined as a category of involution-preserving functors $P \to \mathsf{C}$, where P is an involutive monoid.

*(d) C^P satisfies a universal property concerning the splitting of projectors, and another universal property concerning the epi-mono factorisation of all morphisms: see [G9], Section 3.3.

2.2.7 *Complements on abelian relations*

Many topics of Homological Algebra, from homology to spectral sequences, rely on the study of subquotients and induced relations between them. This study works better if undertaken in $\text{Rel}\,\text{Ab}$ (or the category of relations on any abelian category).

We give here some elementary hints at this topic; the interested reader can find proofs in [G9] or [G10], and their developments in [G9]. This part is not used elsewhere, in this book.

(a) A relation $a\colon A \rightarrowtail B$ in Rel Ab determines two subgroups of A, called *definition* and *annihilator*

$$\begin{aligned}
\operatorname{Def} a &= \{x \in A \mid \exists y \in B\colon (x,y) \in a\}, \\
\operatorname{Ann} a &= \{x \in A \mid (x,0) \in a\},
\end{aligned} \tag{2.23}$$

and two subgroups of B, called *values* and *indeterminacy*

$$\begin{aligned}
\operatorname{Val} a &= \{y \in B \mid \exists x \in A\colon (x,y) \in a\} = \operatorname{Def}(a^{\sharp}), \\
\operatorname{Ind} a &= \{y \in B \mid (0,y) \in a\} = \operatorname{Ann}(a^{\sharp}).
\end{aligned} \tag{2.24}$$

(b) If $K \subset H$ are subgroups of A, the abelian group H/K is called a *subquotient* of A. (The prime example is the homology $\operatorname{Ker} d/\operatorname{Im} d$ of a differential group (A, d).)

The subquotients H/K of A amount to the projectors $e\colon A \rightarrowtail A$ in Rel Ab, i.e. the symmetric idempotent endo-relations of A. The bijective correspondence $H/K \mapsto e \mapsto H/K$ works as follows:

$$\begin{aligned}
e &= \{(x,y) \in A \oplus A \mid x,y \in H, x - y \in K\}, \\
H &= \operatorname{Def} e = \operatorname{Val} e, \qquad K = \operatorname{Ann} e = \operatorname{Ind} e.
\end{aligned} \tag{2.25}$$

(c) A projector $e\colon A \rightarrowtail A$ has two factorisations $e = m p^{\sharp} p m^{\sharp} = q^{\sharp} n n^{\sharp} q$

$$\tag{2.26}$$

where m, n are inclusions of subgroups and p, q are projections on quotients.

A subquotient H/K of A also amounts to a *subobject* $s\colon H/K \rightarrowtail A$ in Rel Ab, where s is the *monorelation* $m p^{\sharp} = q^{\sharp} n$. The projector (2.26) has an epi-mono factorisation $e = s s^{\sharp}\colon A \rightarrowtail A$, with $s^{\sharp} s = 1$.

(Therefore, every projector of Rel Ab splits.)

(d) For $a\colon A \rightarrowtail B$, the support $\underline{e}(a) = a^{\sharp}a\colon A \rightarrowtail A$ and the cosupport $\underline{e}^{*}(a) = a a^{\sharp}\colon B \rightarrowtail B$ correspond to the following subquotients of A and B, respectively

$$\operatorname{Coim} a = \operatorname{Def}(a)/\operatorname{Ann}(a), \qquad \operatorname{Im} a = \operatorname{Val}(a)/\operatorname{Ind}(a). \tag{2.27}$$

(d) The relation a has a canonical factorisation $a = cub$ in Rel Ab

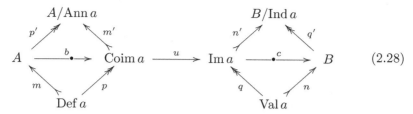

$$(2.28)$$

where $b^\sharp$ and c are monorelations and u is an isomorphism (of Ab and Rel Ab).

Thus Rel Ab has epi-mono factorisations, essentially unique by 2.2.2(b).

(e) All this holds for every category of modules; it can be extended to abelian categories, and much beyond [G9].

2.3 Semigroups, categories, and inverse categories

We begin now the study of the theory of inverse categories, extending that of inverse semigroups. Part of it can be deduced from the latter, via the one-morphism addition $C \mapsto C^\infty$ described in 2.3.1.

We generally prefer to avoid a formal use of this tool, that should nevertheless be kept in mind as a guideline for a conspicuous part of the extended theory.

Another relationship between semigroups and categories has already appeared in 2.2.5: the idempotent completion of a semigroup is a category; other instances will be dealt with in Section 2.8.

2.3.1 Semigroups and categories

There is a strong, two-way relationship between semigroups and categories.

(a) The obvious, well-known part of this relationship, already remarked in Section 1.4, says that a monoid can be viewed as a small category on a single object; a homomorphism of monoids is the same as a functor between the corresponding categories. (The idempotent completion recalled above, in 2.2.5, is a more interesting way of constructing a small category out of a semigroup, unital or not.)

(b) The other way round, from categories to (possibly large) semigroups, there is a less familiar procedure that will be called here the *one-morphism addition*. It should not be overly emphasised because it says nothing about prominent topics of category theory, like limits and adjunctions, and its extension to functors is defective. Nevertheless one can use this procedure to extend the theory of regular, involutive, orthodox and inverse semigroups to categories.

In fact (as already seen in 1.1.6 for semigroups), the class $M = \operatorname{Mor} \mathsf{C}$ of morphisms of the category C can be embedded in a class, written as $\mathsf{C}^\bullet$ or C^∞, by adding an element $\infty \notin M$, that may be called the *illicit morphism*, with no domain nor codomain.

The partial composition of M is extended to a (total) associative multiplication, letting for all $a, b \in M$

$$a.\infty = \infty.a = \infty.\infty = \infty,$$

$$ab = \infty, \quad \text{whenever } ab \text{ is not defined in } \mathsf{C}, \tag{2.29}$$

making ∞ an absorbing element.

Of course a large category C gives a *large semigroup* C^∞, with a proper class of elements.

The identities (and therefore the objects) of C are determined as the elements $e \in \mathsf{C}^\infty$ such that:

(i) $ee = e \neq \infty$,

(ii) for every $a \in \mathsf{C}^\infty$, ea and ae belong to $\{a, \infty\}$.

This complicated, unnatural description explains why all the notions of category theory where objects play an important role, limits for instance, have no reasonable counterpart in semigroup theory.

Given a functor $F \colon \mathsf{C} \to \mathsf{D}$, the extended mapping

$$F^\infty \colon \mathsf{C}^\infty \to \mathsf{D}^\infty, \qquad F^\infty(\infty) = \infty, \tag{2.30}$$

is a semigroup-homomorphism if and only if the functor F is *injective on the objects*, so that it (preserves and) reflects consecutive arrows. Otherwise, two different objects X, X' in C with $F(X) = F(X') = Y$ give

$$F(1_X).F(1_{X'}) = 1_Y, \qquad F^\infty(1_X.1_{X'}) = F^\infty(\infty) = \infty.$$

Here it would be convenient to put an ordering on the semigroup C^∞, where $a \leqslant a \leqslant \infty$ for every $a \in \mathsf{C}^\infty$ (and nothing more). In this way the mapping $F^\infty \colon \mathsf{C}^\infty \to \mathsf{D}^\infty$ becomes a *lax morphism* of ordered semigroups, in the sense that it satisfies the following conditions (for $a, b \in \mathsf{C}^\infty$)

$$F^\infty(a).F^\infty(b) \leqslant F^\infty(ab), \qquad a \leqslant b \implies F^\infty(a) \leqslant F^\infty(b). \tag{2.31}$$

Remarks. (a) According to the context, one may prefer to reverse the order of the extended semigroup $\mathsf{C}^\bullet$, so that the illicit morphism becomes the *minimum*, written perhaps as $-\infty$. Then the extension $F^{\bullet}\colon \mathsf{C}^{\bullet} \to \mathsf{D}^{\bullet}$ of a functor becomes a *colax* morphism of semigroups.

Writing the illicit morphism as 0 can be inconvenient, e.g. for pointed categories.

(b) Large semigroups like Set^{∞} have a relevant role in Lawson's book on inverse semigroups [Ls], where a colax morphism of inverse semigroups is called a 'prehomomorphism' (cf. Section 2.8).

2.3.2 Categories and partial inverses

We say that the morphisms $a\colon X \to Y$ and $b\colon Y \to X$ of the category C are *partial inverses*, or form a *regular pair*, if

$$a = aba, \qquad b = bab. \tag{2.32}$$

This amounts to saying that (a, b) becomes a regular pair in the class-semigroup C^{∞} defined above; the latter has an additional regular pair: (∞, ∞).

The category C will be said to be *regular in the sense of von Neumann*, or *vN-regular*, if every morphism has some partial inverse, or equivalently if C^{∞} is a regular semigroup.

(The term 'regular category' has a different, unrelated meaning in category theory: see [Ba, Bo].)

As for semigroups, each of the relations in (2.32) implies that $ba\colon X \to X$ and $ab\colon Y \to Y$ are idempotents of C (i.e. idempotent endomorphisms). On the other hand, every idempotent endomorphism e is a partial inverse of itself.

Every functor preserves regular pairs of morphisms.

Proposition 1.2.2 extends to categories: the category C is vN-regular if (and only if) for every morphism a there is some morphism b (with reversed domain and codomain, necessarily) such that $a = aba$. The proof is the same.

(One can also use the semigroup C^{∞} to derive the present result from Proposition 1.2.2 — which of course holds as well for large semigroups.)

2.3.3 Definition (Inverse category)

An *inverse category* will be a category K where every morphism $a\colon X \to Y$ has one and only one partial inverse (as defined in 2.3.2), which we write as $a^{\sharp}\colon Y \to X$.

Then K has a *canonical involution* $a \mapsto a^{\sharp}$, which is the unique *regular* one (see 2.2.1). The proof is practically the same as for semigroups (in 1.3.4), and is not rewritten here.

A functor $F \colon \mathsf{K} \to \mathsf{K}'$ between inverse categories automatically preserves partial inverses and commutes with the canonical involutions. An *inverse subcategory* of an inverse category K is defined as a subcategory closed under partial inverses.

If S is an inverse semigroup, then its projector completion S^P (cf. 2.2.6) is an inverse category: indeed, given a morphism $a \colon e \to f$ of S^P, the morphism $a^{\sharp} \colon f \to e$ is a partial inverse of $a \colon e \to f$, and obviously the only one.

A homomorphism $h \colon S \to T$ of inverse semigroups yields a functor between inverse categories

$$H \colon S^P \to T^P, \qquad H(e) = h(e),$$
$$H(a \colon e \to f) = h(a) \colon h(e) \to h(f). \tag{2.33}$$

2.3.4 Idempotents as projectors

In an inverse category K every object X has an inverse monoid $\mathsf{K}(X)$ of endomorphisms, to which we can apply the previous results on inverse semigroups (cf. 1.3.4(b)).

Every idempotent $e \colon X \to X$ is its own partial inverse ($e^{\sharp} = e$), and a *projector* with respect to the canonical involution. These projectors commute and form a 1-semilattice $\mathsf{Prj}\,(X) \subset \mathsf{K}(X)$ (written as $\mathsf{Prj}_{\mathsf{K}}(X)$ when useful)

$$e \leqslant f \;\Leftrightarrow\; e = ef = fe, \qquad e \wedge f = ef = fe. \tag{2.34}$$

Extending 2.2.2, every K-morphism $a \colon X \to Y$ has two associated projectors

$$\underline{e}(a) = a^{\sharp}a \in \mathsf{Prj}\,(X) \qquad\qquad \text{(the *support* of } a\text{)},$$
$$\underline{e}^*(a) = aa^{\sharp} = \underline{e}(a^{\sharp}) \in \mathsf{Prj}\,(Y) \qquad \text{(the *cosupport* of } a\text{)}, \tag{2.35}$$

$$a = a.\underline{e}(a) = \underline{e}^*(a).a = \underline{e}^*(a).a.\underline{e}(a). \tag{2.36}$$

All the idempotents of K turn up as supports and cosupports: indeed an arrow e is an idempotent endomorphism if and only if $e = \underline{e}(e)$, if and only if $e = \underline{e}^*(e)$.

In any category, an *isomorphism* a has precisely one partial inverse: its inverse a^{-1}. Every groupoid (see 1.4.5) is an inverse category.

2.3.5 Partial bijections between sets

We now describe the 'prime example' of an inverse category, namely the category $\mathcal{I}$ of *partial bijections between sets*. It is a subcategory of the category $\mathcal{S}$, of sets and partial mappings (cf. 2.1.3) and contains the inverse monoid $\mathcal{I}(X) = \mathcal{I}(X, X)$ of partial endo-bijections of every set X, already considered above.

An object of $\mathcal{I}$ is a set. A *partial bijection*

$$f = (X_0, Y_0, f_0) \colon X \nrightarrow Y, \tag{2.37}$$

is a bijective mapping $f_0 \colon X_0 \to Y_0$ from a subset $X_0 = \operatorname{Def} f \subset X$ to a subset $Y_0 = \operatorname{Val} f \subset Y$.

Equivalently, f is an *injective* partial mapping, in the category $\mathcal{S}$; but the term 'partial injection' might mask the selfdual character of $\mathcal{I}$.

Given another partial bijection $g \colon Y \nrightarrow Z$, the composite $gf \colon X \nrightarrow Z$ is their composite in $\mathcal{S}$, i.e. the partial bijection

$$x \mapsto g(f(x)),$$

$$\operatorname{Def}(gf) = \{x \mid x \in \operatorname{Def} f, f(x) \in \operatorname{Def} g\}, \tag{2.38}$$

$$\operatorname{Val}(gf) = \{g(f(x)) \mid x \in \operatorname{Def} f, f(x) \in \operatorname{Def} g\}.$$

$\mathcal{I}$ is thus a wide subcategory of $\mathcal{S}$. Its identities are the identity mappings of sets.

Moreover, $\mathcal{I}$ is inverse: the unique partial inverse of $f \colon X \nrightarrow Y$ is the partial bijection $f^\sharp \colon Y \nrightarrow X$ whose associated bijection $f_0^{-1} \colon \operatorname{Val} f \to \operatorname{Def} f$ is the inverse of $f_0 \colon \operatorname{Def} f \to \operatorname{Val} f$.

The support and cosupport of $f \colon X \nrightarrow Y$ are *partial identities* of the sets X and Y

$$\begin{aligned}
\underline{e}(f) &= f^\sharp f \colon X \nrightarrow X, & (\underline{e}(f))_0 &= \operatorname{id}(\operatorname{Def} f), \\
\underline{e}^*(f) &= f f^\sharp \colon Y \nrightarrow Y, & (\underline{e}^*(f))_0 &= \operatorname{id}(\operatorname{Val} f).
\end{aligned} \tag{2.39}$$

The morphism $f \colon X \nrightarrow Y$ is invertible in $\mathcal{I}$ if and only if it is a total bijection, i.e. an isomorphism of Set. An endomorphism $f \colon X \nrightarrow X$ is idempotent if and only if $f = \underline{e}(f)$, if and only if f is a *partial identity* of X over some subset U (and then $\operatorname{Def} f = U = \operatorname{Val} f$).

Exercises and complements. (a) The category $\mathcal{I}$ is equivalent to a wide subcategory of Set$_\bullet$.

*(b) The category $\mathcal{I}$ is a (non-abelian) Puppe-exact category, with distributive lattices of subobjects ([G10], Section 6.3). $\mathcal{I}$ and its extensions play a central role in a construction of universal models for spectral sequences [G9].

*(c) Every inverse category can be embedded in $\mathcal{I}$ [Kas].

2.3.6 Partial homeomorphisms

We are also interested in the inverse category $\mathrm{I}\mathcal{T}$ of *topological spaces and partial homeomorphisms* $f \colon X \nrightarrow Y$ (i.e. homeomorphisms $\mathrm{Def}\, f \to \mathrm{Val}\, f$ between subspaces of X and Y).

In fact, we are more interested in two wide subcategories of $\mathrm{I}\mathcal{T}$:

- $\mathrm{I}\mathcal{C}$, whose arrows $X \nrightarrow Y$ are partial homeomorphisms between *open* subspaces of X and Y,

- $\mathrm{I}\mathcal{D}$, whose arrows $X \nrightarrow Y$ are partial homeomorphisms between *closed* subspaces of X and Y.

We shall see, in 2.4.5, how the categories of partial mappings

$$\mathcal{S}, \quad \mathcal{T}, \quad \mathcal{C}, \quad \mathcal{D}, \quad \mathcal{C}^r,$$

introduced in 2.1.3 and 2.1.4 have an 'inverse core', the categories

$$\mathcal{I} = \mathrm{I}\mathcal{S}, \quad \mathrm{I}\mathcal{T}, \quad \mathrm{I}\mathcal{C}, \quad \mathrm{I}\mathcal{D}, \quad \mathrm{I}\mathcal{C}^r.$$

2.3.7 Theorem (Characterisation of inverse categories)

The following conditions on a category K *are equivalent:*

(i) K *is inverse,*

(ii) K *is vN-regular and its idempotents commute,*

(iii) K *is vN-regular and its monoids* $\mathsf{K}(X) = \mathsf{K}(X, X)$ *are inverse,*

(iv) K *has a regular involution and its idempotents commute,*

(v) K *has a regular involution whose projectors commute,*

(vi) K *has a regular involution whose projectors are closed under composition.*

Note. For a regular involution, a projector is a symmetric idempotent.

Proof It is, essentially, a corollary of the similar characterisations of inverse semigroups, namely Theorem 1.2.6 and Proposition 1.3.5.

(i) $\Rightarrow$ (iii). Obvious.

(iii) $\Rightarrow$ (ii). By Theorem 1.2.6, applied to the semigroups $\mathsf{K}(X)$ of endomorphisms.

(ii) $\Rightarrow$ (i). One can rewrite the first (easy) part of the proof of Theorem 1.2.6, for a morphism $a \colon X \to Y$ with partial inverses $b, c \colon Y \to X$.

(i) $\Rightarrow$ (iv). By 2.3.3, taking into account that we already know that (i) implies (ii).

(iv) $\Rightarrow$ (ii). Obvious.

(iv) $\Leftrightarrow$ (v) $\Leftrightarrow$ (vi). By Proposition 1.3.5, applied to the involutive semi-groups $K(X)$. $\qquad\square$

2.3.8 Lemma (Regularity lemma for categories)

Let the following diagram be given in an arbitrary category C

$$X \mathrel{\substack{a \\ \longrightarrow \\ \longleftarrow \\ b}} Y \mathrel{\substack{a' \\ \longrightarrow \\ \longleftarrow \\ b'}} Z \qquad a = aba, \quad a' = a'b'a'. \tag{2.40}$$

Then the following properties are equivalent:

(i) $a'a = (a'a)(bb')(a'a)$,

(ii) $(b'a')(ab)$ *is an idempotent endomorphism of* Y.

Note. The morphism $(b'a')(ab)$ is a product of idempotents; therefore condition (b) is automatically satisfied when C is an inverse category.

Proof One can rewrite the proof of the Regularity Lemma for semigroups, in 1.2.8. Or apply the one-morphism addition of 2.3.1. $\qquad\square$

2.4 The canonical order of an inverse category

The results of Section 1.3, on the canonical order of inverse semigroups, can be easily extended to inverse categories.

We also introduce pre-inverse categories and their inverse core.

2.4.1 Theorem

Let K *be an inverse category. For two morphisms* a, b *in* K *the following properties are equivalent (and imply that the arrows* a, b *are parallel):*

(i) $\quad a = ab^\sharp a$,

(ii) $\quad a = b(a^\sharp a)$,

(ii*) $\quad a = (aa^\sharp)b$,

(iii) $\quad (aa^\sharp)b(a^\sharp a)$,

(iv) $\quad a = be$, *for some projector* e,

(iv*) $\quad a = fb$, *for some projector* f,

(v) $\quad a = fbe$, *for some projectors* e, f.

Proof One can rewrite the proof of Theorem 1.3.8. Or apply the one-morphism addition of 2.3.1. $\qquad\qquad\square$

2.4.2 Theorem and Definition

(a) The inverse category K *has a* canonical order $a \leqslant b$ *(consistent with composition and partial inverses), defined by the equivalent properties (i)–(v) of the previous theorem.*

(b) This order of K *extends the order of all the meet semilattices* $\mathsf{Prj}\,(X)$ *of projectors, defined in 1.1.7, and is generated by the latter (as an order of categories).*

(c) The projectors $e \in \mathsf{Prj}\,(X)$ *are characterised by the condition* $e \leqslant 1_X$. *Furthermore, the categorical order of* K *is generated by the condition* $e \leqslant 1_X$ *(for every object* X *and every* $e \in \mathsf{Prj}\,(X)$*).*

Proof One could apply the one-morphism addition of 2.3.1 and the corresponding Theorem 1.3.9, for inverse semigroups. But it will be clearer to rewrite the proof, writing as E the class of all projectors of K.

(a) The relation $a \leqslant b$, is reflexive (use property (i)) and transitive (use (iv)); it is also antisymmetric: if $a = be$ and $b = af$ (with $e, f \in E$), then $a = afe = aef = be.ef = af = b$.

The fact that the order is consistent with composition follows from the Regularity Lemma 2.3.8. Indeed, assuming that $a \leqslant b$ and $a' \leqslant b'$, we can apply the lemma to the pairs $(a, b^\sharp)$ and $(a', b'^\sharp)$, concluding that $aa' = (aa')(b'^\sharp b^\sharp)(aa')$, which means that $aa' \leqslant bb'$.

Finally, if $a \leqslant b$, then $a^\sharp \leqslant b^\sharp$.

(b) The first claim is obvious. For the second, suppose we have in K an order of categories $a \prec b$ that extends the order $\leqslant$ of projectors (i.e. idempotents). If the pair a, b satisfies the equivalent properties (i)–(v), then $ab^\sharp$ and $b^\sharp a$ are idempotents and

$$ab^\sharp = (aa^\sharp)(bb^\sharp) \leqslant bb^\sharp, \qquad b^\sharp a = (b^\sharp b)(a^\sharp a) \leqslant b^\sharp b.$$

Therefore, since $\prec$ is an order of categories and extends the order of idempotents:

$$a = ab^\sharp a \prec bb^\sharp a \prec bb^\sharp b = b.$$

(c) The last statement is an obvious consequence of form (iv) in 2.4.1. $\quad\square$

2.4.3 Pre-inverse categories

A *pre-inverse* category C will be an ordered category where

(i) for every object X, the endomorphisms $e \leqslant 1_X$ are idempotent.

We prove below, in Theorem 2.4.4, that such a category has a canonical inverse subcategory.

Exercises and complements. (a) In a pre-inverse category C the endomorphisms $e \leqslant 1_X$ commute and will be called projectors of X; they form a 1-semilattice $\mathsf{Prj}\,(X)$ in the monoid $\mathsf{C}(X)$.

(b) The category $\mathcal{S}$, of sets and partial mappings, is pre-inverse. Every $\mathcal{S}$-concrete category $U \colon \mathsf{C} \to \mathcal{S}$ (see 2.1.3) is also, with respect to the order lifted by U. Various examples are listed in 2.1.4, and others will be in Chapter 4.

(c) In $\mathcal{S}$ an idempotent morphism $e \colon X \nrightarrow X$ need not be $\leqslant 1_X$. (We already know that $e \geqslant 1_X$ means $e = 1_X$.)

2.4.4 Theorem and Definition (The inverse core)

Let C be a pre-inverse category.

(a) There is a wide subcategory $\mathsf{K} = \mathsf{IC}$ *of* C, *formed by the morphisms* $u \colon X \to Y$ *that have a* Morita inverse $u' \colon Y \to X$ *in* C; *the latter is defined (as in [Bir]) by the properties:*

$$\text{(i)} \quad u = uu'u, \qquad u' = u'uu', \qquad u'u \leqslant 1_X, \qquad uu' \leqslant 1_Y.$$

IC *is an inverse category, and will be called the* inverse core *of* C.

The projectors of an object X *in the inverse category* K *(i.e. its idempotent endomorphisms) coincide with the projectors of* X *in* C *(i.e. its endomorphisms* $\leqslant 1_X$*), so that there is no ambiguity in writing* $\mathsf{Prj}\,(X)$. *The partial inverse in* K *is given by the Morita inverse in* C *(which is thus unique).*

For two parallel morphisms u, v *in* K, *the canonical order* $u \leqslant_\mathsf{K} v$ *of inverse categories is defined by the relation* $u = uv^{\sharp}u$ *(see Theorem 2.4.1), and coincides with* $u \leqslant_\mathsf{C} v$.

(b) Every inverse category is pre-inverse, and coincides with its inverse core.

(c) (The universal property) Suppose that $F \colon \mathsf{K}' \to \mathsf{C}$ *is a functor defined on an inverse category, that preserves the orders. Then* F *takes values in the inverse core* IC.

Note. The category of small inverse categories is thus coreflective in the category of small pre-inverse categories, with counit $\varepsilon\colon \mathsf{IC} \to \mathsf{C}$ given by the inclusion: see 1.8.5.

Proof (a) K is a subcategory of C. Indeed, if $u\colon X \to Y$ and $v\colon Y \to Z$ have Morita inverses u' and v', respectively, then $u'v'$ is a Morita inverse of vu, because:

$$vu.u'v'.vu = v.uu'.v'v.u = vv'v.uu'u = vu,$$

$$vu.u'v' = v(uu')v' \leqslant vv' \leqslant 1.$$

The category K is thus vN-regular (in the sense of 2.3.2).

Every idempotent endomorphism e of K satisfies $e \leqslant 1$ in C: indeed, if v is a Morita inverse of e in C then $v = vev = ve.ev \leqslant 1$ is also idempotent and $e = eve = ev.ve \leqslant 1$. As the converse is trivial, the idempotent endomorphisms of K coincide with the endomorphisms $\leqslant 1$ of C, and commute by 2.4.3(a).

The category K is inverse, by Theorem 2.3.7, and the partial inverse of a morphism u in K must coincide with the Morita inverse of u in C.

If $u \leqslant_{\mathsf{K}} v$, then $u = ve$ for a projector $e \leqslant_{\mathsf{C}} 1$ and $u \leqslant_{\mathsf{C}} v$. Conversely, if $u \leqslant_{\mathsf{C}} v$ (with u, v in K), then $uv^{\sharp} \leqslant_{\mathsf{C}} vv^{\sharp} \leqslant_{\mathsf{C}} 1$, whence $uv^{\sharp}$ is idempotent in C. But $uv^{\sharp}$ belongs to K, and is thus a projector of K. Therefore $vu^{\sharp} = uv^{\sharp} \leqslant_{\mathsf{C}} 1$ and $u = uu^{\sharp}u \leqslant_{\mathsf{C}} vu^{\sharp}u \leqslant_{\mathsf{C}} u$, so that $u = v(u^{\sharp}u) \leqslant_{\mathsf{K}} v$.

(b) Follows from 2.4.2(c).

(c) Obvious.   □

2.4.5 Examples and complements

We can now review various pre-inverse and inverse categories, introduced in 2.1.3, 2.1.4, 2.3.5 and 2.3.6.

(a) We already remarked that the ordered category $\mathcal{S}$ of sets and partial mappings is pre-inverse (together with all $\mathcal{S}$-concrete categories). Its inverse core is the ordered subcategory $\mathcal{I} = \mathsf{I}\mathcal{S}$ of sets and partial bijections.

The projectors $e\colon X \nrightarrow X$ of a set X (with respect to $\mathcal{S}$ and $\mathcal{I}$) are the partial identities $e \leqslant 1_X$; they form an ordered set $\mathsf{Prj}\,(X)$ isomorphic to the boolean algebra $\mathcal{P}(X)$ of the subsets of X:

$$\mathsf{Prj}\,(X) \to \mathcal{P}(X), \qquad e \mapsto \mathrm{Def}\,e = \mathrm{Val}\,e.$$

(b) The inverse core of the pre-inverse category $\mathcal{T}$ of topological spaces and partial maps is the category $\mathsf{I}\mathcal{T}$ of spaces and partial homeomorphisms between subspaces of domain and codomain. The projectors $e \leqslant 1_X$ of a space X are its partial identities, and form again an ordered set $\mathsf{Prj}\,(X)$ isomorphic to the boolean algebra $\mathcal{P}(X)$.

(c) Analogously, the inverse core of the category $\mathcal{C}$ (resp. $\mathcal{D}$) of topological spaces and partial continuous mappings defined on open (resp. closed) subspaces is the category $\mathrm{I}\mathcal{C}$ (resp. $\mathrm{I}\mathcal{D}$) of topological spaces and partial homeomorphisms between open (resp. closed) subspaces of domain and codomain.

In $\mathcal{C}$ the projectors $e \leqslant 1_X$ of a space X form an ordered set $\mathsf{Prj}\,(X)$ isomorphic to the 'frame' $\mathcal{O}(X)$ of the open subsets of X (see 6.1.2).

(d) The inverse core of the pre-inverse category $\mathcal{C}^r$ (see 2.1.4) is the wide subcategory $\mathrm{I}\mathcal{C}^r$ (see 2.3.6) of *open euclidean spaces and partial diffeomorphisms of class* C^r, *between open subspaces of domain and codomain.*

(Of course, 'partial C^0-diffeomorphism' means 'partial homeomorphism'.)

2.4.6 Exercises and complements

(a) The ordered category Ord is not pre-inverse.

(b) The ordered category $\mathsf{Rel}\,\mathsf{Set}$ is pre-inverse.

Characterise the endo-relations $e \leqslant 1_X$ and the inverse core $\mathrm{I}(\mathsf{Rel}\,\mathsf{Set})$. Show that an endo-relation $a \geqslant 1_X$ need not be idempotent.

(c) The ordered category $\mathsf{Rel}\,\mathsf{Ab}$ is pre-inverse.

Characterise the endo-relations $e \leqslant 1_A$ and $\mathrm{I}(\mathsf{Rel}\,\mathsf{Ab})$. Show that the condition $a = be$ (for some $e \leqslant 1_A$) is *not* equivalent to $a \leqslant b$. (It is the case in all the categories of 2.4.5.)

2.5 Euclidean spheres and topological groups

After the basic theory of inverse categories, we investigate other prerequisites, about euclidean spheres and topological groups.

The spheres are examined here as topological spaces; they will be dealt with as topological realisations of smooth manifolds in 4.1.2, and as 'directed spaces' in Sections 4.3 and 4.4.

For a reader interested in Topology there are many books, like Munkres [Mu], Kelley [Ke] and Bourbaki [Bou2].

Other references on Algebraic Topology can be found in the next section.

2.5.1 Euclidean spaces

We use the following notation, for spaces which play an important role in Topology and Algebraic Topology.

The space $\mathbb{R}^n$ of n-tuples of real numbers $x = (x_1, ..., x_n)$ is equipped with the euclidean topology, determined by the euclidean norm and the euclidean distance

$$||x|| = (x_1^2 + ... + x_n^2)^{1/2}, \qquad d(x, y) = ||x - y||, \qquad (2.41)$$

or equivalently as a cartesian power of the euclidean line $\mathbb{R}$.

Identifying $\mathbb{R}^n$ with the hyperplane $\mathbb{R}^n \times \{0\} \subset \mathbb{R}^{n+1}$, these spaces form a nested family, and the singleton $\mathbb{R}^0 = \{0\}$ is included in all of them

$$\{0\} = \mathbb{R}^0 \subset \mathbb{R}^1 \subset \mathbb{R}^2 \subset ... \subset \mathbb{R}^n \subset ... \qquad (2.42)$$

A *euclidean space* is any subspace of some $\mathbb{R}^n$, and an *open euclidean space* is any open subspace of some $\mathbb{R}^n$.

$\mathbb{I}$ denotes the standard compact interval $[0, 1]$ of $\mathbb{R}$, on which standard paths and homotopies are parametrised (cf. 4.3.1). Its cartesian power $\mathbb{I}^n$ is the *standard n-cube*, a subspace of $\mathbb{R}^n$.

Inside $\mathbb{R}^{n+1}$, we also have the *standard n-sphere*

$$\mathbb{S}^n = \{x \in \mathbb{R}^{n+1} \mid ||x|| = 1\} \qquad (n \geqslant 0), \qquad (2.43)$$

pointed at $e_1 = (1, 0, ..., 0)$ when useful.

In particular, $\mathbb{S}^0 = \{-1, 1\} \subset \mathbb{R}$ is a discrete space on two points. Adding a lower sphere $\mathbb{S}^{-1} = \emptyset$ as the sphere of $\mathbb{R}^0 = \{0\}$, the spheres form a nested family of subspaces of the family (2.42)

$$\emptyset = \mathbb{S}^{-1} \subset \mathbb{S}^0 \subset \mathbb{S}^1 \subset ... \subset \mathbb{S}^{n-1} \subset ... \qquad (2.44)$$

The convenience of adding $\mathbb{S}^{-1}$ will also appear in several points, below. Its topological dimension should be set at $-\infty$, to preserve the validity of the formula $\dim(X \times Y) = \dim X + \dim Y$.

The standard circle $\mathbb{S}^1$ can also be viewed as a subspace of the complex plane, which topologically is the same as $\mathbb{R}^2$

$$\mathbb{S}^1 = \{z \in \mathbb{C} \mid |z| = 1\}. \qquad (2.45)$$

This gives $\mathbb{S}^1$ the structure of a topological group for complex multiplication, as examined below.

2.5.2 Spheres and cubes

(a) The standard circle $\mathbb{S}^1$ can be obtained as the coequaliser in Top of the following pair of maps

$$\partial^0, \partial^1 \colon \{*\} \rightrightarrows \mathbb{I}, \qquad \partial^0(*) = 0, \quad \partial^1(*) = 1, \tag{2.46}$$

namely the quotient $\mathbb{I}/R$ modulo the equivalence relation that identifies the points of $\partial\mathbb{I} = \{0, 1\}$, the boundary of $\mathbb{I}$ in the line.

In fact, the exponential mapping

$$\varphi \colon \mathbb{I} \to \mathbb{S}^1, \qquad \varphi(t) = (\cos 2\pi t, \sin 2\pi t) = e^{2\pi i t}, \tag{2.47}$$

has equivalence relation $R_\varphi = R$. It induces a bijective map $\mathbb{I}/R \to \mathbb{S}^1$, which is a homeomorphism (since $\mathbb{I}/R$ is a compact space and $\mathbb{S}^1$ is Hausdorff).

The quotient $\mathbb{I}/R$ is often written as $\mathbb{I}/\partial\mathbb{I}$. The quotient X/A, where X is a space and A is a non-empty subset, denotes the topological quotient of X modulo the equivalence relation which collapses the points of A; the quotient can be viewed as a space pointed at the equivalence class A. (See Exercise 2.5.4(c) for the general definition of X/A, without exceptions.)

(b) In general, the n-sphere $\mathbb{S}^n$ can be obtained as the topological quotient of the cube $\mathbb{I}^n$ modulo its boundary $\partial\mathbb{I}^n$ in $\mathbb{R}^n$

$$\mathbb{S}^n \cong \mathbb{I}^n/(\partial\mathbb{I}^n) \qquad (n > 0). \tag{2.48}$$

In fact, starting from a homeomorphism $]0, 1[^n \to \mathbb{R}^n$ and a stereographic projection $\mathbb{R}^n \to \mathbb{S}^n \setminus \{x_0\}$, we get a surjective map $\varphi \colon \mathbb{I}^n \to \mathbb{S}^n$ that takes the boundary $\partial\mathbb{I}^n$ to the point $x_0 \in \mathbb{S}^n$, and is injective out of the boundary. We conclude as in (a): φ induces a homeomorphism from the (compact) space $\mathbb{I}^n/(\partial\mathbb{I}^n)$ to the (Hausdorff) space $\mathbb{S}^n$.

2.5.3 Spheres and discs

The sphere $\mathbb{S}^2$ is the union of two compact hemispheres, which meet at $\mathbb{S}^1$ in the plane $x_3 = 0$

$$\begin{aligned} \mathbb{D}^2_+ &= \{x \in \mathbb{S}^n \mid x_3 \geqslant 0\}, & \mathbb{D}^n_- &= \{x \in \mathbb{S}^n \mid x_3 \leqslant 0\}, \\ \mathbb{D}^2_+ \cup \mathbb{D}^2_- &= \mathbb{S}^2, & \mathbb{D}^2_+ \cap \mathbb{D}^2_- &= \mathbb{S}^1, \end{aligned} \tag{2.49}$$

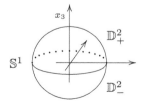

$\mathbb{S}^2$ is thus obtained *pasting* two copies of the compact disc $\mathbb{D}^2$ along their boundary $\mathbb{S}^1$. In other words, we form a commutative diagram

$$\begin{array}{ccc} \mathbb{S}^1 & \xrightarrow{\;i\;} & \mathbb{D}^2 \\ {\scriptstyle i}\downarrow & & \downarrow{\scriptstyle u_1} \\ \mathbb{D}^2 & \xrightarrow[\;u_2\;]{} & \mathbb{S}^2 \end{array} \qquad \mathbb{D}^2 = \{x \in \mathbb{R}^2 \mid ||x|| \leqslant 1\}, \qquad (2.50)$$

where $i\colon \mathbb{S}^1 \to \mathbb{D}^2$ is the inclusion, and $u_1, u_2\colon \mathbb{D}^2 \to \mathbb{S}^2$ embed the disc as the lower or upper closed hemisphere of $\mathbb{S}^2$

$$u_1(x_1, x_2) = (x_1, x_2, -\sqrt{1 - x_1^2 - x_2^2}),$$
$$u_2(x_1, x_2) = (x_1, x_2, \sqrt{1 - x_1^2 - x_2^2}).$$

The sphere has the finest topology making these embeddings continuous, or — in other words — is the quotient of the topological sum $\mathbb{D}^2 + \mathbb{D}^2$ of two discs modulo the equivalence relation that identifies their boundaries.

2.5.4 Topological pushouts

All this is better understood introducing a colimit of topological spaces, that will be dealt with in Section 5.1 for a general category.

We start from a pair of continuous mappings

$$f\colon X_0 \to X_1, \qquad g\colon X_0 \to X_2,$$

with the same domain (replacing two copies of the inclusion $i\colon \mathbb{S}^1 \to \mathbb{D}^2$).

Their *pushout*, or *fibred coproduct*, is a space A equipped with two maps $u_i\colon X_i \to A$ $(i = 1, 2)$ which form a commutative square $u_1 f = u_2 g$, in a universal way

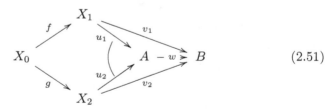

$$(2.51)$$

(i) for every triple (B, v_1, v_2) such that $v_1 f = v_2 g$ there exists a unique map $w\colon A \to B$ such that $w u_1 = v_1$, $w u_2 = v_2$.

(The pushout corner is often marked as above.) The pushout object A is also called a *pasting* over X_0 and written as $X_1 +_{X_0} X_2$. In fact (as proved in Exercise (a), below), the space A can be built as a quotient of the topological sum $X_1 + X_2$. We let $j_i : X_i \to X_1 + X_2$ be the injections of the sum (for $i = 1, 2$) and we let A be the following coequaliser

$$X_0 \underset{j_2 g}{\overset{j_1 f}{\rightrightarrows}} X_1 + X_2 \xrightarrow{\ p\ } A \qquad (2.52)$$

i.e. the quotient of $X_1 + X_2$ modulo the equivalence relation generated by identifying each $j_1 f(x)$ with the corresponding $j_2 g(x)$, for all $x \in X_0$. The maps $u_i : X_i \to A$ of diagram (2.51) are obtained as $u_i = p j_i$.

If $A = \emptyset$, the pushout is the sum $X_1 + X_2$.

If $f = g$ (and $X_1 = X_2$), the pushout $u_1, u_2 : X_1 \rightrightarrows A$ of the diagram $X_1 \leftarrow X_0 \to X_1$ is called the *cokernel pair* of f. The construction of the sphere $\mathbb{S}^2$ in (2.50) is of this type; it is extended to any dimension in Exercise (b).

Exercises and complements. (a) Prove that the construction (2.52), by a sum and a coequaliser, gives indeed the pushout.

(b) Extending the 2-dimensional case, the sphere $\mathbb{S}^n$ is the cokernel pair of the inclusion

$$\mathbb{S}^{n-1} \to \mathbb{D}^n, \qquad \mathbb{D}^n = \{x \in \mathbb{R}^n \mid ||x|| \leqslant 1\} \qquad (n \geqslant 1), \qquad (2.53)$$

of $\mathbb{S}^{n-1}$ into the compact disc $\mathbb{D}^n$ (also called the standard n-dimensional *closed ball*).

This also holds for $n = 0$, if we let $\mathbb{S}^{-1} = \emptyset$, as in (2.44).

(c) The space X/A, defined in 2.5.2(a) for a subset $A \neq \emptyset$ of the space X, can be defined for an arbitrary subset A, as the pushout of the inclusion $i : A \to X$ and the (unique) map $A \to \{*\}$

$$\begin{array}{ccc}
A & \xrightarrow{\ i\ } & X \\
\downarrow & \nearrow & \downarrow \\
\{*\} & \xrightarrow{\ v\ } & X/A
\end{array} \qquad (2.54)$$

In this way, if $A = \emptyset$ we get the topological sum $X + \{*\}$, with an added base-point $v(*)$.

Now the formula $\mathbb{S}^n \cong \mathbb{I}^n/(\partial \mathbb{I}^n)$ of (2.48) also holds for $n = 0$:

$$\mathbb{I}^0/(\partial \mathbb{I}^0) = \{*\}/\emptyset = \{*\} + \{*\} = \mathbb{S}^0. \qquad (2.55)$$

2.5.5 Topological groups

Topological groups, and their actions, will often be used in the next chapters. We only recall some basic facts of this subject, for which we refer to [Bou2], Chapter III.

A *topological group* is a group G equipped with a consistent topology. This means that its structural mappings are continuous: the multiplication $G \times G \to G$ (for the product topology on the domain), and the inversion mapping $G \to G$, $x \mapsto x^{-1}$. The latter is thus an involutive homeomorphism.

It is also important to note that every element $g \in G$ gives a translation map

$$\tau(g) = \varphi(g, -) \colon G \to G, \qquad x \mapsto gx, \qquad (2.56)$$

which is a homeomorphism, with inverse $\tau(g^{-1})$. As $\tau(g)(1_G) = g$, in a topological group all points 'look alike' (topologically, of course).

The morphisms of the category of topological groups are the continuous homomorphisms.

A subgroup of a topological group inherits a consistent topology. The same holds for cartesian products.

Exercises and complements. (a) $\mathbb{R}^n$ is a topological group, for the sum. The non-zero complex numbers form a topological group $\mathbb{C}^*$, for multiplication. $\mathbb{S}^1$ is a topological subgroup of $\mathbb{C}^*$.

(b) The exponential mapping

$$\varphi \colon \mathbb{R} \to \mathbb{C}^*, \qquad \varphi(t) = e^{2\pi i t} = \cos 2\pi t + i.\sin 2\pi t, \qquad (2.57)$$

is a continuous homomorphism of the additive group $\mathbb{R}$ in the multiplicative group $\mathbb{C}^*$. Its kernel is the subgroup $\mathbb{Z} \subset \mathbb{R}$ of the integers; its image is $\mathbb{S}^1$.

(c) The restriction $\varphi' \colon \mathbb{R} \to \mathbb{S}^1$ of φ is an open continuous homomorphism. It induces an isomorphism of topological groups

$$\psi \colon \mathbb{R}/\mathbb{Z} \to \mathbb{S}^1, \qquad (2.58)$$

and an element of $\mathbb{S}^1$ can be viewed as a coset $[t] = t + \mathbb{Z}$ of $\mathbb{R}$ modulo $\mathbb{Z}$. The isomorphism ψ agrees with the homeomorphism $\mathbb{I}/\partial \mathbb{I} \to \mathbb{S}^1$ of 2.5.2(a).

(d) The quotient G/H of a topological group modulo a normal subgroup is the quotient group with the quotient topology: these structures are always consistent.

(e) A topological group is Hausdorff if and only if its unit, or any point, is a closed subset.

(f) A quotient G/H is Hausdorff if and only if H is a closed subgroup.

2.5.6 Continuous actions

Let G be a topological group. A *G-space* is a topological space X equipped with a *left action* of G. This is a continuous mapping

$$\varphi \colon G \times X \to X, \qquad \varphi(g, x) = gx, \qquad (x \in X, g \in G), \qquad (2.59)$$

that satisfies the following axioms (for all $x \in X$ and $g, h \in G$):

(i) $1_G x = x$ *(unitarity)*,

(ii) $h(gx) = (hg)x$ *(compatibility)*.

We also say that G *acts*, or *operates*, on the space X. The elements of G are called *operators*. (In particular, G acts on itself by multiplication.)

The *congruence* $x \equiv_G x'$ is defined in X by the existence of an operator g such that $gx = x'$; an equivalence class $Gx = \{gx \mid g \in G\}$ of this relation is called an *orbit* of the action, and the quotient topological space X/G (modulo this congruence) is called the *orbit space*.

In a G-space X, every operator $g \in G$ gives a translation map

$$\tau(g) = \varphi(g, -) \colon X \to X, \qquad x \mapsto gx, \qquad (2.60)$$

which is a homeomorphism, with inverse $\tau(g^{-1})$. The projection on the orbit space

$$p \colon X \to X/G, \qquad (2.61)$$

is an open map: for every U open in X, the saturated subset $p^{-1}(pU) = \bigcup_g gU$ is open in X, which means that pU is open in X/G.

The action is said to be *transitive* if it has only one orbit:

- for every $x, x' \in X$ there is some $g \in G$ such that $x' = gx$.

The action is *free* (or *fixed-point free*) if it satisfies the following readily equivalent properties:

- for every $x, x' \in X$ there is at most one $g \in G$ such that $x' = gx$,
- for every $g, h \in G$, if $gx = hx$ for some $x \in X$, then $g = h$,
- the only operator $g \in G$ which can have a *fixed point* (some $x \in X$ such that $gx = x$) is 1_G.

The action is transitive and free if and only if, for every $x \in X$, the continuous mapping

$$\varphi_x = \varphi(-, x) \colon G \to X, \qquad g \mapsto gx, \qquad (2.62)$$

is bijective. (It need not be a homeomorphism, as shown by the action of G on the underlying set with the indiscrete topology.)

A *G-morphism* $F\colon X \to Y$ between *G*-spaces is a continuous mapping consistent with the *G*-actions: $F(gx) = g.F(x)$, for all $x \in X$ and $g \in G$. These morphisms form the category $G\mathsf{Top}$ of *G*-spaces.

A *right action* of *G* on *X* is written as xg, and has a 'reversed' compatibility axiom: $(xh)g = x(hg)$, where hg operates *first* by *h* and *then* by *g*. This is equivalent to a left action of the opposite topological group G^{op} on *X*. For a commutative group in additive notation, the action is preferably written on the right, as $x \mapsto x + g$.

Examples and complements. (a) A subgroup *H* of an *abelian* topological group *G* acts on *G* by translations; the orbit space G/H is the quotient topological group. (In the general case, letting *H* operate on the left by multiplication, the elements of the orbit space are the left cosets Hx of *H* in *G*.)

(b) The discrete group $\mathbb{Z}$ acts freely on the euclidean line $\mathbb{R}$, by translations: $x \mapsto x + k$. The orbit space is the topological group $\mathbb{R}/\mathbb{Z}$, homeomorphic to the circle $\mathbb{S}^1$, as proved in Exercise 2.5.5(c).

(c) We already recalled that $\mathbb{S}^1$ is a topological subgroup of the multiplicative group $\mathbb{C}^*$, formed of the unimodular complex numbers $u \in \mathbb{C}$, with $|u| = 1$. It acts on $\mathbb{C}$, by multiplication

$$\mathbb{S}^1 \times \mathbb{C} \to \mathbb{C}, \qquad (u, z) \mapsto uz. \qquad (2.63)$$

The orbits are the circles of $\mathbb{C}$ around the origin, together with the 'trivial circle' [0]. The action is free on $\mathbb{C}^*$, and is not transitive. The orbit space $\mathbb{C}/\mathbb{S}^1$ is homeomorphic to the euclidean half-line $[0, +\infty[$.

2.5.7 *Exercises and complements* (Gluing topological spaces)

These exercises are a preparation for gluing elementary spaces (e.g. the open euclidean spaces), in the topological realisation of a manifold (e.g. a classical differentiable manifold). Solutions can be found below.

(a) For a partial homeomorphism $u\colon X_1 \rightarrowtail X_2$ in $I\mathcal{C}$, one can define a *gluing space* $\mathrm{gl}\,(u)$, with a pair of open embeddings $u^i\colon X_i \rightarrowtail \mathrm{gl}\,(u)$, so that *u* can be viewed as a partial identity of the space $\mathrm{gl}\,(u)$.

(b) Forgetting topologies, a partial bijection $u\colon X_1 \rightarrowtail X_2$ in $\mathcal{I} = I\mathcal{S}$ has a *gluing set* $\mathrm{gl}\,(u)$, where the sets X_i are embedded as subsets, and *u* can be viewed as a partial identity.

(c) The gluing of a partial homeomorphism $u\colon X_1 \rightarrowtail X_2$, mentioned in (a), satisfies a universal property in the *ordered* category $\mathcal{C}$ (a lax colimit, according to 5.6.4)

$$
\begin{array}{c}
X_1 \;-\;-\;\xrightarrow{\;f^1\;}\; Y \\[4pt]
u\downarrow\uparrow u^\sharp \quad \mathrm{gl}\,(u) \;-\;-\;\xrightarrow{\;h\;}\; Y \\[4pt]
X_2 \;-\;-\;\xrightarrow{\;f^2\;}\; Y
\end{array}
\qquad u_2^1 = u, \; u_1^2 = u^\sharp, \qquad (2.64)
$$

- $u^j u_j^i \leqslant u^i$ (for $i \neq j$) in $\mathcal{C}$,

- for every pair of partial maps $f^i \colon X_i \rightarrowtail Y$ in $\mathcal{C}$ such that $f^i u_j^i \leqslant f^j$ ($i \neq j$), there exists a unique morphism $h \colon \mathrm{gl}\,(u) \to Y$ in $\mathcal{C}$ such that $h u^i = f^i$ ($i = 1, 2$),

- if we have a similar pair $g^i \colon X_i \rightarrowtail Y$ with $g u^i = g^i$ and $f^i \leqslant g^i$ (for $i = 1, 2$), then $f \leqslant g$.

(d) The gluing of the partial homeomorphism $u \colon X_1 \rightarrowtail X_2$ can also be obtained as the pushout in Top (defined in 2.5.4) of a suitable pair of topological embeddings f, g, so that $u = g f^\sharp$ in IC

$$
\begin{array}{ccc}
 & X_1 & \\
{\scriptstyle f}\nearrow & & \searrow{\scriptstyle u_1} \\
U & & A \\
{\scriptstyle g}\searrow & & \nearrow{\scriptstyle u_2} \\
 & X_2 &
\end{array}
\qquad (2.65)
$$

   This fact can be extended to the gluing of any intrinsic atlas on $\mathcal{C}$. Yet, as already said in 2.1.4, we prefer to work directly in $\mathcal{C}$ (and other categories of partial mappings) rather than exploiting its presentation in the form $\mathrm{P_M\,Top}$.

Solutions. (a) We take the topological sum $Z = X_1 + X_2$ (cf. 1.7.6). For the sake of simplicity, we suppose that X_1 and X_2 are already disjoint, and embedded as open subspaces of Z. We take on Z the equivalence relation R_u spanned by the relation R:

$$
x \, R \, y \quad \text{whenever } u(x) = y \qquad (x \in \mathrm{Def}\, u, \; y \in \mathrm{Val}\, u). \qquad (2.66)
$$

   We let $\mathrm{gl}\,(u)$ be the topological quotient Z/R_u

$$
X_1 \;\; U \;\; X_2 \qquad (2.67)
$$

The canonical projection $p\colon Z \to Z/R_u$ restricts to two injective mappings $u^i\colon X_i \to \mathrm{gl}\,(u)$, which are homeomorphisms onto their images, open in $\mathrm{gl}\,(u)$. The open subspaces $\mathrm{Def}\,u \subset X_1$ and $\mathrm{Val}\,u \subset X_2$ are identified in $\mathrm{gl}\,(u)$, as an open subspace U. Letting $e\colon \mathrm{gl}\,(u) \rightarrowtail \mathrm{gl}\,(u)$ be the partial identity on U, the commutative diagram of $I\mathcal{C}$

$$
\begin{array}{ccc}
X_1 & \overset{u}{\dashrightarrow} & X_2 \\[2pt]
{\scriptstyle u^1}\big\downarrow & & \big\downarrow{\scriptstyle u^2} \\[2pt]
\mathrm{gl}\,(u) & \underset{e}{\dashrightarrow} & \mathrm{gl}\,(u)
\end{array}
\qquad\qquad U = p(\mathrm{Def}\,u) = p(\mathrm{Val}\,u), \tag{2.68}
$$

shows that u is 'represented' by the partial identity e.

(c) If $x \in \mathrm{Def}\,u$, $u^2 u(x) = u^1(x)$, whence $u^2 u \leqslant u^1$. Similarly, taking $y \in \mathrm{Def}\,u^\sharp = \mathrm{Val}\,u$ and $x = u^\sharp(y)$ we have $u^1 u^\sharp(y) = u^2(x)$, whence $u^1 u^\sharp \leqslant u^2$.

Supposing we have two partial maps $f^i\colon X_i \rightarrowtail Y$ in $\mathcal{C}$ such that $f^i u^i_j \leqslant f^j$ $(i \neq j)$, we can define a partial map f in $\mathcal{C}$ letting

$$
f\colon \mathrm{gl}\,(u) \rightarrowtail Y, \qquad \mathrm{Def}\,f = u^1(\mathrm{Def}\,f^1) \cup u^2(\mathrm{Def}\,f^2),
$$

$$
f(u^i(x)) = f^i(x) \quad \text{if } x \in \mathrm{Def}\,f^i \quad (i = 1, 2).
$$

This is legitimate, because $u^1(x) = u^2(y)$ implies $u(x) = y$, and then $f^2(y) = f^2(u(x)) = f^1(x)$. Moreover f is continuous on $\mathrm{Def}\,f$, because it is on each subspace $u^i(\mathrm{Def}\,f^i)$, which are open in $\mathrm{gl}\,(u)$. The final point is obvious.

(d) The partial homeomorphism $u\colon X_1 \rightarrowtail X_2$ is a topological embedding $g\colon U \to X_2$, where $U = \mathrm{Def}\,u$ is an open subspace of X_1.

Letting $f\colon U \to X_1$ be the inclusion, the pushout (2.65) is constructed in (2.52) as the quotient of the space $X_1 + X_2$ described in (a).

2.6 *Complements on topology

We review here some topological constructions and properties that are not technically used in this book, but may nevertheless clarify our analysis of manifolds. In particular, one cannot deal with topological manifolds without thinking of topological dimension.

2.6.1 *Alexandrov compactification*

Every space X can be embedded in a compact space $X^\bullet$, called the *Alexandrov compactification*, or *one-point compactification*, of X. It is a beautiful construction, which requires some non-trivial work on the properties of compactness, without being really difficult. (A reader might like to work it out, on the basis of the following brief outline.)

As a set, we let

$$X^{\bullet} = X \cup \{\infty\}, \tag{2.69}$$

adding a point $\infty \notin X$. The topology of $X^{\bullet}$ contains open sets of two distinct kinds (and nothing more):

- the open sets U of X,
- the complements $V_K = X^{\bullet} \setminus K$ of the compact closed subsets K of X.

(If X is Hausdorff, a compact subset of X is necessarily closed in X, which simplifies the second point above.)

The space $X^{\bullet}$ is compact, and contains X as an open subspace.

(If X is not compact, then it is dense in $X^{\bullet}$; otherwise the added point is open in $X^{\bullet}$ — but it is still convenient to add it, to give a uniform construction.)

The construction is particularly interesting when the space X is Hausdorff, and locally compact: then $X^{\bullet}$ is Hausdorff. Moreover, the construction is determined by being *the* compact Hausdorff space containing X and one more point (up to the name of the latter, of course).

Complements and examples. (a) We recall that a space X is locally compact if any point has a fundamental system of compact neighbourhoods. If X is Hausdorff, this is equivalent to saying that any point has a compact nbd.

(b) The sphere $\mathbb{S}^n$ is (homeomorphic to) the one-point compactification of $\mathbb{R}^n$ (for $n \geqslant 0$).

It is sufficient to note that $\mathbb{S}^n$ is a compact Hausdorff space, and taking out any point we get a space homeomorphic to $\mathbb{R}^n$ (by stereographic projection).

(c) The discrete space $\mathbb{N}$ is a locally compact Hausdorff space. Its *Alexandrov compactification* $\mathbb{N}^{\bullet}$ adds a point ∞ adherent to $\mathbb{N}$. A limit of a sequence $(x_k)_{k \in \mathbb{N}}$ in a space X is the same as a topological limit of the mapping $x \colon \mathbb{N} \to X$ at ∞, as a point of $\mathbb{N}^{\bullet}$ adherent to $\mathbb{N}$ (independently of the uniqueness of limits in X).

2.6.2 Suspension

The *suspension* ΣX of a (non-empty) space X is defined as the topological quotient of the cylinder $X \times \mathbb{I}$ modulo the equivalence relation which collapses the lower basis $X \times \{0\}$ to a point v_0, and the upper basis $X \times \{1\}$ to another point v_1.

This construction gives $\Sigma \mathbb{S}^n \cong \mathbb{S}^{n+1}$ (for $n \geqslant 0$), and produces all spheres starting from $\mathbb{S}^0$

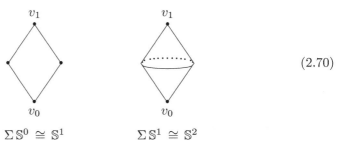

$$\Sigma \mathbb{S}^0 \cong \mathbb{S}^1 \qquad\qquad \Sigma \mathbb{S}^1 \cong \mathbb{S}^2$$

(2.70)

Exceptions should be avoided, in order to get a suspension functor Σ: Top $\to$ Top. The suspension ΣX of a general topological space (possibly empty) is defined as the colimit of the solid diagram at the left, where ∂^0 and ∂^1 are the faces $X \to X \times \mathbb{I}$ of the cylinder (as in (1.49))

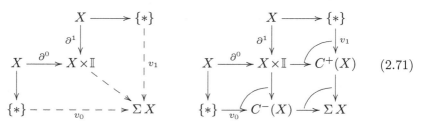

(2.71)

The colimit can be obtained by three pushouts, as in the right diagram above:

- $C^+(X)$ is the upper cone on the space X, obtained by collapsing the upper basis $X \times \{1\}$ of the cylinder $X \times \mathbb{I}$ to a point v_1 (or adding this point, if $X = \emptyset$),

- $C^-(X)$ is the lower cone,

- ΣX is the 'common quotient' of both cones.

In this way we get the previous result when $X \neq \emptyset$. In addition, we get $\Sigma \emptyset = \mathbb{S}^0$, which is consistent with letting $\mathbb{S}^{-1} = \emptyset$.

2.6.3 Hints at Algebraic Topology

Algebraic Topology can clarify many points we are considering. The core of this discipline is constructing *functors from a category of topological spaces to a category of algebraic structures*, and using them to reduce (difficult) topological problems to (simpler) algebraic ones.

We have thus:

- the sequence of *singular homology functors* $H_n \colon \mathsf{Top} \to \mathsf{Ab}$,

- the *fundamental group functor* $\pi_1 \colon \mathsf{Top}_\bullet \to \mathsf{Gp}$,

- the *fundamental groupoid functor* $\Pi_1 \colon \mathsf{Top} \to \mathsf{Gpd}$ (whose construction is sketched in 4.3.1(b)),

- the higher homotopy functors, etc.

For this fascinating field there are elementary textbooks, like Vick [Vi] and Massey [Mas], and more advanced ones, like Hilton–Wylie [HiW], Spanier [Sp] and Hatcher [Hat]; the last is freely downloadable.

Some results. (a) The singular homology functors give, on the standard spheres $\mathbb{S}^n$

$$H_0(\mathbb{S}^0) \cong \mathbb{Z}^2, \quad H_k(\mathbb{S}^0) = 0 \qquad\qquad (k > 0),$$

$$H_0(\mathbb{S}^n) \cong H_n(\mathbb{S}^n) \cong \mathbb{Z} \qquad\qquad (n > 0), \qquad\qquad (2.72)$$

$$H_k(\mathbb{S}^n) = 0 \qquad\qquad (n > 0,\ 0 \neq k \neq n),$$

as one can find in any book on Algebraic Topology. (The trivial group is written as 0.) This has many important consequences.

(b) For instance, these results prove that the sphere $\mathbb{S}^n$ is not homeomorphic to any other sphere $\mathbb{S}^m$ (for $m \neq n$), because any functor preserves isomorphisms (as remarked in 1.5.5(d)).

(c) (*Topological Dimension Theorem*) As a consequence, if the euclidean spaces $\mathbb{R}^m$ and $\mathbb{R}^n$ are homeomorphic, the same is true of their one-point compactifications $\mathbb{S}^m$ and $\mathbb{S}^n$, *and we conclude that $m = n$.*

Let us note that one cannot get this result applying the homology functors, directly, to $\mathbb{R}^m$ and $\mathbb{R}^n$: these spaces are contractible, i.e. homotopy equivalent to a point. Therefore each functor H_k has the same result on all of them

$$H_0(\mathbb{R}^n) \cong \mathbb{Z}, \qquad H_k(\mathbb{R}^n) = 0 \qquad (n \geqslant 0,\ k > 0).$$

(More generally, a homotopy-invariant functor cannot distinguish homotopy equivalent spaces.)

Loosely speaking, *the n-dimensional character of the topology of $\mathbb{R}^n$ appears more clearly in its Alexandrov compactification.*

(d) Applying (2.69) we also conclude that the sphere $\mathbb{S}^n$ (for $n > 0$) is not a retract of $\mathbb{R}^{n+1}$, or of any contractible space, because $H_n(\mathbb{S}^n) \cong \mathbb{Z}$ is not a retract of the trivial group. This fact is also the source of important consequences, for instance the well-known Brouwer fixed-point theorem.

(For $n = 0$ the argument can be adapted: $H_0(\mathbb{S}^0) \cong \mathbb{Z}^2$ is not a retract of $H_0(\mathbb{R}) \cong \mathbb{Z}$; but the Intermediate Value Theorem already proves that $\mathbb{S}^0$ is not a retract of $\mathbb{R}$.)

2.6.4 The dimension of open euclidean spaces

Open euclidean spaces can have disparate forms.

While considering n-dimensional manifolds it is important to know that a non-empty open euclidean space has a topological dimension, *invariant up to homeomorphism* — although we shall not use this result, in a technical sense.

This fact comes out of a theorem, generally proved in Algebraic Topology, that implies the Topological Dimension Theorem of 2.6.3(c), and is far more difficult to prove.

(a) (*Brouwer Invariance of Domain Theorem*) If U is an open subspace of a euclidean space $\mathbb{R}^m$ and $f \colon U \to \mathbb{R}^m$ is a topological embedding, then its homeomorphic image $V = f(U)$ is open in $\mathbb{R}^m$ (cf. [Sp], Theorem 4.7.16).

(b) As a consequence, if two non-empty open subspaces U of $\mathbb{R}^m$ and V of $\mathbb{R}^n$ are homeomorphic, then $m = n$.

In fact, we can suppose $n \leqslant m$ and $\mathbb{R}^n \subset \mathbb{R}^m$. A homeomorphism $U \to V$ gives a topological embedding $f \colon U \to \mathbb{R}^m$; applying (a), the image $V = f(U)$ is also open in $\mathbb{R}^m$. As $V \neq \emptyset$, there are open balls of $\mathbb{R}^m$ contained in $V \subset \mathbb{R}^n$, which rules out the case $n < m$.

2.7 *The transfer functor of inverse categories

Inverse categories have a peculiar form of direct and inverse images, which works on projectors instead of subobjects.

More precisely, for every inverse category K we construct a 'transfer functor' Prj : K → Slt, with values in an inverse category of semilattices and pairs of mappings. It was introduced in [G1].

This section is of interest in the theory of inverse categories, but of a marginal use in this book.

2.7.1 The inverse category of 1-semilattices and transfer pairs

Unital meet semilattices have a greatest element 1, that is the identity of the meet operation, and are called 1-semilattices (as in 1.1.2). They form the category 1Slh *of 1-semilattices and (unitary) homomorphisms*.

But here we are more interested in the category Slt *of 1-semilattices and transfer pairs* (introduced in [G1]), that we construct now on the same objects, showing that it is inverse and plays a structural role for all inverse categories.

A *transfer pair* $a = (a_\bullet, a^\bullet) \colon X \rightarrowtail Y$ between the 1-semilattices X, Y is a pair of mappings *in opposite directions* that satisfy the following conditions

(i) $a_\bullet \colon X \to Y$ and $a^\bullet \colon Y \to X$ are product-preserving mappings,

(ii) $a^\bullet a_\bullet(x) = x.a^\bullet(1_Y)$, $\quad a_\bullet a^\bullet(y) = y.a_\bullet(1_X)$ $\quad$ (for $x \in X$, $y \in Y$).

The composite of $a \colon X \rightarrowtail Y$ with a consecutive morphism $b = (b_\bullet, b^\bullet) \colon Y \rightarrowtail Z$ is obviously defined as the pair

$$ba = (b_\bullet a_\bullet, a^\bullet b^\bullet) \colon X \rightarrowtail Z. \tag{2.73}$$

This is legitimate: for all $x \in X$ we have

$$\begin{aligned} a^\bullet b^\bullet b_\bullet a_\bullet(x) &= a^\bullet(b^\bullet b_\bullet(a_\bullet(x))) = a^\bullet(a_\bullet(x).b^\bullet(1_Z)) \\ &= a^\bullet a_\bullet(x).a^\bullet b^\bullet(1_Z) = x.a^\bullet(1_Y).a^\bullet b^\bullet(1_Z) \\ &= x.a^\bullet(1_Y.b^\bullet(1_Z)) = x.a^\bullet b^\bullet(1_Z), \end{aligned}$$

and similarly $b_\bullet a_\bullet a^\bullet b^\bullet(z) = z.b_\bullet a_\bullet(1_X)$, for $z \in Z$.

The category Slt has an obvious involution, with $(a_\bullet, a^\bullet)^\sharp = (a^\bullet, a_\bullet)$. This involution is regular, since

$$aa^\sharp a = (a_\bullet a^\bullet a_\bullet, a^\bullet a_\bullet a^\bullet) \colon X \rightarrowtail Y,$$

and for all $x \in X$

$$a_\bullet a^\bullet a_\bullet(x) = a_\bullet(x.a^\bullet(1_Y)) = a_\bullet(x).a_\bullet a^\bullet(1_Y) = a_\bullet(x).1_Y.a_\bullet(1_X) = a_\bullet(x).$$

For a symmetric idempotent $e = (e_\bullet, e_\bullet) \colon X \rightarrowtail X$, the mapping $e_\bullet = e^\bullet$ is idempotent and

$$e_\bullet(x) = e_\bullet e_\bullet(x) = e^\bullet e_\bullet(x) = x.e_\bullet(1).$$

Therefore a projector of X amounts to the multiplication by a fixed element $e_0 = e_\bullet(1)$ of the semilattice X. Such mappings commute, whence Slt is an inverse category, by Theorem 2.3.7.

We prove now that a transfer pair $(a_\bullet, a^\bullet)$ is determined by its 'covariant part' $a_\bullet$.

The downward-closed subset generated by an element $x_1 \in X$ is written as $\downarrow x_1 = \{x \in X \mid x \leqslant x_1\}$, as in 1.1.2.

2.7.2 Proposition

Let $(a_\bullet, a^\bullet)\colon X \rightarrowtail Y$ be a transfer pair of 1-semilattices.

(a) This pair of mappings restricts to an isomorphism of 1-semilattices

$$a_1\colon X_1 \rightleftarrows Y_1 : a^1,$$

$$X_1 = {\downarrow}x_1 = a^\bullet(Y) \subset X, \qquad x_1 = a^\bullet(1_Y), \qquad (2.74)$$

$$Y_1 = {\downarrow}y_1 = a_\bullet(X) \subset Y, \qquad y_1 = a_\bullet(1_X).$$

(b) The subset $X_1 = {\downarrow}x_1$ and its unit $x_1 = a^\bullet(1_Y)$ are determined by $a_\bullet\colon X \to Y$ as:

$$x_1 = \max\{x \in X \mid \forall\, x' \in X\colon x' < x \Rightarrow a_\bullet(x') < a_\bullet(x)\}. \qquad (2.75)$$

Dually, the set $Y_1 = {\downarrow}y_1$ and the element $y_1 = a_\bullet(1_X)$ are determined by $a^\bullet\colon Y \to X$.

(c) The pair $a = (a_\bullet, a^\bullet)$ is determined by its covariant part $a_\bullet$ (and by its contravariant part as well); it is also determined by the partial bijection $(X_1, Y_1, a_1)\colon X \rightarrowtail Y$.

Proof (a) The inclusion $a^\bullet(Y) \subset \{x \in X \mid x \leqslant a^\bullet(1_Y)\}$ is obvious; conversely, if $x \leqslant a^\bullet(1_Y)$, then $x = x.a^\bullet(1_Y) = a^\bullet a_\bullet(x) \in a^\bullet(Y)$.

The mappings a_1, a^1 are inverse to each other: if $x = a^\bullet(y)$, then

$$a^\bullet a_\bullet(x) = a^\bullet a_\bullet a^\bullet(y) = a^\bullet(y) = x.$$

Since these mappings preserve the product (by definition of a transfer pair), they form an isomorphism of 1-semilattices.

(b) We know by (a) that $a_\bullet$ is injective on ${\downarrow}a^\bullet(1_Y)$. On the other hand, if we have an element $x \in X$ such that $x' < x$ implies $a_\bullet(x') < a_\bullet(x)$ (for all $x' \in X$), then $x \leqslant a^\bullet(1_Y)$, by the following argument:

- trivially $x.a^\bullet(1_Y) \leqslant x$,

- suppose, for a contradiction, that $x.a^\bullet(1_Y) < x$; then $a_\bullet(x) = a_\bullet a^\bullet a_\bullet(x) = a_\bullet(x.a^\bullet(1_Y)) < a_\bullet(x)$, which is absurd.

(c) The mapping $a_\bullet$ determines X_1 by (b), and $Y_1 = a_\bullet(X)$ (trivially). By (a), it determines the restricted bijection $a_1\colon X_1 \to Y_1$ and its inverse $a^1\colon Y_1 \to X_1$. It follows that $a^\bullet(y) = a^\bullet(a_\bullet a^\bullet(y)) = a^1(y.a_\bullet(1_X))$ is determined by $a_\bullet$.

Finally, from the partial bijection $(X_1, Y_1, a_1)\colon X \rightarrowtail Y$ we recover $x_1 = \max(X_1)$ and $a_\bullet(x) = a_1(x.x_1)$. $\qquad\square$

2.7.3 Concreteness

The category Slt is made concrete by the functor

$$U \colon \mathsf{Slt} \to \mathsf{Set}, \qquad U(X) = |X|, \qquad U(a_\bullet, a^\bullet) = a_\bullet, \qquad (2.76)$$

where $|X|$ is the underlying set of the 1-semilattice X; the faithfulness of U has been proved in the previous proposition. Of course, the selfdual category Slt is also made co-concrete by the faithful functor $\mathsf{Slt}^{op} \to \mathsf{Set}$ taking $(a_\bullet, a^\bullet)$ to $a^\bullet$.

The previous proposition also yields a faithful functor with values in the inverse category of partial bijections

$$\begin{aligned} W \colon \mathsf{Slt} \to \mathcal{I}, \qquad W(X) = |X|, \\ W(a_\bullet, a^\bullet) = (X_1, Y_1, a_1) \colon |X| \rightharpoonup |Y|. \end{aligned} \qquad (2.77)$$

2.7.4 Theorem and Definition (The transfer functor)

Every inverse category K *has a* functor of projectors

$$\begin{aligned} \mathsf{Prj} \colon \mathsf{K} \to \mathsf{Slt}, \qquad X \mapsto \mathsf{Prj}\, X, \\ (a \colon X \to Y) \mapsto ((a_P, a^P) \colon \mathsf{Prj}\, X \rightharpoonup \mathsf{Prj}\, Y), \qquad (2.78) \\ a_P(e) = aea^\sharp, \qquad a^P(f) = a^\sharp f a \qquad (e \in \mathsf{Prj}\, X, \ f \in \mathsf{Prj}\, Y). \end{aligned}$$

This is also called the transfer functor *of* K, *and written as* Prj_K *when useful. We say that* K *is a* transfer inverse category *when this functor* Prj *is faithful.*

Proof First, for $e, e' \in \mathsf{Prj}\, X$ we have that:

$$a_P(e).a_P(e') = aea^\sharp.ae'a^\sharp = a.e(a^\sharp a)e'.a^\sharp = a(a^\sharp a)ee'a^\sharp = a_P(ee'),$$

which also shows that $a_P(e).a_P(e) = a_P(e)$. Therefore a_P takes values in $\mathsf{Prj}\, Y$ and preserves the product. The same holds for $a^P = (a^\sharp)_P$.

The pair (a_P, a^P) is a transfer pair (as defined in 2.7.1), because:

$$a^P a_P(e) = a^\sharp(aea^\sharp)a = e.(a^\sharp a) = e.a^P(1), \qquad a_P a^P(f) = f.a_P(1).$$

Finally, to show that Prj is a functor, a composite $ba \colon X \to Y \to Z$ in K gives, for $e \in \mathsf{Prj}\, X$

$$b_P a_P(e) = b(aea^\sharp)b^\sharp = (ba)e(ba)^\sharp = (ba)_P(e).$$

Similarly $a^P b^P(e'') = (ba)^P(e'')$, for $e'' \in \mathsf{Prj}\, Z$. $\qquad \square$

2.7.5 Exercises and complements

(a) (*A solved exercise*) We begin by showing that Slt is a transfer inverse category.

We have already seen, at the end of 2.7.1, that in the inverse category Slt a transfer pair $e = (e_\bullet, e^\bullet) \colon X \to X$ is a projector if and only if the mapping $e_\bullet = e^\bullet \colon X \to X$ is the multiplication $x \wedge - \colon y \mapsto xy$ by a fixed element $x = e_\bullet(1)$.

The transfer functor of Slt is thus isomorphic to the identity functor, via the transformation

$$\iota \colon \mathrm{id} \to \mathrm{Prj} \colon \mathsf{Slt} \to \mathsf{Slt}, \qquad \iota X \colon X \to \mathrm{Prj}_{\mathsf{Slt}}(X),$$

$$(\iota X)_\bullet(x) = (x \wedge -, x \wedge -) \colon X \to X, \qquad (\iota X)^\bullet(e) = e_\bullet(1) \in X, \tag{2.79}$$

whose naturality is easily checked. Therefore the transfer functor is faithful.

(b) Prove that $\mathcal{I}$ is also a transfer inverse category. *Hints.* Use the functor

$$\mathcal{P} \colon \mathcal{I} \to \mathsf{Slt}, \qquad X \mapsto \mathcal{P}X, \qquad f \mapsto \mathcal{P}f,$$

$$(\mathcal{P}f)_\bullet(A) = f_0(A \cap X_0) \qquad \text{(for } A \subset X\text{)}, \tag{2.80}$$

$$(\mathcal{P}f)^\bullet(B) = f_0^{-1}(B \cap Y_0) \qquad \text{(for } B \subset Y\text{)},$$

that sends the set X to its boolean algebra of subsets $\mathcal{P}X$, and a partial bijection $f = (X_0, Y_0, f_0) \colon X \to Y$ to the transfer pair $\mathcal{P}f = ((\mathcal{P}f)_\bullet, (\mathcal{P}f)^\bullet)$ defined by direct and inverse images along f.

(c) The categories I$\mathcal{T}$, I$\mathcal{C}$ and I$\mathcal{D}$ are also transfer inverse categories.

2.7.6 Functors and projectors

A functor $F \colon \mathsf{K} \to \mathsf{L}$ between inverse categories gives, for every object A of K, a *homomorphism of 1-semilattices*

$$\mathrm{Prj}_F(X) \colon \mathrm{Prj}_\mathsf{K}(X) \to \mathrm{Prj}_\mathsf{L}(FX), \qquad e \mapsto F(e). \tag{2.81}$$

Moreover, a morphism $a \colon X \to Y$ gives two commutative squares

$$\begin{array}{ccc}
\mathrm{Prj}_\mathsf{K}(X) & \xrightarrow{\ \mathrm{Prj}_F(X)\ } & \mathrm{Prj}_\mathsf{L}(FX) \\
{\scriptstyle a_P}\big\uparrow\big\downarrow{\scriptstyle a^P} & & {\scriptstyle (Fa)_P}\big\uparrow\big\downarrow{\scriptstyle (Fa)^P} \\
\mathrm{Prj}_\mathsf{K}(Y) & \xrightarrow[\ \mathrm{Prj}_F(Y)\ ]{} & \mathrm{Prj}_\mathsf{L}(FY)
\end{array} \tag{2.82}$$

$$(Fa)_P \, \mathrm{Prj}_F(X) = \mathrm{Prj}_F(Y).a_P \colon e \mapsto F(aea^\sharp),$$

$$(Fa)^P \, \mathrm{Prj}_F(Y) = \mathrm{Prj}_F(X).a^P \colon e \mapsto F(a^\sharp ea).$$

Let us note that this family ($\mathsf{Prj}_F(X))_A$ of *homomorphisms* does *not* form a natural transformation between the functors Prj_K and $\mathsf{Prj}_L.F$, because these functors take values in the category Slt of 1-semilattices and *transfer pairs*.

*A reader acquainted with the theory of 'double categories' will note that this family forms a *horizontal transformation of vertical functors*

$$\mathsf{Prj}_F: \mathsf{Prj}_K \to \mathsf{Prj}_L.F: \mathsf{K} \to \mathbb{S}\mathsf{lht},$$

with values in a *double category* $\mathbb{S}\mathsf{lht}$; the latter consists of 1-semilattices, homomorphisms and transfer pairs, with double cells as in (2.82): see [G1, GP1].

Here we shall not use this formal aspect, whose bases are only mentioned in 5.6.5.*

2.8 *Global orders and Ehresmann's pseudogroups

Ehresmann's analysis of local structures is based on 'inductive groupoids', with a global order on all morphisms. These groupoids have an extended composition, Ehresmann's pseudoproduct (recalled in Section 0.6 of the general Introduction), which makes them into 'pseudogroups', a special form of inverse semigroups.

We examine here an extension of these groupoids, called a 'functorially ordered groupoid' by Ehresmann, an 'inductive groupoid' by M. Lawson, and a 'semi-inductive groupoid' here. The previous correspondence makes them equivalent to inverse semigroups: this result is called the Ehresmann–Schein–Nambooripad Theorem, in Lawson's book [Ls].

We follow Lawson's analysis of this topic; the main proofs can be found there. Ehresmann's 'inductive sets' and 'inductive categories' will be considered in Section 6.7.

2.8.1 Global orders on categories

A *globally ordered category* will be a category C equipped with a *global order*. This is an order relation on the whole class $\mathrm{Mor}\,\mathsf{C}$ of its morphisms, written here as $f' \subset f$, that satisfies the following conditions:

(i) $f' \subset f \Rightarrow (\mathrm{Dom}\,f' \subset \mathrm{Dom}\,f \text{ and } \mathrm{Cod}\,f' \subset \mathrm{Cod}\,f)$,

(ii) $(f' \subset f,\ g' \subset g,\ \mathrm{Cod}\,f = \mathrm{Dom}\,g,\ \mathrm{Cod}\,f' = \mathrm{Dom}\,g') \Rightarrow g'f' \subset gf$,

(iii) $(f' \subset f,\ \mathrm{Dom}\,f' = \mathrm{Dom}\,f,\ \mathrm{Cod}\,f' = \mathrm{Cod}\,f) \Rightarrow f' = f$,

where the relation $X' \subset X$ on objects means that $1_{X'} \subset 1_X$.

For Ehresmann an 'ordered category' means a category equipped with such a relation ([E6], p. 374).

The prime example is now the category Set of sets and (total) mappings, where the relation $X' \subset X$ on objects is (ordinary) weak inclusion and the relation $f' \subset f$, for mappings $f\colon X \to Y$ and $f'\colon X' \to Y'$, means that f' is a restriction of f:

$$X' \subset X, \qquad Y' \subset Y, \qquad \text{for all } x \in X',\ f'(x) = f(x). \qquad (2.83)$$

(This implies that $f(X') \subset Y'$.) The relationship of this structure with $\mathcal{S}$ will be examined in Section 6.7.

For Top we are mainly interested in the global order of 'open inclusions', which is related to manifolds (and to our category $\mathcal{C}$). Here the relation $X' \subset X$ means that

$$X' \text{ is an open subspace of the topological space } X, \qquad (2.84)$$

and the relation $f' \subset f$ is defined as in (2.83).

The subcategory $C^r\mathsf{Top}$ of open euclidean spaces and (total) C^r-mappings inherits a global order (corresponding to our category $\mathcal{C}^r$).

2.8.2 Remarks

(a) The category Top has other global orders of interest (related to $\mathcal{T}$ and $\mathcal{D}$), not used here, where the relation $X' \subset X$ means that

- X' is a subspace of the topological space X,
- X' is a closed subspace of the topological space X.

(b) Ehresmann and Lawson use the single-sorted form of a category, mentioned in 1.4.5(c). Here we translate their notation in the usual two-sorted version, with objects and morphisms.

(c) Ehresmann worked on local structures in many papers and books, like [E1]–[E5] and many others, and in some case modified his terminology. We mostly refer to Volume [E6] ('Structures locales') of his collected works, containing Andrée C. Ehresmann's helpful comments.

(d) With respect to the local order of the category $\mathcal{S}$ of sets and partial mappings, the global order on Set (or other concrete categories) makes a deeper use of set-theory, based on the notion of inclusion *between arbitrary sets* — rather than *between subsets of a given set*.

Marginally, we also note that the restriction of the local order $f' \leqslant f$ of $\mathcal{S}$ to the subcategory Set is the discrete order $f' = f$, and has nothing to do with $f' \subset f$.

*(e) The formal aspects of this alternative local/global will be made clear in Chapter 5: a category equipped with a local order amounts to a category

enriched on Ord (cf. 5.5.4(b)), while a (small) category with a global order amounts to an *internal category in* Ord (cf. 5.6.7(b)), that satisfies the additional condition 2.8.1(iii).

2.8.3 Semi-inductive groupoids

We consider now a structure introduced by Ehresmann as a 'functorially ordered groupoid' ([E6], p. 391), and called here a *semi-inductive groupoid*.

This is a groupoid G equipped with a global order $f' \subset f$ (as defined above) that also satisfies the following conditions:

(iv) $f \subset g \Rightarrow f^{-1} \subset g^{-1}$,

(v) if $f \colon X \to Y$ is in G and $X' \subset X$ (i.e. $1_{X'} \subset 1_X$), there is a unique $f' \subset f$ with Dom $f' = X'$, called the *restriction* $f|^{X'}$ *of* f *on* X' (or from X'),

(vi) the class Ob G, ordered by the restricted relation $X' \subset X$, is a meet-semilattice.

This notion is selfdual: using (iv), one can show that (v) is equivalent to its dual (v*)

(v*) if $f \colon X \to Y$ is in G and $Y' \subset Y$ (i.e. $1_{Y'} \subset 1_Y$), there is a unique $f'' \subset f$ with Cod $f'' = Y'$, that will be called the *restriction* $f|_{Y'}$ *of* f *to* Y' (or with values in Y').

For this structure, Lawson uses the term 'inductive groupoid' ([Ls], Section 4.1). In Ehresmann's terminology, an 'inductive groupoid' is also assumed to have all joins of upper-bounded families of maps, or equivalently all meets of non-empty families ([E6], p. 363).

The main examples are now groupoids of bijective mappings, like:

(a) Bij = Iso(Set) = Iso($\mathcal{S}$), the groupoid of sets and bijections,

(b) Iso(Top) = Iso($\mathcal{C}$), the groupoid of topological spaces and homeomorphisms,

(c) Iso(C^rTop) = Iso($\mathcal{C}^r$), the groupoid of open euclidean spaces and C^r-diffeomorphisms.

They are equipped with a global order, inherited from Set, or Top, or C^rTop, respectively. (Again, we note that this has nothing to do with the restriction of the local orders of $\mathcal{S}$, $\mathcal{C}$ and $\mathcal{C}^r$, which is discrete.)

A functor $F \colon G \to G'$ between semi-inductive groupoids will be said to be *monotone* if it preserves the global orders, and a *semi-inductive functor* if it also preserves the binary meet of objects.

2.8.4 From semi-inductive groupoids to inverse semigroups

We now observe that the class Mor Bij of all bijections of sets becomes a (large) semigroup $\underline{S}(\mathsf{Bij})$ by means of the following *extended composition* $g \cdot f$, or (Ehresmann's) *pseudoproduct*

- for two bijections $f \colon X \to Y$ and $g \colon Y' \to Z$, the bijection $g \cdot f \colon X' \to Z'$ is defined by:

$$X' = f^{-1}(Y \cap Y'), \qquad Z' = g(Y \cap Y'),$$
$$(g \cdot f)(x) = g(f(x)) \qquad\qquad (x \in X'). \tag{2.85}$$

This composition is associative but has no identity: $\underline{S}(\mathsf{Bij})$ is a semigroup, and actually a large inverse semigroup.

In general, every semi-inductive groupoid G has an *associated inverse semigroup* $\underline{S}(\mathsf{G})$ consisting of the class Mor G equipped with the extended product, or pseudoproduct; the latter is defined by means of the restrictions of 2.8.3

$$b \cdot a = b|^H . a|_H, \qquad H = \mathrm{Cod}\,(a) \cap \mathrm{Dom}\,(b). \tag{2.86}$$

This is indeed an associative product; $\underline{S}(\mathsf{G})$ is an inverse semigroup with partial inverses given by the inverses in G, and canonical order equal to the global order of G (as proved in [Ls], Section 4.1, Proposition 7).

A semi-inductive functor $F \colon \mathsf{G} \to \mathsf{G}'$ between semi-inductive groupoids (see 2.8.1) preserves the pseudoproduct; more generally a monotone functor gives a *colax morphism* of inverse semigroups (called a prehomomorphism in [Ls])

$$F(b \cdot a) \leqslant F(b) \cdot F(a). \tag{2.87}$$

Remarks. Ehresmann constructs a C^r-manifold with an atlas of charts in the globally ordered groupoid Iso(Top) of topological spaces and (total) homeomorphisms, by imposing that the transition maps belong to the subgroupoid Iso(C^rTop) of open euclidean spaces and (total) diffeomorphisms of class C^r.

Composition uses the pseudoproduct of these homeomorphisms, as recalled in Section 0.6 of the general Introduction. The inverse semigroups obtained in this way are called *pseudogroups of transformations* ([E6], p. 363).

2.8.5 From inverse semigroups to semi-inductive groupoids

The other way round, let S be a (possibly large) inverse semigroup.

Its idempotent completion S^E (defined in 2.2.5) is a category and gives us the *associated groupoid* $\mathsf{G} = \underline{G}(S) = \mathrm{Iso}(S^E)$, where:

- an object is any idempotent of S,
- if e, f are idempotents of S, then

$$\begin{aligned}\mathsf{G}(e, f) &= \{a \in S \mid a^\sharp a = e,\ aa^\sharp = f\} \\ &= \{a \in S \mid \underline{e}(a) = e,\ \underline{e}^*(a) = f\},\end{aligned} \tag{2.88}$$

- the composite of $a\colon e \to f$ and $b\colon f \to g$ is the product ba in S.

This is legitimate because in any category isomorphisms compose; or directly because

$$\underline{e}(ba) = a^\sharp b^\sharp ba = a^\sharp fa = a^\sharp a = \underline{e}(a), \qquad \underline{e}^*(ba) = \underline{e}^*(b).$$

(In the one-sorted version of a groupoid where G consists of its morphisms alone — as used by Ehresmann and Lawson — the previous construction amounts to taking the same set as S with a *restricted composition* $b \circ a = ba$, which is defined precisely when $\underline{e}^*(a) = \underline{e}(b)$.)

$\underline{G}(S)$ is equipped with a *global order* $a \subset b$ defined on all pair of morphisms by the canonical order of S (see 1.3.9), and is a semi-inductive groupoid, as defined above.

If $S = \mathcal{I}(X)$ is the inverse semigroup of partial endo-bijections of the set X (see 1.2.4), then $\underline{G}(S)$ is the groupoid of total bijections between subsets of X.

For every inverse semigroup S and every semi-inductive groupoid G we have:

$$\underline{S}\,\underline{G}(S) = S, \qquad \underline{G}\,\underline{S}(\mathsf{G}) = \mathsf{G}. \tag{2.89}$$

In Lawson [Ls], Section 4.1, Theorem 8, all this is fixed in the following (more complete) result, called the Ehresmann–Schein–Nambooripad Theorem:

The category of semi-inductive groupoids and monotone (resp. semi-inductive) functors is isomorphic to the category of inverse semigroups and colax (resp. ordinary) homomorphisms.

(Again, a semigroup can be large.)

2.8.6 Comments

To formalise the general morphisms of C^r-manifolds, pseudogroups are not sufficient: we are to come back to the category Top equipped with the global order of open inclusions, described in 2.8.1, and to its globally ordered subcategory $C^r\mathsf{Top}$.

This approach will be further examined in Section 6.7, and compared to the present one.

Outside of these comparisons, we shall not use large semigroups of mappings multiplied by the pseudoproduct, nor globally ordered categories.

This makes appeal to inclusion and intersection of arbitrary sets — which we prefer to avoid. Also Ehresmann would often restrict his analysis to a given space and its subspaces (cf. [E6], p. 8); a restriction which can be managed for differentiable manifolds, but would exclude other local structures like general fibre bundles.

3

Cohesive categories and manifolds

We develop here our formal approach to manifolds, presented in the general Introduction. Cohesive categories are studied in Sections 3.1–3.4, and their manifolds in Sections 3.5–3.6.

The leading examples are the ordered categories of partial mappings $\mathcal{S}$, $\mathcal{C}$, $\mathcal{C}^r$, $\mathcal{D}$ introduced in 2.1.4, and their inverse cores $\mathcal{I} = \mathrm{I}\mathcal{S}$, $\mathrm{I}\mathcal{C}$, $\mathrm{I}\mathcal{C}^r$, $\mathrm{I}\mathcal{D}$ (see 2.4.5).

A central role in this analysis is played by the 'projectors' (restrictions of identities): to define a manifold on these categories we only need the order relation, but we need the projectors and their structure to define the general morphisms of manifolds (in Section 3.6).

Theoretically, the main results are the Gluing Completion Theorems (Sections 3.5 and 3.6, proved in Section 3.9) and the Cohesive Completion Theorems (Sections 3.7 and 3.8).

3.1 Introducing cohesive structures

This section introduces, in an informal way, 'cohesive structures' in the category $\mathcal{S}$ of sets and partial mappings, in other $\mathcal{S}$-concrete categories and in other contexts.

Cohesive categories will be defined in Section 3.2 and cohesive sets in Section 6.6.

3.1.1 Cohesion in the category of partial mappings

The category $\mathcal{S}$ of sets and partial mappings was defined in 2.1.3. Recall that, for a partial mapping $f \colon X \to Y$, we write as Def f the subset of X on which f is defined.

We also recall, from 2.1.3, that $\mathcal{S}$ is an ordered category: for $f, g\colon X \twoheadrightarrow Y$ we say that $f \leqslant g$, when f is a restriction of g, that is

(i) Def $f \subset$ Def g, $f(x) = g(x)$ for all $x \in$ Def f.

We now introduce a *proximity relation* $f\,!\,g$, in $\mathcal{S}$ (i.e. a binary relation between parallel maps, which is reflexive, symmetric and consistent with composition), that will be called the *linking relation*, or *compatibility*

(ii) $f\,!\,g$ if f and g are parallel maps and coincide on Def $f \cap$ Def g.

Order and linking are closely related. For instance, a set $\varphi \subset S(X, Y)$ of parallel partial mappings that has an upper bound (i.e. there is some $g\colon X \twoheadrightarrow Y$ such that $f \leqslant g$, for all $f \in \varphi$) is obviously a *linked set* of parallel maps (i.e. $f\,!\,f'$ for all $f, f' \in \varphi$).

Conversely, if φ is a linked set, the least upper bound $f_1 = \bigvee \varphi$ exists, and is given by the set-theoretical union of the graphs of all $f \in \varphi$. It will be called a *linked join*; let us note that this join is *compositive*, i.e. preserved by composition.

On the other hand, the greatest lower bound $f_0 = \bigwedge \varphi$ always exists if $\varphi \neq \emptyset$, and is given by the intersection of the graphs of all $f \in \varphi$. But it is easy to verify that *this meet is preserved by composition in $\mathcal{S}$ if and only if the set φ is linked*; it will then be called a *linked meet*.

The linking relation has a trivial restriction to total mappings: $f\,!\,g$, if and only if $f = g$. This is also true of the ordering.

3.1.2 Cohesion and projectors

In the pre-inverse category $\mathcal{S}$, the projectors $e\colon X \twoheadrightarrow X$ of a set X are the partial identities, characterised by the property $e \leqslant 1_X$ (cf. 2.4.3); they form an ordered set $\mathsf{Prj}\,(X)$ isomorphic to the boolean algebra $\mathcal{P}(X)$ of the subsets of X, via $e \mapsto$ Def e.

The projectors of $\mathcal{S}$ are thus determined by the order; conversely, they determine the order and the linking relation

(i) $f \leqslant g$ $\Leftrightarrow$ $f = ge$, for some projector e,

(ii) $f\,!\,g$ $\Leftrightarrow$ $f = fe,\ g = ge',\ fe' = ge$, for projectors e, e'.

A pair (e, e') satisfying (ii) will be called a *resolution* of f and g; it gives the meet $f \wedge g = fe' = ge$.

Furthermore, every partial mapping $f\colon X \twoheadrightarrow Y$ has a least projector $e \in \mathsf{Prj}\,X$ such that $f = fe$, namely the partial identity on Def f; it will be written as $\underline{e}(f)$ and called the *support* of f.

Again, supports determine the order and the linking relation

(iii) $f \leqslant g \quad \Leftrightarrow \quad f = g\underline{e}(f),$

(iv) $f \, ! \, g \quad \Leftrightarrow \quad f\underline{e}(g) = g\underline{e}(f).$

Extending these facts, we shall introduce in Sections 3.2 and 3.3 the notions of 'cohesive category' $(\mathsf{C}, \leqslant, !)$, of 'prj-cohesive category' $(\mathsf{C}, \mathsf{Prj})$, and of 'e-cohesive category' $(\mathsf{C}, \underline{e})$.

Every prj-cohesive category C has an associated cohesive structure defined as above, in (i)–(ii). Every e-cohesive category C has an associated cohesive structure, defined as in (iii)–(iv), and an associated prj-cohesive structure with projectors determined by the condition $e = \underline{e}(e)$.

3.1.3 Cohesion for continuous partial mappings

Dealing with topological spaces, our basic examples are the $\mathcal{S}$-concrete categories $\mathcal{C}$ and $\mathcal{C}^r$.

If C is any of them, the (faithful) forgetful functor $\mathsf{C} \to \mathcal{S}$ 'lifts' to C an e-cohesive structure provided with *arbitrary linked joins* and *binary linked meets* (as in 3.1.1), that are distributive on the former. This structure is determined by the projectors $e \leqslant 1_X$ of each object X; as we have seen in 2.4.5, these projectors form an ordered set $\mathsf{Prj}\,(X)$, isomorphic to the frame $\mathcal{O}(X)$ of the open sets of X.

Similarly, the category $\mathcal{D}$, has an e-cohesive structure, lifted from $\mathcal{S}$, with *finite linked joins* and *binary linked meets*.

These cohesive structures are studied below. Their manifolds will be studied in Chapter 4, with other examples, related to foliations and fibre bundles.

3.1.4 Cohesion for inverse categories

Inverse categories, studied in Chapter 2, have a cohesive structure which works in a partially different way.

Now the prime example is the category $\mathcal{I} = \mathsf{I}\mathcal{S}$ of sets and partial bijections.

As a subcategory of $\mathcal{S}$, it is an ordered category: $f \leqslant g$, means that the partial bijection f is a restriction of g. Note that $f \leqslant g$, is equivalent to the condition $f^\sharp \leqslant g^\sharp$ (about their partial inverses), and coincides with the canonical order of $\mathcal{I}$ as an inverse category, expressed in 2.4.1 by many equivalent properties, like $f = (ff^\sharp)g(f^\sharp f)$.

On the other hand, the linking relation $f \, !_\mathcal{S} \, g$ of 3.1.1(ii) is not invariant under the involution $(-)^\sharp$ of $\mathcal{I}$. The linking relation $f \,!\, g$ of interest in $\mathcal{I}$ is defined by the following equivalent conditions (for parallel morphisms):

(i) f and g have a common upper bound in $\mathcal{I}$,

(ii) $f \,!_\mathcal{S}\, g$ and $f^\sharp \,!_\mathcal{S}\, g^\sharp$,

(iii) the partial bijections f and g coincide where they are both defined, and the same holds of $f^\sharp$ and $g^\sharp$.

As a general fact, we shall see that an inverse category K has a *canonical cohesive structure*, that extends the canonical order $a \leqslant b$, studied in the previous chapter (see 2.4.1, 2.4.2)

$$a \leqslant b \iff a = ab^\sharp a \iff a = b.a^\sharp a \iff a = a^\sharp a.b$$
$$\iff a = (aa^\sharp)b(a^\sharp a) \dots \tag{3.1}$$

$$a \,!\, b \iff (a.b^\sharp b = b.a^\sharp a \text{ and } bb^\sharp.a = aa^\sharp.b). \tag{3.2}$$

This structure is not prj-cohesive, in the previous sense: its linking relation is described by *double resolutions*, on domain and codomain, or equivalently by *supports* (on domain) and *cosupports* (on codomain): $\underline{e}(a) = a^\sharp a$ and $\underline{e}^*(a) = aa^\sharp$. The canonical structure of inverse categories will be studied in Section 3.4, and its gluing completion in Section 3.5.

We shall also see, in 3.4.6, that *every prj-cohesive category* C *has an inverse core* IC (in the sense of Theorem 2.4.4).

3.1.5 Quantaloids and upper bounded subsets

As a structure related to these topics, a *quantaloid* is an ordered category C where all hom-sets $C(X, Y)$ are complete lattices, and composition distributes over arbitrary joins.

This important notion will have a secondary role here, in Section 4.6 and Chapter 6.

In fact, the $\mathcal{S}$-concrete categories that we are considering, or will be, are not quantaloids: a subset of parallel morphisms has a join if and only if it is upper bounded, which need not be the case (see Exercise (c), below).

We might be lead to introduce a generalisation of quantaloids based on this fact, and define an *upper-conditioned quantaloid*, or *uc-quantaloid*, as an ordered category C where every upper bounded subset of parallel morphisms has a join, preserved by composition.

Unfortunately, this simple setting seems to be insufficient for our goals: the morphisms of manifolds on C, defined in Section 3.6, will rely on projectors and resolutions, or we would not be able to compose them.

Exercises. (a) The category Rel Set is a quantaloid.

(b) The category Rel Ab is not a quantaloid. *Hints:* use the projectors $e_H \colon A \twoheadrightarrow A$ associated to the subgroups H of an abelian group A.

(c) The ordered categories $\mathcal{S}, \mathcal{T}, \mathcal{C}, \mathcal{D}, \mathcal{C}^r$ are uc-quantaloids, and are not quantaloids.

(d) In $\mathcal{S}, \mathcal{C}$ and $\mathcal{C}^r$ a set φ of parallel morphisms is upper bounded if (and only if) it is pairwise upper bounded. In $\mathcal{D}$ this holds for a finite set φ.

3.1.6 *Cohesion for measurable functions*

We end this section with other examples of cohesive structures, that will not be used in this book.

Let X be a measurable space and Y a normed vector space on the real field. The *set* Y, pointed at 0, has a very simple cohesive structure:

$$
\begin{aligned}
y \leqslant y' &\Leftrightarrow (y = 0 \text{ or } y = y'), \\
y \,!\, y' &\Leftrightarrow (y \leqslant y' \text{ or } y' \leqslant y) \Leftrightarrow (y = 0 \text{ or } y' = 0 \text{ or } y = y').
\end{aligned}
\tag{3.3}
$$

The linked subsets of Y are of the form $\emptyset$, $\{y\}$, $\{0, y\}$; each of them has an obvious join.

The normed space $L^\infty(X, Y)$ of bounded measurable mappings from X to Y inherits a pointwise cohesive structure

$$
\begin{aligned}
f \leqslant f' &\Leftrightarrow (\forall x \in X \colon fx \leqslant f'x) \Leftrightarrow (\forall x \colon fx = 0 \text{ or } fx = f'x), \\
f \,!\, f' &\Leftrightarrow (\forall x \in X \colon (fx)\,!\,(f'x)) \\
&\Leftrightarrow (\forall x \in X \colon fx = 0 \text{ or } f'x = 0 \text{ or } fx = f'x).
\end{aligned}
\tag{3.4}
$$

This structure is *finitely cohesive*, i.e. provided with finite linked joins. It is easy to guess that the universal completion of $L^\infty(X, Y)$ with respect to countable joins of linked sets is the vector space $M(X, Y)$ of all measurable mappings from X to Y.

Indeed any such map $f \colon X \to Y$ is the linked join of the increasing sequence of bounded measurable mappings $f_n = e_n.f$, where $e_n \colon Y \to Y$ is the following measurable (non-linear) mapping: $e_n(y) = y$ if $||y|| \leqslant n$, $e_n(y) = 0$ otherwise.

We already noted (in 2.1.5) that $\mathcal{S}$ is equivalent to the category Set, of pointed sets and pointed (everywhere defined) mappings. This equivalence gives to Set, a cohesive structure described as in (3.3), which can be lifted from pointed sets to vector spaces.

3.1.7 *Cohesion for operators*

Starting from the category Ban of Banach spaces and continuous linear mappings (see 1.4.1), the category $L^\infty(\mathbf{a}, \mathsf{Ban})$ of *bounded measurable operators* on a boolean σ-algebra $\mathbf{a}$ was studied in [G5], in 1990.

It has for objects all the pairs (X, E) where X is a Banach space and

$$E\colon \mathbf{a} \to \mathsf{Ban}(X)$$

is a (bounded) σ-additive spectral measure with values in X (cf. Dunford–Schwartz [DS], XV.2.3–4). A morphism $S\colon (X, E) \to (Y, F)$ is a bounded linear mapping $S\colon X \to Y$ that commutes with the measures E, F, in the sense that $S.E(a) = F(a).S$, for all $a \in \mathbf{a}$.

The category $L^\infty(\mathbf{a}, \mathsf{Ban})$ has a natural prj-cohesive structure, defined as in 3.1.2(i), (ii): the projectors of the object (X, E) are the endomorphisms $E(a)$, for $a \in \mathbf{a}$. The structure is not e-cohesive, nor complete with regard to linked joins: its σ-cohesive completion 'is' the category $M(\mathbf{a}, \mathsf{Ban})$ of closed, densely defined, measurable operators, as shown in [G5].

3.2 Cohesive categories

Abstracting the properties that we have examined in the previous section, we give an axiomatic definition of a cohesive category, which comprises both aspects we are interested in: the $\mathcal{S}$-concrete case (where the linking relation is controlled on domains) and the case of inverse categories (where the linking relation involves both domains and codomains).

3.2.1 *Compositive joins and meets*

Let C be an ordered category. We begin with a notion which will be frequently used.

We say that a family $(a_i\colon X \to Y)_{i \in I}$ of parallel maps has a *compositive join* $a = \bigvee_{i \in I} a_i$ if a is the least upper bound of the family in the ordered set $\mathsf{C}(X, Y)$, and is *preserved by composition*. In other words:

(i) for all $x\colon X' \to X$ and $y\colon Y \to Y'$ we have $yax = \bigvee_i ya_i x$, in the ordered set $\mathsf{C}(X', Y')$.

Compositive meets are defined in the same way. A *meet-semilatticed* category will be an ordered category with *compositive binary meets* (of all pairs of parallel maps):

$$y(a \wedge a')x = yax \wedge ya'x. \tag{3.5}$$

The stronger *cartesian compositive property* (or middle-four interchange for meet and composition)

$$(b \wedge b').(a \wedge a') = ba \wedge b'a', \tag{3.6}$$

will hold for linked meets in prj-cohesive categories (in Theorem 3.3.3).

(Both notions come out of enrichment of C, over the category of meet-semilattices equipped with different monoidal structures: the obvious tensor product and the cartesian product, see 5.5.6.)

Exercises and complements. Compositive joins have the following elementary properties, left to the reader. We assume that $a = \bigvee_{i \in I} a_i$ is a compositive join in $C(X, Y)$. Compositive meets enjoy dual properties.

(a) (*Associativity*) If, for every i, $a_i = \bigvee_j a_{ij}$ ($j \in J_i$) is a compositive join, then $a = \bigvee_{ij} a_{ij}$ (for $i \in I$, $j \in J_i$) is also.

(b) (*Composition*) For $x \colon X' \to X$ and $y \colon Y \to Y'$, $yax = \bigvee_i ya_i x$ is a compositive join.

(c) (*Expansion*) If $a_i \leqslant a'_i \leqslant a$, for every $i \in I$, the element a is also the compositive join $\bigvee_i a'_i$.

(d) (*Extension*) For a family $(a_i)_{i \in J}$ of morphisms $a_i \leqslant a$ that extends $(a_i)_{i \in I}$ to a set $J \supset I$, the element a is also the compositive join $\bigvee_{i \in J} a_i$.

3.2.2 Definition (Cohesive category)

A *cohesive category* will be a category C provided with two binary relations on parallel morphisms, the *order* $\leqslant$ and the *linking* (or *compatibility*) relation !, that satisfy the following four axioms:

(CH.1) the relation $\leqslant$ is an order of categories: reflexive, transitive, antisymmetric and consistent with composition,

(CH.2) the relation ! is reflexive, symmetric and consistent with composition: if $a \,!\, a'$ and $b \,!\, b'$ are consecutive pairs of morphisms, then $ba \,!\, b'a'$,

(CH.3) if $a \leqslant a'$, $b \leqslant b'$ and $a' \,!\, b'$, then $a \,!\, b$,

(CH.4) if $a \,!\, b$, the meet $a \wedge b$ exists and is preserved by composition. It will be called a *linked meet*.

As a consequence, if $a, b \leqslant c$ then $a \,!\, b$ (by (CH.2, 3)). We say that the cohesive category C is *link-filtered* if the converse holds too

(i) $a \,!\, b$ if and only if a and b have a common upper bound,

in which case the linking relation is determined by the order. (See 1.1.2 for filtered ordered sets.)

A link-filtered cohesive category is the same as an ordered category provided with binary *upper-bounded meets*, consistent with composition (obviously, one defines $a\,!\,b$, as above, in (i)).

Every category has a *discrete* cohesive structure where both relations $a \leqslant b$, and $a\,!\,b$, coincide with $a = b$. On the other hand, a cohesive category with *locally indiscrete linking* ($a\,!\,a'$ if and only if a and a' are parallel morphisms) is the same as a meet-semilatticed category, as defined in 3.2.1.

The notion of cohesive category is selfdual.

A *cohesive functor* $F\colon \mathsf{C} \to \mathsf{D}$ is a functor between cohesive categories which preserves order, linking and linked binary meets.

3.2.3 Linked joins of morphisms

Let C be a cohesive category.

A *linked* (or *compatible*) *set* α of C is any set of parallel morphisms such that $a\,!\,a'$ for all $a, a' \in \alpha$. If also β is a linked set, $\alpha\,!\,\beta$ will mean that α and β are parallel sets and $a\,!\,b$, for all $a \in \alpha$, $b \in \beta$; or equivalently, that $\alpha \cup \beta$ is a linked set of C. Any set of parallel morphisms which has an upper bound is linked.

We say that a set α of parallel morphisms of C has a *linked join* if:

(i) α has a compositive join $\bigvee \alpha$ (in particular, it is a linked set),

(ii) for each linked morphism b (i.e. $b\,!\,a$, for all a in α), $(\bigvee \alpha)\,!\,b$ and moreover $(\bigvee \alpha) \wedge b$ is the compositive join of the set $\{a \wedge b \mid a \in \alpha\}$ (which is linked, by (CH.3)).

It is easy to see that linked joins satisfy properties similar to those considered in 3.2.1(a)–(d) for compositive joins. Moreover, property (ii) will follow from (i), in convenient hypotheses: see Theorem 3.3.3(b).

In order to formulate hypotheses on these joins, we shall use a 'cardinal bound' ρ as a general substitute for the terms:

- *finite*, or finitely,
- *countable*, or countably, or σ-,
- *arbitrary*, or totally.

Thus, a ρ-*set* $\alpha \subset \mathsf{C}(X, Y)$ of parallel morphisms of C is a finite, or countable, or arbitrary subset, according to the 'value' of our cardinal bound ρ. The finite case includes the empty subset, and the countable case includes the finite one. (In the papers [G4, G5] we used a more general cardinal bound, for which we have no real use here.)

Considering a ρ-set $\alpha \subset \mathsf{C}(X,Y)$ or a family $(a_i)_{i \in I}$ in $\mathsf{C}(X,Y)$ indexed by a ρ-set are equivalent things, as far as we are considering the order and linking relation, joins and meets, upper bounds (or any notion invariant up to repeating issues).

Exercises and complements. (a) For a set $\alpha \subset \mathsf{C}(X,Y)$, the down-closed subset $\downarrow\alpha$ generated by α is linked if and only if α is. If $a = \bigvee \alpha$ is a linked join, also $a = \bigvee \downarrow\alpha$ is.

(b) If $\hat{a} = \bigvee \alpha$ and $\hat{b} = \bigvee \beta$ are linked joins of two linked sets $\alpha \,!\, \beta$, then $\hat{a} \,!\, \hat{b}$ and $\hat{a} \wedge \hat{b}$ is the linked join of the set $\{a \wedge b \mid a \in \alpha,\, b \in \beta\}$.

3.2.4 Definition (Totally cohesive categories)

A *ρ-cohesive category* will be a cohesive category C where every linked ρ-set of parallel morphisms has a linked join.

In particular we have the notions of *finitely cohesive*, *σ-cohesive* and *totally cohesive* category according to the 'value' of the cardinal bound ρ, in 3.2.3. Each of these conditions is stronger than the previous one.

The cohesive category C is ρ-cohesive if and only if:

(CH.5ρ) every linked ρ-set $\alpha \subset \mathsf{C}(X,Y)$ has a join $\bigvee \alpha$, compositive in C; linked binary meets distribute over joins of linked ρ-sets

$$(\bigvee \alpha) \wedge b = \bigvee_{a \in \alpha} (a \wedge b) \qquad \text{(if } \alpha \text{ is a linked } \rho\text{-set and } \alpha \,!\, b). \qquad (3.7)$$

The necessity of (CH.5ρ) being obvious, we assume that it holds and $\alpha!b$. Then the set $\beta = \alpha \cup \{b\}$ is a linked ρ-set, whence $\bigvee \beta$ is the upper bound of the morphisms $\bigvee \alpha$ and b, and $(\bigvee \alpha)!b$. Moreover the meets $a \wedge b$ ($a \in \alpha$) form a linked ρ-set, by (CH.3), and their join is compositive.

A *ρ-cohesive functor* $F \colon \mathsf{C} \to \mathsf{D}$ will be a functor between ρ-cohesive categories which preserves order, linking, linked binary meets and linked ρ-joins.

In a ρ-cohesive category, a ρ-set of parallel morphisms is linked if and only if it is *pairwise* upper bounded. A totally cohesive category is a stronger notion than a uc-quantaloid, even forgetting about all conditions on binary meets: see Exercise 3.2.6(g).

We shall see, in Section 3.7, that every cohesive category has a totally cohesive completion (together with a finitely cohesive and a σ-cohesive completion). However, most of the cohesive category used in this book already are totally cohesive.

3.2.5 Lifted cohesive structures

Let $U \colon \mathsf{C} \to \mathsf{D}$ be a faithful functor with values in a cohesive category. Then C has an order and a linking relation lifted by U: if a, b are parallel morphisms in C we let:

$$a \leqslant b \iff Ua \leqslant Ub \text{ in } \mathsf{D},$$
$$a \,!\, b \iff Ua \,!\, Ub \text{ in } \mathsf{D}.$$
$$(3.8)$$

The axioms (CH.1–3) are automatically satisfied in C. (CH.4) need not be, but we can test it using the isotone embeddings

$$U_{XY} \colon \mathsf{C}(X, Y) \to \mathsf{D}(UX, UY) \qquad (X, Y \text{ in } \mathsf{C}). \qquad (3.9)$$

(a) We say that C *has a cohesive structure lifted from* D, or that it is *cohesive over* D, if the following equivalent conditions are fulfilled:

(i) for all $a \,!\, b$, in $\mathsf{C}(X, Y)$ the meet $Ua \wedge Ub$ in D lifts to C, in the sense that there is a (unique) $c \in \mathsf{C}(X, Y)$ such that $Uc = Ua \wedge Ub$ (and then $c = a \wedge b$ in C),

(i′) C is cohesive and U is a cohesive functor.

(b) Let D be ρ-cohesive. We say that C *has a ρ-cohesive structure lifted from* D if the following equivalent conditions are also fulfilled:

(ii) for every linked ρ-set α of morphisms $X \to Y$ in C, the join of $U(\alpha)$ in D lifts to C, in the sense that there is a (unique) $c \in \mathsf{C}(X, Y)$ such that $Uc = \bigvee U(\alpha)$ (and then $c = \bigvee \alpha$),

(ii′) C is ρ-cohesive and U is a ρ-cohesive functor.

(c) If the functor U is full, all embeddings (3.9) are isomorphisms of ordered sets, and the lifting conditions above are automatically fulfilled.

3.2.6 Exercises and complements

The following results on the cohesive structure of $\mathcal{S}$-concrete and $\mathcal{I}$-concrete categories will be frequently used. The easy proofs, essentially sketched in Section 3.1, can be found in Chapter 7.

(a) The cohesive structure of $\mathcal{S}$, described in 3.1.1, is totally cohesive.

(b) The cohesive structures of $\mathcal{C}$ and $\mathcal{C}^r$, described in 3.1.3, are totally cohesive and lifted from $\mathcal{S}$. The structure of $\mathcal{C}^r$ is also lifted from $\mathcal{C}$.

(c) The cohesive structure of $\mathcal{D}$, described in 3.1.3, is finitely cohesive; it is lifted from $\mathcal{S}$ and from $\mathcal{C}$.

(d) The inverse category $\mathcal{I}$ has a link-filtered cohesive structure, described in 3.1.4. We have seen that it is not lifted from the structure of $\mathcal{S}$: two partial bijections $f, g\colon X \nrightarrow Y$ are linked in $\mathcal{I}$ when $f \,!_{\mathcal{S}}\, g$ and $f^\sharp \,!_{\mathcal{S}}\, g^\sharp$.

However, $\mathcal{I}$ is totally cohesive, and its linked joins and linked binary meets are the same as in $\mathcal{S}$.

(e) The inverse categories $\mathrm{I}\mathcal{C}$ and $\mathrm{I}\mathcal{C}^r$ have a totally cohesive structure, lifted from $\mathcal{I} = \mathrm{I}\mathcal{S}$. The category $\mathrm{I}\mathcal{D}$ has a finitely cohesive structure, lifted from $\mathcal{I} = \mathrm{I}\mathcal{S}$ and from $\mathrm{I}\mathcal{C}$.

(f) The wide subcategory $\mathcal{S}'$ of $\mathcal{S}$ consisting of those partial mappings whose definition-set has no more than (say) five elements lifts from $\mathcal{S}$ a cohesive structure that is not link-filtered, nor finitely cohesive.

(g) We have seen that the category $\mathcal{T}$ of topological spaces and partial maps is a uc-quantaloid (in 3.1.5). Letting $f \,!\, g$, when this pair of partial maps is upper bounded in $\mathcal{T}$, the structure is finitely cohesive, but not σ-cohesive.

This linking relation is not lifted from $\mathcal{S}$: if the partial maps f, g are set-theoretically compatible, their join *in $\mathcal{S}$* need not be continuous on its definition-set.

*(h) The category $L^\infty(a, \mathsf{Ban})$ is finitely cohesive; as recalled in 3.1.7, its σ-cohesive completion is important [G5].

3.2.7 Elementary properties

Let C be a ρ-cohesive category. A non-empty ρ-set $\alpha \subset \mathsf{C}(X, Y)$ of parallel morphisms is linked if and only if it has some upper bound (e.g. $\bigvee \alpha$).

If α and β are parallel linked ρ-sets of morphisms and $\alpha \,!\, \beta$, then $\bigvee \alpha \,!\, \bigvee \beta$ and

$$(\bigvee \alpha) \wedge (\bigvee \beta) = \bigvee a \wedge b \qquad (a \in \alpha, \, b \in \beta). \tag{3.10}$$

Furthermore, if α and γ are consecutive linked ρ-sets of morphisms, the set $\gamma\alpha = \{ca \mid a \in \alpha, \, c \in \gamma\}$ is again a linked ρ-set, by (CH.2), and

$$\bigvee (\gamma\alpha) = \bigvee \gamma . \bigvee \alpha. \tag{3.11}$$

3.2.8 Exercises

Let C be a cohesive category.

(a) C is finitely cohesive if and only if it satisfies the following two conditions, for all objects X, Y:

(CH.5a) the ordered set $C(X, Y)$ has a least element $0_{XY} : X \to Y$ (called a *zero* morphism), that is compositive in C (composed with any other gives a zero morphism),

(CH.5b) every linked pair $a \,!\, b$ in $C(X, Y)$ has a join $a \vee b$, compositive in C; linked binary meets distribute over joins of linked pairs.

(b) C is σ-cohesive if and only if it *also* satisfies:

(CH.5c) every increasing sequence (a_n) in $C(X, Y)$, obviously linked, has a join $\bigvee a_n$, compositive in C; linked binary meets distribute over increasing countable joins.

(c) Let X be an ordered category, equipped with the linking relation 3.2.2(i), where $a \,!\, b$ means that a and b have a common upper bound. Then X is ρ-cohesive if and only if:

(C.1) X has compositive upper-bounded binary meets,

(C.2ρ) ρ-sets of parallel maps, filtered in X, have a compositive join; meets of upper-bounded pairs of maps distribute over these joins.

3.2.9 Cohesive functors and transformations

(a) A ρ-cohesive functor $F : C \to D$ between ρ-cohesive categories has been defined in 3.2.4: it has to preserve order, linking, linked binary meets and linked ρ-joins. In the finitary and countable case there are characterisations of such functors, similar to those of 3.2.8.

A *ρ-cohesive transformation*

$$\varphi : F \to G : C \to D$$

will be a natural transformation between ρ-cohesive functors.

(b) A *ρ-cohesive subcategory* of the ρ-cohesive category C is any subcategory C' which is closed under linked binary meets and linked ρ-joins; then C', equipped with the induced order and linking relation, is ρ-cohesive, and the inclusion $C' \to C$ is also.

A *ρ-cohesive embedding* $F : C \to D$ will be a ρ-cohesive functor that is injective on objects and reflects the order and linking relations. Then F is faithful and $F(C)$ is a ρ-cohesive subcategory of D, isomorphic to C.

(c) Taking out all conditions on linked ρ-joins, we have *cohesive functors* between cohesive categories (already defined in 3.2.2), *cohesive transformations*, and so on.

3.3 Prj-cohesive and e-cohesive categories

As suggested in Section 3.1, cohesive structures are often defined by assigning, for each object X, a set of *commuting idempotent endomorphisms* of X, its 'projectors'. This yields the notions of prj-cohesive and e-cohesive category, the latter being stronger than the former.

These notions are not selfdual. At the end of this section we also consider the dual cases, of prj*-cohesive and e*-cohesive category.

Let us note that, in this book, we are not directly interested in prj-cohesive categories (which were necessary for our work on measurable operators, in [G5]). We could restrict our study to the e-cohesive case, but the exposition would result less clear: most properties of e-cohesive categories are better analysed using general 'resolutions' rather than the minimal resolutions given by supports.

We also note that, in prj-cohesive (and e-cohesive) categories, the existence of compositive joins ensures that binary linked meets distribute over them: see Theorem 3.3.3(b).

Related structures, like dominical categories [Di, He, DiH], p-categories [Rs, RoR] and restriction categories [CoL], are examined in 3.3.9(c), (d).

3.3.1 *Definition* (Prj-category)

A *prj-cohesive category*, or *prj-category*, will be a category C equipped, for every object X, with a set $\mathsf{Prj}\,(X) \subset \mathsf{C}(X, X)$ of endomorphisms of X, called the *projectors* of X, so that:

(PCH.1) every identity is a projector, and every projector is idempotent; if e and f are projectors of the same object, then $ef = fe$ is a projector,

(PCH.2) if $a \colon X \to Y$ is in C and $f \in \mathsf{Prj}\,(Y)$, there is some $e \in \mathsf{Prj}\,(X)$ such that $fa = ae$.

Thus $\mathsf{Prj}\,(X)$ is a commutative idempotent submonoid of $\mathsf{C}(X, X)$ and a 1-semilattice in its own right, with unit 1_X and:

$$e \wedge f = ef = fe, \qquad e \leqslant f \Leftrightarrow e = ef = fe. \qquad (3.12)$$

A *prj-cohesive functor* $F \colon \mathsf{C} \to \mathsf{D}$, or *prj-functor*, is a functor between prj-categories which preserves projectors.

Every category has a *discrete* prj-structure, where the only projectors are the identities.

3.3.2 *The cohesive structure*

A prj-category C has the following *associated* order and linking relations (that make C into a cohesive category, as proved below)

(i) $a \leqslant b$ $\Leftrightarrow$ $a = be$, for some projector e (and then $ae = a$),

(ii) $a \,!\, b$ $\Leftrightarrow$ $a = ae$, $b = be'$, $ae' = be$, for some projectors e, e'.

In the second case we say that the pair (e, f) is a *resolution* of the *linked pair* (a, b).

The order extends the canonical order of projectors: indeed, if $e = fe'$ in $\mathsf{Prj}\, X$, then $ef = fe'f = fe' = e$. An endomorphism $a \in \mathsf{C}(X, X)$ is a projector if and only if $a \leqslant 1_X$; therefore, all the existing joins and non-empty meets of projectors, in $\mathsf{Prj}\, X$ or $\mathsf{C}(X, X)$, are the same.

The identity 1_X is the greatest element (and empty meet) of $\mathsf{Prj}\, X$. The ordered set $\mathsf{C}(X, X)$ need not have a greatest element (as we have seen in 3.1.5(c)), but the identity 1_X is a maximal element in $\mathsf{C}(X, X)$: if $a \geqslant 1_X$ then $1 = ae$ (for some projector e), hence $e = aee = ae = 1$ and $a = 1$.

Examples and remarks. (a) The categories $\mathcal{S}, \mathcal{C}, \mathcal{C}^r$ and $\mathcal{D}$ are prj-cohesive, with projectors given by their partial identities. All of them are totally cohesive, except $\mathcal{D}$ which is finitely cohesive (see 3.2.6).

(b) We shall see that the structure of a prj-category is determined by the order, *as a pre-inverse category satisfying* (i) (in 3.3.4). The inverse core IC will be studied in 3.4.6–3.4.8.

(c) The projector e in (i) and (ii) can be replaced with any projector $e' \leqslant e$ such that $ae' = a$.

(d) The cohesive structure of a discrete prj-category is discrete: the relations $a \leqslant b$, and $a \,!\, b$, are both equivalent to $a = b$.

3.3.3 *Theorem*

(a) A prj-category C, equipped with the associated order and linking relation, is a cohesive category (as defined in 3.2.2).

If (e, e') is a resolution of the linked pair (a, b), then

$$a \wedge b = ae' = be. \tag{3.13}$$

If $a, b\colon X \to Y$ and $c, d\colon Y \to Z$ are linked pairs, the cartesian *compositive property (3.6) holds:*

$$X \underset{b}{\overset{a}{\rightrightarrows}} Y \underset{d}{\overset{c}{\rightrightarrows}} Z \qquad ca \wedge db = (c \wedge d).(a \wedge b). \tag{3.14}$$

Every upper bounded set of parallel morphisms is linked. Every set ε of parallel projectors is linked; the linked meet of two parallel projectors is their meet in $\mathsf{Prj}\,X$, $e \wedge f = ef = fe$, which is thus compositive in C.

(b) A prj-category C is ρ-cohesive if (and only if) every linked ρ-set of parallel morphisms has a join, preserved by composition: this join is necessarily a linked join.

(c) A functor between prj-cohesive categories is cohesive if and only if it is prj-cohesive, if and only if it preserves the order.

Note. A stronger form of point (b) will be given in Corollary 3.8.6.

Proof The symbols $e, f, e', f' \ldots$ always denote projectors.

(a) For the first two axioms (CH.1, 2) the only non-trivial verification concerns the composition. Let the arrows of diagram (3.14) be given in C.

If $a \leqslant b$, and $c \leqslant d$, let $a = be$, $c = df$; by (PCH.2) there is a projector e' such that $fa = ae'$, and $db.ee' = dae' = dfa = ca$, so that $ca \leqslant db$.

On the other hand, if $a\,!\,b$, and $c\,!\,d$, let

$$a = ae, \qquad b = be', \qquad ae' = be,$$
$$c = cf, \qquad d = df', \qquad cf' = df. \tag{3.15}$$

By (PCH.2) there are projectors e_1, e_2 such that $fa = ae_1$ and $f'b = be_2$; we want to show that $(ee_1, e'e_2)$ is a resolution of the pair (ca, db), which will prove that these morphisms are linked.

Indeed $ca.ee_1 = cae_1 = cfa = ca$, and analogously $db.e'e_2 = db$. Last

$$ca.e'e_2 = cbee_2 = cbe_2e = cf'be = dfbe = dfae'$$
$$= dae_1e' = dae'e_1 = db.ee_1. \tag{3.16}$$

As to (CH.3), if $a \leqslant a'$, $b \leqslant b'$ and $a'\,!\,b'$, let $a = a'e$, $b = b'f$ and take a resolution (e', f') of (a', b'). Then (ee', ff') is a resolution of (a, b):

$$a.ee' = a, \qquad b.ff' = b, \qquad a.ff' = a'eff' = b'efe' = b.ee'.$$

Next we prove (CH.4) and property (3.13). Let $a\,!\,b$, with resolution (e, e'): $a = ae$, $b = be'$, $ae' = be$; we have to show that $h = ae' = be$ is the meet of a and b. Clearly $h \leqslant a, b$; if $k \leqslant a, b$ then $k = af = bf'$ and $k = ae.f = bf'.e \leqslant h$.

It is now easy to deduce the cartesian property (3.14): with the hypothesis $a\,!\,b$, $c\,!\,d$ and the notation of (3.15), we have from (3.16) that:

$$ca \wedge db = ca.e'e_2 = df.be = (c \wedge d).(a \wedge b).$$

(b) Suppose we have a ρ-join $a = \bigvee a_i$, preserved by composition, and take a parallel map b linked with all a_i. We have a resolution of the set formed by all a_i and b

$$a_i e_i = a_i, \qquad\qquad b e_b = b,$$
$$a_i e_j = a_j e_i, \qquad a_i e_b = b e_i = a_i \wedge b \qquad (i \in I).$$

We let $e = \bigvee_i e_i$, a compositive join by hypothesis. Now the pair (e, e_b) is a resolution of the pair (a, b), and shows that it is linked

$$ae = \bigvee_j a e_j = \bigvee_{ij} a_i e_j = \bigvee_i a_i = a, \qquad b e_b = b,$$
$$a e_b = \bigvee_i a_i e_b = \bigvee_i b e_i = be.$$

The distributivity of $- \wedge b$ on the given join follows: computing binary meets by resolutions, as in (3.13), we get a compositive join

$$a \wedge b = a e_b = (\bigvee_i a_i) e_b = \bigvee_i a_i e_b = \bigvee_i (a_i \wedge b).$$

(c) The last points are trivial; in particular a functor between prj-cohesive categories preserves the order if and only if it preserves the projectors, in which case it also preserves resolutions, whence the linking relation and also binary linked meets, because of (3.13). $\square$

3.3.4 Proposition and Definition

For an ordered category C *the following conditions are equivalent.*

(i) C *has a prj-cohesive structure, with projectors defined by*

$$\mathsf{Prj}\,(X) = \{a \in \mathsf{C}(X, X) \mid a \leqslant 1_X\}. \tag{3.17}$$

(ii) C *has a cohesive structure* $(\leqslant, !)$*, with the order and linking relation given by 3.3.2(i)–(ii), for the projectors defined as above.*

(iii) C *is pre-inverse (i.e. $a \leqslant 1$ implies $aa = a$), and $a \leqslant b$ if and only if there is some $e \leqslant 1$ such that $a = be$.*

The ordered category (resp. the cohesive category) C *will be said to be* prj-cohesive *when (iii) is met: the prj-cohesive structure described above is the only possible one.*

Proof We know that (i) $\Rightarrow$ (ii), by 3.3.2; (ii) $\Rightarrow$ (iii) is obvious. Assuming (iii), we define the projectors by (3.17), making C into a prj-cohesive category.

As to axiom (PCH.1), we have already seen that in a pre-inverse category the endomorphisms $e \leqslant 1_X$ form a 1-semilattice (in 2.4.3(a)). Then

(PCH.2) follows from the characterisation of the ordering: if f is a projector, the relation $fa \leqslant a$ implies the existence of a projector e such that $fa = ae$.

The rest is obvious: a prj-cohesive structure is determined by its projectors, which are determined by the condition $e \leqslant 1_X$ (for some object X of the category). $\qquad\square$

3.3.5 Definition (e-category)

An *e-cohesive category*, or *e-category*, will be a category C equipped, for every object X, with a projector-set $\mathsf{Prj}(X) \subset \mathsf{C}(X, X)$ satisfying three axioms (the first is the same as axiom (PCH.1) of 3.3.1)

(ECH.0) every identity is a projector, and every projector is idempotent; if e and f are projectors of the same object, then $ef = fe$ is a projector,

(ECH.1) for every $a\colon X \to Y$ in C, the set of projectors e of X such that $ae = a$ has a least element $\underline{e}(a)$, called the *support* of a,

(ECH.2) for every $a\colon X \to Y$ and $b\colon Y \to Z$ in C, we have $\underline{e}(b).a = a.\underline{e}(ba)$.

There are elementary properties, for consecutive morphisms $a\colon X \to Y$ and $b\colon Y \to Z$, and projectors $e \in \mathsf{Prj}(X)$, $f \in \mathsf{Prj}(Y)$

$$\underline{e}(e) = e, \tag{3.18}$$

$$\underline{e}(ba) \leqslant \underline{e}(a), \tag{3.19}$$

$$fa = a.\underline{e}(fa), \tag{3.20}$$

$$\underline{e}(ae) = e.\underline{e}(ae) = \underline{e}(a).e, \tag{3.21}$$

$$\text{if } a \text{ is a monomorphism}, \underline{e}(a) = 1. \tag{3.22}$$

In particular, (3.20) shows that axiom (PCH.2) is also satisfied: *every e-cohesive category is prj-cohesive.* Therefore, it has a cohesive structure: the order $a \leqslant b$, and the linking relation $a \, ! \, b$, are defined as in 3.3.2(i), (ii). They can also be characterised as below, in (3.24)–(3.25).

The discrete prj-structure is e-cohesive: the support of a morphism $a\colon X \to Y$ is 1_X.

An *e-cohesive functor* $F\colon \mathsf{C} \to \mathsf{D}$, or *e-functor*, will be a functor between e-categories which preserves supports

$$F(\underline{e}(a)) = \underline{e}(Fa) \qquad \text{(for } a \text{ in } \mathsf{C}\text{)}. \tag{3.23}$$

By (3.18) it also preserves projectors, hence it is prj-cohesive and cohesive (by Theorem 3.3.3).

3.3.6 Exercises and complements (The e-cohesive structure)

Let C be an e-cohesive category.

(a) The order and linking relations of C are characterised as follows:

$$a \leqslant b \iff a = b.\underline{e}(a), \tag{3.24}$$

$$\begin{aligned} a \,!\, b \quad &\iff \quad a.\underline{e}(b) = b.\underline{e}(a) \\ &\iff \quad (\underline{e}(a), \underline{e}(b)) \text{ is a resolution of } (a, b), \tag{3.25} \\ &\iff \quad a.\underline{e}(b) \leqslant b \text{ and } b.\underline{e}(a) \leqslant a. \end{aligned}$$

(b) If $a \leqslant b$, then $\underline{e}(a) \leqslant \underline{e}(b)$.

(c) If $a \,!\, b$, then

$$a \wedge b = a.\underline{e}(b) = b.\underline{e}(a), \qquad \underline{e}(a \wedge b) = \underline{e}(a) \wedge \underline{e}(b). \tag{3.26}$$

(d) Assuming that C is ρ-cohesive and α is a linked ρ-set:

$$\underline{e}(\vee \alpha) = \vee_{a \in \alpha} \underline{e}(a). \tag{3.27}$$

(e) (*Lifted e-structure*) Let $U \colon C \to D$ be a faithful functor with values in an e-cohesive category. We define the projectors of C as the endomorphisms e such that $U(e)$ is a projector in D.

We say that C *has an e-cohesive structure lifted from* D if the following equivalent conditions are fulfilled:

(i) for every a in $C(X, Y)$ the support $\underline{e}(Ua)$ in D lifts to C, in the sense that there is a (unique) $e \in C(X, X)$ such that $Ue = \underline{e}(Ua)$ (and then $e = \underline{e}(a)$ in C),

(i′) C is e-cohesive (with the previous projectors) and U is an e-cohesive functor.

(f) (*Lifted e-structure, continued*) Furthermore, if D is ρ-cohesive, we say that C *has a* ρ-*cohesive e-structure lifted from* D if the following equivalent conditions are also fulfilled:

(ii) for every linked ρ-set α of morphisms $X \to Y$ in C, the join of the set $U(\alpha)$ in D lifts to C, in the sense that there is a (unique) $c \in C(X, Y)$ such that $Uc = \vee U(\alpha)$ (and then $c = \vee \alpha$),

(ii″) C is a ρ-cohesive e-category and U is a ρ-cohesive e-functor.

3.3.7 Exercises and complements (Inverse images of projectors)

Let C be an e-category. Every morphism $a\colon X \to Y$ in C determines a mapping

$$a^P \colon \mathsf{Prj}\,(Y) \to \mathsf{Prj}\,(X), \qquad a^P(f) = \underline{e}(fa). \qquad (3.28)$$

(a) Prj becomes thus a contravariant functor

$$\mathsf{Prj} : \mathsf{C}^{\mathrm{op}} \to \mathsf{1Slh}, \qquad (3.29)$$

with values in the category 1Slh of 1-semilattices and their homomorphisms (introduced in 2.7.1).

This means that, for $a\colon X \to Y$, $b\colon Y \to Z$ and $f, f' \in \mathsf{Prj}\,(Y)$

$$(ba)^P = a^P b^P, \qquad 1^P = 1,$$
$$a^P(ff') = a^P(f).a^P(f'). \qquad (3.30)$$

(b) Other properties, for a morphism $a\colon X \to Y$ and projectors $e, e' \in \mathsf{Prj}\,(X)$, $f \in \mathsf{Prj}\,(Y)$:

$$e^P(1) = e, \qquad (3.31)$$

$$fa = a.a^P(f), \qquad (3.32)$$

$$e^P(e') = ee'. \qquad (3.33)$$

*(c) Conversely, if C satisfies (ECH.0, 1) and the mappings (3.28) are given, satisfying (3.30)–(3.32), then C is e-cohesive, with $\underline{e}(a) = a^P(1)$.

3.3.8 Examples and remarks

(a) The categories $\mathcal{S}, \mathcal{C}, \mathcal{C}^r, \mathcal{D}$ of 2.1.3 and 2.1.4 are e-cohesive: the support $\underline{e}(f)$ of the partial mapping f is the partial identity on Def f.

In each of these categories the set $\mathsf{Prj}\,(X)$ consists of those endomorphisms of X that are partial identities (on a subset, or an open subspace, or a closed subspace of X, respectively).

$\mathcal{C}$ and $\mathcal{C}^r$ have a totally cohesive e-structure, and $\mathcal{D}$ a finitely cohesive e-structure, all of them lifted from $\mathcal{S}$, in the sense of 3.3.6(f).

Many other examples will be dealt with in Chapter 4.

(b) An e-cohesive category need not be link-filtered, as shown by the category $\mathcal{S}' \subset \mathcal{S}$ considered in 3.2.6(f).

*(c) The category $L^\infty(a, \mathsf{Ban})$ described in 3.1.7 is prj-cohesive and not e-cohesive [G5].

3.3.9 Complements

(a) (*Cartesian products*) The cartesian product $C = \Pi_i\, C_i$ (see 1.4.6) of a family of cohesive categories $(C_i)_{i \in I}$ is cohesive, with the product order relation and the product linking:

$$(a_i) \leqslant (b_i) \quad \Leftrightarrow \quad \text{for all indices } i,\ a_i \leqslant b_i \text{ in } C_i,$$
$$(a_i)\,!\,(b_i) \quad \Leftrightarrow \quad \text{for all indices } i,\ a_i\,!\,b_i \text{ in } C_i. \qquad (3.34)$$

If all the factors C_i are prj-cohesive, so is the product C, with

$$\mathsf{Prj}_C((X_i)_{i \in I}) = \Pi_i\, \mathsf{Prj}\,(X_i).$$

If all the factors C_i are e-cohesive, so is the product C, with

$$\underline{e}((a_i)_{i \in I}) = (\underline{e}(a_i))_{i \in I}.$$

(b) (*Duality*) A *prj*-cohesive category* will be a pair (C, Prj) satisfying (PCH.1) and

(PCH.2*) for all $a \colon X \to Y$ and $e \in \mathsf{Prj}\,(X)$ there is some $f \in \mathsf{Prj}\,(Y)$ such that $ae = fa$.

The associated cohesion structure has $a \leqslant b$ if there is some projector f such that $a = fb$; analogously, the linking is determined by *coresolutions* of pairs of morphisms.

An *e*-cohesive category* will be a pair (C, Prj) that satisfies the axioms (ECH.0, 1*, 2*). Every morphism $a \colon X \to Y$ has then a *cosupport* $\underline{e}^*(a) \in \mathsf{Prj}\,(Y)$.

These notions will be useful in the next section, dealing with inverse categories.

(c) (*Dominical categories and p-categories*) Some related structures predate the articles [G3, G4] on e-categories.

A *dominical category*, as defined by R.A. Di Paola and A. Heller [Di, He, DiH], is a category C equipped with a functor $\times \colon C \times C \to C$ and suitable further structure, meant to formalise the categories of partial mappings when equipped with the obvious product of a pair of partial maps

$$a \times a' \colon X \times X' \to Y \times Y', \qquad \mathrm{Def}\,(a \times a') = (\mathrm{Def}\,a) \times (\mathrm{Def}\,a').$$

A *p-category*, as defined by G. Rosolini and E. Robinson [Rs, RoR], is a generalisation of the previous structure, still equipped with a product as above. In both cases the partial identity on Def a (written here as $\underline{e}(a)$) is denoted as dom a.

Every p-category C (a fortiori every dominical category) is e-cohesive, with $\underline{e}(a) = \operatorname{dom} a$ and:

$$\begin{aligned}
\operatorname{Prj}(X) &= \{\operatorname{dom} a \mid a \in C(X, X)\} \\
&= \{e \in C(X, X) \mid \exists\, a \text{ in } C\colon e = \operatorname{dom} a\} \qquad (3.35) \\
&= \{e \in C(X, X) \mid e = \operatorname{dom} e\}.
\end{aligned}$$

This follows from the following properties of domains, proved in [Rs] 2.1.4–5 (or [RoR], 1.4) for morphisms $a\colon X \to Y$, $b\colon Y \to Z$, $a'\colon X \to Y'$

- (i) $\operatorname{dom} 1_X = 1_X$,
- (ii) $\operatorname{dom}(ba) = \operatorname{dom}((\operatorname{dom} b)a)$,
- (iii) $(\operatorname{dom} b).a = a.\operatorname{dom}(ba)$,
- (iv) $(\operatorname{dom} a).(\operatorname{dom} a') = (\operatorname{dom} a').(\operatorname{dom} a)$,
- (v) $a.\operatorname{dom} a = a$,
- (vi) $\operatorname{dom}(\operatorname{dom} a) = \operatorname{dom} a$,
- (vii) $(\operatorname{dom} a).(\operatorname{dom} a) = (\operatorname{dom} a)$,
- (viii) $\operatorname{dom}((\operatorname{dom} a).(\operatorname{dom} a')) = (\operatorname{dom} a).(\operatorname{dom} a')$.

Indeed the second and third equalities in (3.35) come from (vi). The axiom (ECH.0) follows from (i), (vii), (iv) and (viii), while (ECH.2) coincides with (iii). As to (ECH.1): if $a\colon X \to Y$, then $a = a.\operatorname{dom} a$, by (v); on the other hand, if $a = ae$ and $e \in \operatorname{Prj}(X)$, then $e \leqslant \operatorname{dom} a$, since (by (ii) and (viii))

$$\begin{aligned}
\operatorname{dom} a = \operatorname{dom}(ae) &= \operatorname{dom}((\operatorname{dom} a)e) = \operatorname{dom}((\operatorname{dom} a)(\operatorname{dom} e)) \\
&= (\operatorname{dom} a).(\operatorname{dom} e) = (\operatorname{dom} a).e.
\end{aligned}$$

(d) *(Restriction categories)* Other related structures came after [G3, G4]: restriction categories, introduced in 2002 by R. Cockett and S. Lack [CoL], are equivalent to e-categories.

A *restriction category* is equipped with an operation assigning to each map $f\colon A \to B$ a map $\underline{e}(f)\colon A \to A$ (written as $\bar{f}$), under four axioms:

- (R1) $f\underline{e}(f) = f$,   for all $f\colon A \to B$,
- (R3) $\underline{e}(f)\underline{e}(g) = \underline{e}(g)\underline{e}(f)$,   for all $f\colon A \to B$ and $g\colon A \to C$,
- (R3) $\underline{e}(g\underline{e}(f)) = \underline{e}(g)\underline{e}(f)$,   for all $f\colon A \to B$ and $g\colon A \to C$,
- (R4) $\underline{e}(g)f = f\underline{e}(gf)$,   for all $f\colon A \to B$ and $g\colon B \to C$.

If these axioms hold, it is easy to see that the set $\operatorname{Prj}(A) \subset C(A, A)$ of all supports $\underline{e}(f)$ of maps defined on A is a commutative idempotent submonoid. (ECH.1) is fulfilled: if $g\underline{e}(f) = g$, then

$$\underline{e}(g) = \underline{e}(g\underline{e}(f)) = \underline{e}(g)\underline{e}(f) \leqslant \underline{e}(f),$$

in the semilattice $\operatorname{Prj}(A)$. Finally, (ECH.2) is the same as (R4).

Conversely, if C is e-cohesive, (R1) holds by (ECH.1), (R2) by (ECH.0), (R3) by (3.21), and (R4) by (ECH.2).

3.4 Inverse categories and cohesion

Inverse categories have been studied in Chapter 2.

We now show that an inverse category has a canonical cohesive structure, invariant under the involution, which combines an e-cohesive and an e*-cohesive structure without belonging to any of these kinds.

Inverse categories will supply the intrinsic atlas of a generalised manifold; for instance, the intrinsic atlas of a C^r-manifold is formed of open euclidean spaces and partial C^r-diffeomorphisms between open subsets, and lives in the inverse core IC^r of the cohesive category C^r.

K is always an inverse category.

3.4.1 Supports and cosupports

Let us recall that, in the inverse category K, we have a unique regular involution, that sends a morphism a to its unique partial inverse $a^\sharp$ (see Section 2.3).

Every object X has a 1-semilattice $\mathsf{Prj}(X)$ (cf. 2.3.4) consisting of its projectors, the idempotent endomorphisms $e: X \to X$, which are automatically symmetric with respect to the canonical involution: $e = e^\sharp$.

Every morphism a has two associated projectors (as in 2.3.4):

$$\underline{e}(a) = a^\sharp a \in \mathsf{Prj}(X) \qquad \text{(the support of } a\text{)},$$
$$\underline{e}^*(a) = aa^\sharp \in \mathsf{Prj}(Y) \qquad \text{(the cosupport of } a\text{)}. \tag{3.36}$$

Moreover, K has a canonical order (of categories) $a \leqslant b$, characterised by the following equivalent properties (from Theorems 2.4.1 and 2.4.2):

(i) $a = ab^\sharp a$,

(ii) $a = b.\underline{e}(a)$,

(ii*) $a = \underline{e}^*(a).b$,

(iii) $a = \underline{e}^*(a).b.\underline{e}(a)$,

(iv) $a = be$, for some projector e,

(iv*) $a = fb$, for some projector f,

(v) $a = fbe$, for some projectors e, f.

Note. Using the transfer of projectors (in 2.7.4), the support and cosupport of a morphism a are expressed as:

$$\underline{e}(a) = a^P(1_Y), \qquad \underline{e}^*(a) = a_P(1_X).$$

3.4.2 Two preliminary cohesive structures

We use the supports and cosupports of the inverse category K to define two cohesive structures, related by the involution.

First, the projectors of K satisfy the axioms (ECH.0–2) of 3.3.5 (see Exercise (a)), and define a *left cohesive structure* K_L, which is e-cohesive and determined by the supports $\underline{e}(a) = a^\sharp a$

$$K_L = (K, \leqslant, !'),$$

$$a \leqslant b \iff a = b\underline{e}(a), \tag{3.37}$$

$$a \,!'\, b \iff a\underline{e}(b) = b\underline{e}(a) \iff ba^\sharp \in \mathsf{Prj}\,(Y),$$

for $a, b \in K(X, Y)$.

Note that the order relation coincides with the canonical order $\leqslant$ of K, recalled above. (The last equivalence in the definition of $a \,!'\, b$ is verified in Exercise (b).)

Applying the involution of K, we also have a right cohesive structure K_R, which is e*-cohesive and determined by the cosupports $\underline{e}^*(a) = aa^\sharp$

$$K_R = (K, \leqslant, !''),$$

$$a \leqslant b \iff a = \underline{e}^*(a)b, \tag{3.38}$$

$$a \,!''\, b \iff \underline{e}^*(b)a = \underline{e}^*(a)b \iff b^\sharp a \in \mathsf{Prj}\,(X),$$

for $a, b \in K(X, Y)$, again.

The two order relations coincide, but the linking relations are related by the involution

$$a \,!'\, b \iff a^\sharp \,!''\, b^\sharp, \tag{3.39}$$

and generally different (as already said in 3.1.4). The reader is invited to work out Exercise (c), describing these relations in the inverse category $\mathcal{I}$ of partial bijections of sets.

Exercises. (a) The projectors of the inverse category K define an e-cohesive structure with supports $\underline{e}(a) = a^\sharp a$.

(b) For $a, b \colon X \to Y$: $a(b^\sharp b) = b(a^\sharp a)$ if and only if $ba^\sharp \in \mathsf{Prj}\,(Y)$.

(c) Characterise the relations $a \,!'\, b$ and $a \,!''\, b$ in $\mathcal{I}$, showing that they are different. The cohesive structures $\mathcal{I}_L$ and $\mathcal{I}_R$ are not even finitely cohesive. *Hints:* use the embedding of $\mathcal{I}$ in $\mathcal{S}$.

3.4.3 The cohesive structure of an inverse category

The *canonical cohesive structure* $(\leqslant, !)$ of the inverse category K will be given by its canonical order $\leqslant$ together with the following linking relation, that is *invariant under the involution* of K

$$a \,!\, b \quad \Leftrightarrow \quad a \,!'\, b \text{ and } a \,!''\, b$$

$$\Leftrightarrow \quad a(b^{\sharp} b) = b(a^{\sharp} a) \text{ and } (bb^{\sharp})a = (aa^{\sharp})b, \qquad (3.40)$$

$$\Leftrightarrow \quad ba^{\sharp} \in \mathsf{Prj}\,(Y) \text{ and } b^{\sharp} a \in \mathsf{Prj}\,(X).$$

The axioms (CH.1–4) of Definition 3.2.2 are satisfied, since the two cohesive structures defined above, in 3.4.2, have the same order relation. The binary linked meets are computed below, in (3.41).

In the inverse category $\mathcal{I}$ of partial bijections of sets, the relation $a \,!\, b$ means that the partial bijections a, b have an upper bound (or equivalently a join) in $\mathcal{I}$ (see Exercise 34.2(c)).

The inverse category K need not be link-filtered: e.g. consider the inverse subcategory of $\mathcal{I}$ formed by those partial bijections whose definition-set has no more than five elements.

3.4.4 Proposition

(a) In the cohesive structure of the inverse category K, if $a \,!\, b$, in $\mathsf{K}(X, Y)$, $e \in \mathsf{Prj}\,(X)$ and $f \in \mathsf{Prj}\,(Y)$, we have:

$$a \wedge b = ab^{\sharp} b = ba^{\sharp} a = bb^{\sharp} a = aa^{\sharp} b = ab^{\sharp} a = ba^{\sharp} b, \qquad (3.41)$$

$$(a \wedge b)_P(e) = a_P(e) \wedge b_P(e), \qquad (a \wedge b)^P(f) = a^P(f) \wedge b^P(f), \qquad (3.42)$$

$$\underline{e}(a \wedge b) = \underline{e}(a) \wedge \underline{e}(b) = b^{\sharp} a = a^{\sharp} b,$$
$$\underline{e}^*(a \wedge b) = \underline{e}^*(a) \wedge \underline{e}^*(b) = ba^{\sharp} = ab^{\sharp}. \qquad (3.43)$$

Meets satisfy the cartesian compositive property (3.6).

(b) A set α of parallel morphisms in K is linked if and only if it has a double resolution (e_a), (f_a) of projectors, satisfying (for $a, b \in \alpha$)

$$a = a.e_a = f_a.a, \qquad a.e_b = b.e_a, \qquad f_b.a = f_a.b. \qquad (3.44)$$

The least double resolution is given by the families $e_a = \underline{e}(a)$, $f_a = \underline{e}^*(a)$, of supports and cosupports of all $a \in \alpha$.

(c) K is ρ-cohesive if and only if it has compositive joins of linked ρ-sets of parallel morphisms (which are linked joins).

Proof (a) The axioms (CH.1–4) of Definition 3.2.2 are a straightforward consequence of the definition in 3.4.3, since the e- and $\underline{e}^*$-structures (defined in 3.4.2) are cohesive structures with the same order relation.

Linked meets can be computed according to the first structure, which is e-cohesive, or according to the second one, $\underline{e}^*$-cohesive, giving

$$a \wedge b = a(b^\sharp b) = b(a^\sharp a), \qquad a \wedge b = (bb^\sharp)a = (aa^\sharp)b.$$

The last two expressions in (3.41) follow at once from $ba^\sharp \in \mathsf{Prj}\,(Y)$ and $b^\sharp a \in \mathsf{Prj}\,(X)$.

The cartesian compositive property of meets follows from Theorem 3.3.3 (applied to the first cohesive structure of K). For (3.42):

$$(a \wedge b)_P(e) = (a \wedge b).e.(a \wedge b)^\sharp = (ae \wedge be).(a^\sharp \wedge b^\sharp)$$
$$= aea^\sharp \wedge beb^\sharp = a_P(e) \wedge b_P(e).$$

(b) Obvious. (c) From 3.3.3(b). $\qquad\qquad\square$

3.4.5 Complements and remarks

(a) (*Cohesive functors*) Every functor $F \colon \mathsf{K} \to \mathsf{K}'$ between inverse categories is automatically a cohesive functor (as defined in 3.2.2). In fact, it preserves supports and cosupports, and therefore the canonical order and linking relations; it also preserves linked binary meets, by (3.41).

If K and K' are ρ-cohesive inverse categories, saying that F is a ρ-cohesive functor (cf. 3.2.4) simply means that it preserves all the existing ρ-joins of parallel morphisms (which are linked).

(b) A faithful functor between inverse categories also reflects the order and the linking relation. An $\mathcal{I}$-concrete inverse category always has the cohesive structure lifted from $\mathcal{I}$.

(c) A full and faithful cohesive functor $F \colon \mathsf{K} \to \mathsf{K}'$ between inverse categories gives local isomorphisms of ordered sets $\mathsf{K}(X, Y) \to \mathsf{K}'(FX, FY)$; we also know that it preserves and reflects the linking relation. Therefore, if K' is ρ-cohesive, also K and F are.

(d) In an inverse category, the condition $a \,!'\, b$ (resp. $a \,!''\, b$) is sufficient to ensure that a and b have a compositive meet with respect to the canonical order, using the left e-cohesive structure (resp. the $\underline{e}^*$-cohesive structure).

However, these 'lateral meets' need not satisfy property (3.42), nor (3.43) (see Exercise 3.4.2(c)). The good notion is linked meets, in the present symmetric sense.

3.4.6 The inverse core of a prj-category

Let C be a prj-category. We have already remarked that C is pre-inverse (in 3.3.4).

Its inverse core $\mathsf{K} = \mathsf{IC}$ is the wide subcategory formed of those morphisms $u \colon X \to Y$ that have a Morita inverse $u' \colon Y \to X$ in C

$$u = uu'u, \quad u' = u'uu', \quad u'u \leqslant 1_X, \quad uu' \leqslant 1_Y. \tag{3.45}$$

Let us recall that K is an inverse category whose projectors (i.e. idempotent endomorphisms) coincide with the assigned projectors of C (i.e. the endomorphisms $\leqslant 1$). The partial inverse in K is given by the Morita inverse in C.

The canonical order $u \leqslant v$ in K as an inverse category is defined by the relation $u = ve$ (for some projector e), and coincides with the restriction of the order of C as a prj-category. The linking relation $u \,!_\mathsf{K}\, v$ (see 3.4.3) is defined by supports and cosupports

$$u \,!_\mathsf{K}\, v \quad \Leftrightarrow \quad (u \,!\, v \text{ and } u^\sharp \,!\, v^\sharp \text{ in } \mathsf{C}). \tag{3.46}$$

Therefore the embedding $\mathsf{IC} \to \mathsf{C}$ preserves the cohesive structure, and need not reflect the linking relation (see the examples in 3.4.8). The inverse-core construction clearly applies to prj-cohesive functors.

3.4.7 Theorem

Let C *be a prj-category and* $\mathsf{K} = \mathsf{IC}$ *its inverse core, with its canonical e-cohesive structure.*

(a) If C *is ρ-cohesive, so is* K*; the embedding* $\mathsf{K} \to \mathsf{C}$ *is a ρ-cohesive functor.*

(b) If C *is ρ-cohesive over* $\mathcal{S}$*, then* K *is a ρ-cohesive e-category over* $\mathcal{I}$*.*

Proof (a) If φ is a linked ρ-set in K, so is $\varphi^\sharp = \{u^\sharp \mid u \in \varphi\}$. Both φ and $\varphi^\sharp$ are also linked in C, with resolutions $e_u = u^\sharp u$ and $e_{u^\sharp} = uu^\sharp$ (for $u \in \varphi$).

Letting $e = \vee e_u$ (in C), the join $\vee \varphi$ in C belongs to K (where it is obviously a compositive join)

$$(\vee \varphi^\sharp)(\vee \varphi) = \vee_{u,v \in \varphi} u^\sharp v = \vee e_u e_v = \vee e_u = e \leqslant 1, \tag{3.47}$$

$$(\vee \varphi)(\vee \varphi^\sharp)(\vee \varphi) = (\vee \varphi).e = \vee_{u,v} u.e_v = \vee \varphi. \tag{3.48}$$

(b) We have a ρ-cohesive functor $U \colon \mathsf{C} \to \mathcal{S}$ that reflects order and linking. U is faithful, and the associated functor $IU \colon \mathsf{IC} \to \mathcal{I}$ is also, as a subfunctor of U.

The functor IU preserves and reflects the order and the linking relation, as already observed in 3.4.5(a), (b). It preserves linked binary meets (by 3.4.5(a)) and all linked ρ-joins (by the previous point). $\qquad\square$

3.4.8 Examples and complements

We can now review the e-cohesive structure of the inverse categories $\mathcal{I}$, $I\mathcal{C}$, $I\mathcal{D}$ and $I\mathcal{C}^r$ of 2.4.5.

(a) $\mathcal{I} = I\mathcal{S}$ is the inverse category of sets and partial bijections. It is a totally cohesive e-category, by 3.4.7. We have already remarked, in (3.46) (and concretely in 3.4.2), that its linking relation is a symmetric version of the linking relation lifted from $\mathcal{S}$.

(b) $I\mathcal{C}$ (resp. $I\mathcal{D}$) is the inverse category of topological spaces and partial homeomorphisms between open (resp. closed) subspaces. $I\mathcal{C}$ is a totally cohesive e-category, while $I\mathcal{D}$ is finitely cohesive; in both cases, their structures are lifted from $\mathcal{I}$.

(c) $I\mathcal{C}^r$ is the inverse category of open euclidean spaces and partial C^r-diffeomorphisms between open subspaces. It is a totally cohesive subcategory of $\mathcal{C}$, and its structure is lifted from $\mathcal{I}$.

*(d) Let C be a prj-cohesive category. A reader interested in category theory can note that an internal adjunction $u \dashv v$ in the ordered category C is defined by the following inequalities (see 6.3.4)

$$vu \geqslant 1, \qquad uv \leqslant 1,$$

which imply $vu = 1$ (by the maximality of identities). It amounts thus to a split monomorphism u of the inverse core $\mathsf{K} = I\mathsf{C}$ (with $v = u^\sharp$).

In $\mathcal{S}$ this means an everywhere defined partial bijection, i.e. an injective mapping. In $\mathcal{C}$ (resp. in $\mathcal{D}$) this means a topological embedding with an open (resp. a closed) image (see 1.4.9(a)).

3.5 Manifolds and gluing completion for inverse categories

We are now ready to study manifolds on an inverse category K; the reader can think of $I\mathcal{C}^r$, the category of open euclidean spaces and partial diffeomorphisms of class C^r between open subspaces. (In Section 3.6 the basis will be an e-category, like $\mathcal{C}^r$.)

K is always an inverse category, with its canonical cohesive structure, and ρ is a cardinal bound in the sense of 3.2.3: a wild card for *finite* (and finitely), or *countable* (and countably), or *arbitrary* (and totally).

A manifold on K is defined by an 'intrinsic atlas' in K. If K is ρ-cohesive, we form the category ρIMf K of ρ-manifolds and *symmetric profunctors*, that is proved to be the inverse ρ-gluing completion of K, in Theorem 3.5.8. The long proof is deferred to Section 3.9.

If K is indeed IC^r, we can interpret the total gluing completion IMf K as the category of C^r-manifolds and partial diffeomorphisms of class C^r, as we shall see in Section 4.1. Other structures defined by 'local models', as fibre bundles, vector bundles, foliations and simplicial complexes, will also be handled in this way, in Chapter 4. Paracompact C^r-manifolds will correspond to the σ-gluing completion.

Categorically, our manifolds are symmetric enriched categories. The term 'profunctor' comes from the theory of enriched categories, reviewed in Chapter 6. The basic case, profunctors between categories, can be found in 6.5.1.

3.5.1 Manifolds on inverse categories

A (symmetric) *manifold* on the inverse category K, indexed by a set I, will be a diagram U of K consisting of

- objects U_i, for $i \in I$ (the *charts*),

- morphisms $u^i_j \colon U_i \to U_j$, for $i, j \in I$ (the *transition morphisms*),

satisfying three axioms (for $i, j, k \in I$):

(i) $u^i_i = 1_{U_i}$ (*identity law*),

(ii) $u^j_k . u^i_j \leqslant u^i_k$ (*composition law*, or *triangle inequality*),

(iii) $u^j_i = (u^i_j)^\sharp$ (*symmetry law*).

The diagram U will often be written as $((U_i), (u^i_j))_I$, or as U_I. We say that U is a ρ-*manifold* if the set I is a ρ-set: finite, countable or arbitrary, according to the case.

The *gluing* of the manifold U in K (if it exists) will be an object $X = \text{gl}\, U$ provided with a family of morphisms $u^i \colon U_i \to X$ $(i \in I)$, such that

$$u^j . u^i_j \leqslant u^i, \qquad u^{j\sharp} . u^i \leqslant u^i_j \qquad \text{(for all } i, j \in I\text{)}, \tag{3.49}$$

and universal in the obvious sense: for every family $y^i \colon U_i \to Y$ $(i \in I)$ satisfying (3.49) (replacing u^i with y^i, of course), there is a unique morphism $f \colon X \to Y$ in K such that $y^i = f u^i$ (for $i \in I$).

The inverse category K will be said to be ρ-*gluing* (as an inverse category) if it is ρ-cohesive and every ρ-manifold has a gluing. In particular, *gluing inverse category* will mean *totally gluing*.

From now on, in this section, *we assume that the inverse category* K *is* ρ-*cohesive.*

Remarks. (a) From a formal point of view, we shall see in Chapter 6 that the manifold U is a small category 'symmetrically' enriched on the involutive ordered category K.

(b) The term 'triangle inequality' comes from Lawvere's interpretation of generalised metric spaces as enriched categories, reviewed in 6.2.5.

(c) The identity law for enrichment on an ordered category would simply give $1_{U_i} \leqslant u_i^i$, but this is equivalent to (i) in any inverse category, or prj-category.

3.5.2 Proposition

With the previous notation, a family of morphisms $u^i \colon U_i \to X$ ($i \in I$) in K *is the gluing of the manifold U if and only if (for all $i, j \in I$):*

   (i) $u^{j\sharp}u^i = u_j^i$,

   (ii) $1_X = \mathsf{V}_i\, u^i u^{i\sharp}$.

Condition (i) can be replaced with the conjunction of the following properties (for all $i, j \in I$):

   (i') $u^j u_j^i \leqslant u^i$, $u^{j\sharp}u^i \leqslant u_j^i$,

   (i'') $u^{i\sharp}u^i = \operatorname{id} U_i$.

If all these conditions hold, and $y^i \colon U_i \to Y$ ($i \in I$) is any family of morphisms satisfying (3.49), the unique morphism $f \colon X \to Y$ such that $y^i = fu^i$ is given by a linked join:

$$f = \mathsf{V}_i\, y^i u^{i\sharp}. \tag{3.50}$$

Every ρ-cohesive functor between ρ-cohesive inverse categories preserves the existing gluings of ρ-manifolds.

Proof First we assume that X is the gluing of U; then (i') holds by definition. To prove (ii), we consider the projector $e = \mathsf{V}_i\, u^i.u^{i\sharp} \colon X \to X$. Clearly $eu^i = u^i$, for all i; by the uniqueness in the universal property of the gluing, it follows that $e = 1$.

For (i''), we fix some $h \in I$ and consider the family of morphisms $z^i = u_h^i \colon U_i \to U_h$ (for $i \in I$); the family satisfies the conditions (3.49)

$$z^j.u_j^i = u_h^j u_j^i \leqslant u_h^i = z^i, \qquad z^{j\sharp}.z^i = u_j^h u_h^i \leqslant u_j^i.$$

There is then a unique morphism $z \colon X \to U_h$ such that $z^i = z.u^i$ for all i; in particular $z.u^h = z^h = u^h_h = \operatorname{id} U_h$, whence u^h is a monomorphism and $u^{h\sharp}.u^h = \operatorname{id} U_h$ (cf. 2.2.2).

Secondly, the conjunction of (i') and (i'') implies (i):

$$u^i_j = 1_{U_i}.u^i_j = u^{j\sharp}.u^j.u^i_j \leqslant u^{j\sharp}.u^i \leqslant u^i_j.$$

Last, if (i) and (ii) hold, the family $u^i \colon U_i \to X$ ($i \in I$) satisfies (3.49). Let $y^i \colon U_i \to Y$ ($i \in I$) be any family of morphisms that also does. The morphisms $y^i.u^{i\sharp} \colon X \to Y$ defined in (3.50) form a linked ρ-set

$$u^j y^{j\sharp}.y^i u^{i\sharp} \leqslant u^j u^i_j u^{i\sharp} \leqslant u^i u^{i\sharp} \leqslant 1,$$
$$y^j u^{j\sharp}.u^i y^{i\sharp} \leqslant y^j u^i_j y^{i\sharp} \leqslant y^i y^{i\sharp} \leqslant 1.$$

Therefore their linked join $f \colon X \to Y$ exists in K, and it is easy to verify that $y^i = f u^i$, for all indices i.

The last claim on ρ-cohesive functors is now obvious. $\qquad\square$

3.5.3 Symmetric profunctors

K is always a ρ-cohesive inverse category. We now form the category $\rho\mathsf{IMf}\,\mathsf{K}$ of ρ-manifolds on K and 'symmetric profunctors' between them; we show below that this category is the inverse ρ-gluing completion of K.

A *symmetric* (or *bilinked*) *profunctor* between ρ-manifolds

$$a = (a^i_h)_{I,H} \colon U \to V,$$
$$U = (U_i, u^i_j)_I, \qquad V = (V_h, v^h_k)_H, \tag{3.51}$$

will be a family of K-morphisms $a^i_h \colon U_i \to \mathsf{V}_h$, satisfying the following inequalities (for $i, j \in I$ and $h, k \in H$):

(i) $\quad a^j_h u^i_j \leqslant a^i_h, \qquad v^h_k a^i_h \leqslant a^i_k, \qquad\qquad$ (*profunctor laws*),

(ii') $\quad a^i_k a^{i\,\sharp}_h \leqslant v^h_k \qquad\qquad\qquad\qquad\quad$ (*left symmetry law*),

(ii'') $\quad a^{j\,\sharp}_h a^i_h \leqslant u^i_j \qquad\qquad\qquad\qquad\quad$ (*right symmetry law*).

The left symmetry law relies on supports (by Exercise (a)), and therefore on the left cohesive structure K_L of (3.37), while the right law can be expressed with cosupports.

All this is equivalent to extending U and V to a new symmetric manifold on the disjoint union $I + H$, with new transition morphisms a^i_h and a^h_i

$$a^i_h \colon U_i \to \mathsf{V}_h, \qquad a^h_i = (a^i_h)^\sharp \colon V_h \to U_i,$$
$$a^h_j a^i_h \leqslant u^i_j, \qquad a^i_k a^h_i \leqslant v^h_k \qquad\qquad (i, j \in I,\ h, k \in H). \tag{3.52}$$

The symmetry laws of the profunctor are thus rewritten as laws of the extended symmetric manifold. (For the sake of simplicity, we are assuming that I and H are disjoint.)

We shall see, in Chapter 6, that (i) is the usual condition for a profunctor $a: U \to V$ between categories enriched on an ordered category [Bet, Wal].

The symmetry laws are added for two reasons: first, arbitrary profunctors would *not* compose (unless K has arbitrary joins); second, we want to construct an inverse category, where the profunctor a has a (unique) partial inverse.

The *matrix composition* of $a: U_I \to V_H$ with a consecutive profunctor $b = (b_m^h): V_H \to W_M$ is given by

$$ba = (c_m^i): U_I \to W_M, \qquad c_m^i = \bigvee\nolimits_h b_m^h\, a_h^i. \tag{3.53}$$

(The join is indeed a linked ρ-join in K, and gives a symmetric profunctor ba, as verified in Exercise (b).)

It is easy to see that $\rho\mathrm{IMf}\,\mathsf{K}$ is indeed a category. The identity of $U = (U_i, u_j^i)$ is the symmetric endo-profunctor $1_U = (u_j^i)_{I,I}$.

The ρ-cohesive inverse category K has an obvious full embedding

$$J: \mathsf{K} \to \rho\mathrm{IMf}\,\mathsf{K}, \qquad X \mapsto (X, 1_X), \tag{3.54}$$

that identifies the object X with the manifold $(X, 1_X)$, indexed by the singleton.

Varying the cardinal bound ρ, we write as $\mathrm{IMf}\,\mathsf{K}$ the 'total' case, namely the category of all manifolds on K and their symmetric profunctors. We write as $\mathrm{fIMf}\,\mathsf{K}$ the full subcategory of finitely indexed manifolds on K, and as $\sigma\mathrm{IMf}\,\mathsf{K}$ that of σ-indexed manifolds.

Exercises and complements. (a) The left and right symmetry laws (ii) can equivalently be expressed using supports and cosupports:

(iii′) $a_k^i\, \underline{e}(a_h^i) = v_k^h a_h^i$ (*left symmetry law*),

(iii″) $\underline{e}^*(a_h^j)\, a_h^i = a_h^j u_j^i$ (*right symmetry law*).

(b) Verify that, in (3.53), the join is legitimate and gives a symmetric profunctor.

(c) The *left linked profunctors* (only supposed to satisfy condition (ii′)) would not compose: the left cohesive structure K_L is not ρ-cohesive, generally. For the inverse category $\mathcal{I}$, we would only be able to compose such profunctors in the category $\mathrm{Mf}\,\mathcal{S}$, introduced in Section 3.6.

3.5.4 *Theorem* (The inverse structure)

The category $\mathsf{M} = \rho\mathrm{IMf}\,\mathsf{K}$ *(on a ρ-cohesive inverse category* K*) is inverse. The partial inverse of the symmetric profunctor* $a = (a_h^i)_{I,H} \colon U_I \to V_H$ *is*

$$a^\sharp = (a_i^h)_{H,I} \colon V_H \to U_I, \tag{3.55}$$

where $a_i^h = (a_h^i)^\sharp$ *is the extended component introduced in (3.52), a partial inverse in* K.

For a ρ-manifold $U = (U_i, u_j^i)_I$, *the following conditions on a family* $e = (e_j^i \colon U_i \to U_j)_{i,j\in I}$ *of morphisms of* K *are equivalent (leaving the quantifier 'for all i, j' understood):*

(i) e *is a projector of* U *in* M,

(ii) e *is an endomap of* U *and* $e_j^i \leqslant u_j^i$,

(iii) e *is an endomap of* U, $e_i = e_i^i \in \mathsf{Prj}\,U_i$ *and* $e_j^i = u_j^i e_i = e_j u_j^i$,

(iv) $e = (u_j^i e_i)$, *with* $e_i \in \mathsf{Prj}\,U_i$ *and* $u_i^j e_j u_j^i \leqslant e_i$.

When they hold, $e = ee = e^\sharp$ *and* $e_i^j = (e_j^i)^\sharp$ *(for all i, j).*

Note. The components e_i^i are projectors of K, but the other components $e_j^i \colon U_i \to U_j$ (for $i \neq j$) are not even endomorphisms, generally.

Proof See 3.9.1. (The proof is straightforward, and can be written as an exercise.)   □

3.5.5 *Theorem* (The cohesive structure)

In the inverse category $\mathsf{M} = \rho\mathrm{IMf}\,\mathsf{K}$ *(on a ρ-cohesive inverse category* K*), we fix two parallel morphisms:*

$$a, b \colon U \to V, \qquad U = (U_i, u_j^i)_I, \qquad V = (V_h, v_k^h)_H.$$

(a) *For* $e = (u_j^i e_i) \in \mathsf{Prj}\,(U)$ *and* $f = (v_k^h f_h) \in \mathsf{Prj}\,(V)$

$$(fae)_h^i = f_h . a_h^i . e_i \qquad (\text{for } i \in I,\, h \in H). \tag{3.56}$$

(b) *The cohesive structure of the inverse category* M *works as follows:*

$$a \leqslant b \;\;\Leftrightarrow\;\; a_h^i \leqslant b_h^i \text{ in } \mathsf{K} \quad (\text{for all } i \in I,\, h \in H), \tag{3.57}$$

$$a \,!\, b \;\;\Leftrightarrow\;\; b_k^h a_h^i \leqslant u_j^i \text{ and } b_k^i a_i^h \leqslant v_k^h \quad (\text{for all } i, j \text{ and } h, k),$$
$$\;\;\Leftrightarrow\;\; a_h^i \,!\, b_h^i \text{ (for all } i, h) \text{ and } (a_h^i \vee b_h^i)_{I,H} \text{ belongs to } \mathsf{M}, \tag{3.58}$$

$$a \wedge b = (a_h^i \wedge b_h^i)_{I,H}, \qquad a \vee b = (a_h^i \vee b_h^i)_{I,H} \qquad (\text{for } a \,!\, b). \tag{3.59}$$

(c) The inverse category $M = \rho\text{IMf}\,K$ *is* ρ-*cohesive and the embedding* $J: K \to M$ *is a* ρ-*cohesive functor.*

(Actually M is ρ-gluing, as we prove in Theorem 3.5.8.)

(d) If $e_i \in \text{Prj}\,U_i$ *(*$i \in I$*) is an arbitrary family of projectors of the charts of* U, *there exists a least projector* $\hat{e} = (e^i_j)_{I,I}$ *of the manifold* U, *with* $\hat{e}_i \geqslant e_i$ *for all* i, *and is given by*

$$e^i_j = \bigvee{}_h u^h_j e_h u^i_h. \tag{3.60}$$

(e) A ρ-*cohesive functor* $F: K \to L$ *between* ρ-*cohesive inverse categories has an obvious extension*

$$\underline{F} = \rho\text{IMf}\,F: \rho\text{IMf}\,K \to \rho\text{IMf}\,L,$$
$$\underline{F}(U) = (FU_i, Fu^i_j)_I, \quad \underline{F}((a^i_h)_{ih}) = ((Fa^i_h): \underline{F}(U) \to \underline{F}(V)). \tag{3.61}$$

Proof See 3.9.2.   □

3.5.6 Lemma

K *is always a* ρ-*cohesive inverse category, and* $U_I = (U_i, u^i_j)_I$ *is a* ρ-*manifold on* K, *written as* U *when viewed as an object of* $M = \rho\text{IMf}\,K$.

(a) A family of K-*morphisms* $u = (u^i: U_i \to X)_{i\in I}$ *satisfying condition* (3.49) *is the same as a morphism in* M *(a symmetric profunctor)*

$$u = (u^i): U \to X, \tag{3.62}$$

from U *to the trivial manifold* $J(X) = (X, 1_X)$ *on the singleton.*

(b) The pair $(X, (u^i))$ *is the gluing of* U_I *in* K *if and only if* $u: U \to X$ *is an isomorphism in* M.

(c) Using the extension $\underline{J}: M \to \rho\text{IMf}\,M$ *defined in* (3.61), *the* M-*object* U *is the gluing of the manifold* $\underline{J}(U)$ *in* M, *via the* M-*isomorphism*

$$u = (u^i: JU_i \to U)_{i\in I}, \qquad u^i = (u^i_j)_{j\in I}: JU_i \to U, \tag{3.63}$$

which is an 'interpretation' of the identity of the manifold U.

Proof (a) The family (u^i) is a symmetric profunctor $U \to X$ if and only if

$$u^i \leqslant u^i, \qquad u^j u^i_j \leqslant u^i \qquad \text{(profunctor laws)},$$
$$u^{j\sharp} u^i \leqslant u^i_j, \qquad u^i u^{i\sharp} \leqslant 1_X \qquad \text{(symmetry laws)},$$

and these conditions are trivially true or equivalent to (3.49).

(b) Computing the components of the endomorphisms $u^\sharp u \colon U \to U$ and $uu^\sharp \colon X \to X$

$$(u^\sharp u)^i_j = u^{j\sharp}.u^i, \qquad uu^\sharp = \bigvee_i u^i.u^{i\sharp},$$

we see that the conditions $u^\sharp u = 1_U$ and $uu^\sharp = 1_X$ are respectively equivalent to 3.5.2(i) and 3.5.2(ii).

(c) Obvious: the only problem is distinguishing the different roles. $\qquad\square$

3.5.7 *Lemma* (The gluing property)

K *is always a ρ-cohesive inverse category.*

(a) K *is ρ-gluing if and only if the canonical embedding* $J \colon \mathsf{K} \to \rho\mathrm{IMf}\,\mathsf{K}$ *(in (3.54)) is an equivalence of categories.*

(b) In this case, the pseudo inverse of J is the functor of gluings

$$\mathrm{gl} \colon \rho\mathrm{IMf}\,\mathsf{K} \to \mathsf{K}, \qquad U \mapsto \mathrm{gl}\,U, \qquad a \mapsto \mathrm{gl}\,a, \qquad (3.64)$$

determined, up to natural isomorphism, by choosing a gluing $(\mathrm{gl}\,U, (u^i))$, *for every manifold U_I on* K.

Note. For the sake of simplicity, it can be convenient to constrain the choice of gluings, so that the gluing of a trivial manifold $(X, 1_X)$ is the object X (instead of some isomorphic object of K). Then $\mathrm{gl}\,.J = \mathrm{id}\,\mathsf{K}$. (Of course we still have $J.\mathrm{gl} \cong \mathrm{id}\,\mathsf{M}$.)

Proof (a) If K is ρ-gluing, the full embedding $\mathsf{K} \to \rho\mathrm{IMf}\,\mathsf{K}$ is essentially surjective on the objects, by the previous lemma, and therefore an equivalence of categories, by 1.6.3.

Conversely, if this functor is an equivalence of ρ-cohesive inverse categories, then it preserves and reflects the canonical orders, ρ-joins (by 3.4.5(c)) and ρ-gluings (by 3.5.2), so that K is also ρ-gluing.

(b) In this case, the pseudo inverse $\mathrm{gl} \colon \rho\mathrm{IMf}\,\mathsf{K} \to \mathsf{K}$ is constructed as specified in the statement (see 1.6.5). $\qquad\square$

3.5.8 *Theorem and Definition* (The inverse gluing completion)

If K *is a ρ-cohesive inverse category, the category* $\rho\mathrm{IMf}\,\mathsf{K}$ *is the* inverse ρ-gluing completion *of* K.

More precisely, we have the following facts, for every ρ-cohesive functor $F \colon \mathsf{K} \to \mathsf{L}$ *with values in a ρ-gluing inverse category.*

(a) The category $\mathsf{M} = \rho\mathrm{IMf}\,\mathsf{K}$ *is ρ-gluing and the canonical embedding* $J\colon \mathsf{K} \to \mathsf{M}$ *(in 3.5.5(c)) is ρ-cohesive.*

(b) (Universal property) The functor F can be extended to a ρ-cohesive functor $G\colon \mathsf{M} \to \mathsf{L}$, determined up to functorial isomorphism

$$\mathsf{K} \xrightarrow{\ J\ } \rho\mathrm{IMf}\,\mathsf{K}$$

$$F = GJ. \qquad (3.65)$$

with F (down arrow) from K to L and G the diagonal.

(c) If F is faithful, so is G. If F is full and faithful, so is G.

(d) (Manifold realisation) For a ρ-manifold U_I on K, $G(U_I) = \mathrm{gl}\,\underline{F}(U)$ is the gluing in L of the manifold $\underline{F}(U)$, defined in (3.61).

Furthermore, if F is faithful, G is also and will be called the realisation in L of the manifolds of K. *Now a morphism $a\colon U_I \to V_H$ of the category $\rho\mathrm{IMf}\,\mathsf{K}$ is determined as a morphism $\hat{a} = G(a)\colon \mathrm{gl}\,\underline{F}(U) \to \mathrm{gl}\,\underline{F}(V)$ of L whose 'components'*

$$\hat{a}_h^i = (v^h)^\sharp\,\hat{a}\,u^i\colon FU_i \to FV_h \qquad (i \in I,\, h \in H), \qquad (3.66)$$

'belong' to K, in the sense that there is a (unique) $a_h^i\colon U_i \to V_h$ in K such that $F(a_h^i) = \hat{a}_h^i$.

(e) (Gluing density) If the functor $F\colon \mathsf{K} \to \mathsf{L}$ is a ρ-cohesive embedding, then F satisfies the previous universal property (and L is the inverse ρ-gluing completion of K) if and only if K is gluing ρ-dense in L: every ρ-manifold on K has a gluing in L.

Proof See 3.9.3. $\qquad\qquad\qquad\qquad\qquad\qquad\qquad\qquad\qquad\qquad\square$

3.5.9 Exercises and complements

These examples are important. The non-obvious solutions are below.

(a) The inverse category $\mathcal{I} = \mathrm{I}\mathcal{S}$ of small sets and partial bijections is gluing: the gluing of the manifold $(U_i, u_j^i)_I$ is the set

$$X = \mathrm{gl}\,U = (\textstyle\sum_i U_i)/R, \qquad (3.67)$$

where R is the equivalence relation on the disjoint union $\sum_i U_i$ that identifies every $x \in \mathrm{Def}\,u_j^i \subset U_i$ with $u_j^i(x) \in U_j$. The partial bijections $u^i\colon U_i \to X$ are obvious; they are injective and everywhere defined.

(b) Similarly, the inverse category $\mathrm{I}\mathcal{C}$ is totally gluing. *Hints:* we proceed as above, adding topologies. The partial bijections $u^i \colon U_i \to X$ are now open topological embeddings.

Exercise 2.5.7(c) amounts to the gluing of a manifold indexed by $\{1, 2\}$.

(c) The inverse category $\mathbb{I}C^r$ is totally cohesive and not even finitely gluing: its (total) gluing completion $\mathrm{IMf}\,(\mathbb{I}C^r)$ 'is' (i.e. can be interpreted as) the category of C^r-manifolds and partial C^r-diffeomorphisms. (We recall that 'C^0-diffeomorphism' means homeomorphism.)

(d) In the inverse category $\mathcal{I}$ of partial bijections, the symmetry laws (ii') and (ii'') of 3.5.3 on a profunctor $a \colon U_I \to V_H$ have a clear meaning.

- The condition $a^i_k a^{i\,\sharp}_h \leqslant v^h_k$ means that, if the elements $y \in V_h$ and $y' \in V_k$ are the image of the same element of U_i, then they are related by the change of charts v^h_k; globally, these conditions amount to saying that the relation $\mathrm{gl}\,U \twoheadrightarrow \mathrm{gl}\,V$ produced by the profunctor a is single-valued.

- The condition $a^j_h {}^\sharp a^i_h \leqslant u^i_j$ means that, if the elements $x \in U_i$ and $x' \in U_j$ have the same image in V_h, then they are related by the change of charts u^i_j; globally, these conditions amount to saying that the relation $\mathrm{gl}\,U \twoheadrightarrow \mathrm{gl}\,V$ produced by the profunctor a is injective.

Together, these families of conditions say that the profunctor a gives a partial bijection $\mathrm{gl}\,U \twoheadrightarrow \mathrm{gl}\,V$. We have a similar interpretation in all the inverse categories examined above, which are concrete on $\mathcal{I}$.

Solutions. (b) We proceed as in (a), putting on the gluing-set X the quotient topology of the topological sum $\sum_i U_i$. It is also the finest topology that makes all the mappings $u^i \colon U_i \to X$ continuous.

The projection $p \colon \sum_i U_i \to (\sum_i U_i)/R$ *is an open mapping.* In fact, if W is open in U_j, the R-saturated subset $p^{-1}(pW) = \sum_i u^j_i(W)$ is open in $\sum_i U_i$, and $p(W)$ is open in the quotient. Therefore all u^i are open topological embeddings, i.e. everywhere defined partial homeomorphisms with open image.

(c) The inclusion $F \colon \mathbb{I}C^r \to \mathbb{I}C$ extends, by the universal property of the gluing completion, to a faithful totally cohesive functor

$$G \colon \mathrm{IMf}\,(\mathbb{I}C^r) \to \mathbb{I}C \qquad \text{(the *topological realisation*),} \qquad (3.68)$$

that sends the manifold $U = (U_i, u^i_j)_I$ to the topological space $X = \mathrm{gl}\,U$, the gluing of U in $\mathcal{C}$ examined in (b).

This space X is locally euclidean (with a locally constant dimension): every partial homeomorphism $u^i \colon U_i \to X$ gives a homeomorphism from an open euclidean space U_i to an open subspace $\mathrm{Val}\,u^i$ of X. The latter cover X, and every transition morphism $u^i_j = u^{j\,\sharp}.u^i \colon U_i \to U_j$ *belongs to* $\mathbb{I}C^r$.

This allows us to reconstruct the manifold in the usual form, recalled at the beginning of the general Introduction (in Section 0.1).

Applying Theorem 3.5.8(d), a symmetric profunctor $a\colon U_I \to V_H$ in IMf (I$\mathcal{C}^r$) amounts to its gluing $\hat{a}\colon \mathrm{gl}\, U \twoheadrightarrow \mathrm{gl}\, V$ in I$\mathcal{C}$, a partial homeomorphism between open subspaces whose components

$$a_h^i = v^{h\sharp}\,\hat{a}\,u^i \colon U_i \to V_h$$

are C^r-diffeomorphisms.

3.6 Manifolds and gluing completion for prj-categories

In this section, C is always a ρ-cohesive prj-category and $\mathsf{K} = \mathrm{IC}$ its ρ-cohesive inverse core (cf. 3.4.7). The more particular case of a ρ-cohesive *e-category* C is treated in 3.6.8. (We are not able to define linked profunctors for general ρ-cohesive categories.)

Manifolds on C will amount to manifolds on the inverse core K, but the linked profunctors between ρ-manifolds on C are more general than symmetric profunctors on K: they form a category $\rho\mathrm{Mf}\,\mathsf{C}$ whose inverse core is $\rho\mathrm{IMf}\,\mathsf{K}$.

If C is the totally cohesive e-category $\mathcal{C}^r$ of euclidean open sets and partial C^r-mappings (defined on an open subspace), $\mathrm{Mf}\,\mathcal{C}^r$ will be the category of all C^r-manifolds and partial C^r-mappings. Its inverse core $\mathrm{IMf}\,\mathcal{C}^r$ has the same objects, with partial C^r-diffeomorphisms.

3.6.1 Manifolds and gluing

A (symmetric) *manifold* on C will be a diagram $U = (U_i, u_j^i)_I$ in C, whose transition morphisms $u_j^i \colon U_i \to U_j$ $(i, j \in I)$ satisfy:

(i) $u_i^i = 1_{U_i}$ *(identity law)*,

(ii) $u_k^j.u_j^i \leqslant u_k^i$ *(composition law, or triangle inequality)*,

(iii) $u_j^i = u_j^i u_i^j u_j^i$ *(symmetry law)*.

Since $u_i^j u_j^i \leqslant u_i^i = 1_{U_i}$ and because of the symmetry property (iii), all the morphisms u_j^i are actually in the inverse core $\mathsf{K} = \mathrm{IC}$, and satisfy $(u_j^i)^\sharp = u_i^j$. In other words, the manifolds of C *are precisely those of* K. Also here we say that U is a ρ-*manifold* if the set of indices I is a ρ-set.

The *gluing* $X = \mathrm{gl}\, U$ of the manifold U in C is defined as its *lax colimit* in the ordered category C (see 5.6.4), that is an object X provided with a 'universal lax cocone' $u^i \colon U_i \to X$ $(i \in I)$ in C.

This means that

(a) $u^j.u^i_j \leqslant u^i$, for all i, j,

(b) for every lax cocone $y^i: U_i \to Y$ (satisfying $y^j.u^i_j \leqslant y^i$), there is a unique morphism $y: X \to Y$ in C such that $y^i = y.u^i$ (for $i \in I$),

(c) if $y', y'': X \to Y$ and $y'.u^i \leqslant y''.u^i$ (for $i \in I$), then $y' \leqslant y''$.

(Property (c) implies the uniqueness of the morphism y in (b).) We show in the theorem below that *this notion is equivalent to the gluing of U in* K, as defined in 3.5.1.

A prj-category will be said to be *ρ-gluing* if it is ρ-cohesive and every ρ-manifold has a gluing.

Note. Manifolds and gluings make sense in any ordered category C; if C is pre-inverse, a manifold in C amounts to a manifold in $\mathsf{K} = \mathsf{IC}$.

3.6.2 Theorem (Gluings)

C *is always a ρ-cohesive prj-category. Let $U = (U_i, u^i_j)_I$ be a ρ-manifold on C (and K), and $u^i: U_i \to X$ ($i \in I$) a family of morphisms in C.*

Then (X, u^i) is the gluing of U in C if and only if it is in K. If this is the case, the morphisms u^i are monomorphisms of K, and for every lax cocone $y^i: U_i \to Y$ in C, the corresponding morphism $y: X \to Y$ is given by the following linked join in C

$$y = \bigvee_i y^i u^{i\sharp}. \tag{3.69}$$

The ρ-cohesive prj-category C is ρ-gluing if and only if its inverse core IC is (as a ρ-cohesive inverse category).

Every ρ-cohesive functor between ρ-cohesive prj-categories preserves the existing gluings of ρ-manifolds.

Proof If (X, u^i) is the gluing of U in K (characterised in Proposition 3.5.2) the formula (3.69) concerns the join of a linked ρ-set in $\mathsf{C}(X, Y)$, with resolution $e_i = u^i u^{i\sharp} \in \mathsf{Prj}\, X$ ($i \in I$)

$$
\begin{aligned}
y^i u^{i\sharp}.e_i &= y^i, \\
y^i u^{i\sharp}.e_j = y^i.u^{i\sharp}u^j u^{j\sharp} &= y^i u^i_j u^{j\sharp} \leqslant y^j u^{j\sharp}.
\end{aligned}
\tag{3.70}
$$

It is now easy to check, as in 3.5.2, the properties 3.6.1(b), (c) in C.

Conversely, we assume that (X, u^i) is the gluing of U in C. We fix an index $h \in I$ and consider, as in the proof of 3.5.2, the morphisms of C

$$z^i = u^i_h: U_i \to U_h \quad (i \in I).$$

They form a lax cocone from U, hence there is one morphism $z \colon X \to U_h$ of C such that $z^i = zu^i$ (for $i \in I$). In particular, $zu^h = 1$; moreover

$$(u^h z).u^i = u^h z^i = u^h u_h^i \leqslant u^i \quad (i \in I),$$

so that $u^h z \leqslant 1$ (by 3.6.1(c)); therefore u^h is in IC, with partial inverse $(u^h)^\sharp = z$.

It is now sufficient to verify the conditions 3.5.2(i), (ii); the relation

$$u^{h\sharp}.u^i = zu^i = z^i = u_h^i, \tag{3.71}$$

gives the first, by the arbitrariness of $h \in I$. For the second, the relation

$$(\bigvee_i u^i u^{i\sharp}).u^j = u^j, \tag{3.72}$$

and the uniqueness property in 3.6.1(b) give $\bigvee_i (u^i u^{i\sharp}) = 1$.

The last claim of the statement follows now from the last claim in 3.5.2.

$\square$

3.6.3 Linked profunctors

We now define the category $\rho\mathrm{Mf}\,\mathsf{C}$ of ρ-manifolds on C and linked profunctors between them. (C is always a ρ-cohesive prj-category.)

A *profunctor*

$$a = (a_h^i)_{I,H} \colon (U_i, u_j^i)_I \to (V_h, v_k^h)_H, \tag{3.73}$$

is a family of C-morphisms $a_h^i \colon U_i \to V_h$ such that, for all $i, j \in I$ and $h, k \in H$

(i) $a_h^j u_j^i \leqslant a_h^i, \qquad v_k^h a_h^i \leqslant a_k^i$ *(profunctor laws)*.

It will be said to be *linked* (or *compatible*) if it has a *resolution* $e_{ih} \in \mathrm{Prj}\,U_i$ ($i \in I$, $h \in H$), defined by the property

(ii) $a_k^i e_{ih} = v_k^h a_h^i$ *(left linking law)*.

The concrete meaning of this condition has already been discussed, in Exercise 3.5.3(c).

Again, the composition of linked profunctors is matrix-like: given a as above and a consecutive linked profunctor

$$b = (b_m^h)_{H,M} \colon (V_h, v_k^h)_H \to (W_m, w_n^m)_M,$$

the composite $c = ba$ is defined as:

$$c = (c_m^i)_{I,M}, \qquad c_m^i = \bigvee_h (b_m^h a_h^i). \tag{3.74}$$

We prove below that the composition is well defined, and $\rho\mathrm{Mf}\,\mathsf{C}$ is a category. The identity of $U = (U_i, u^i_j)$ is the symmetric endo-profunctor $1_U = (u^i_j)_{I,I}$ (as in 3.5.3).

We shall also prove that $\rho\text{IMf}\,\mathsf{K}$ *coincides with the inverse core* $\text{I}(\rho\text{Mf}\,\mathsf{C})$, in 3.6.6.

Complements and remarks. (a) Varying the cardinal bound ρ, as in 3.5.3, we write as Mf C the category of all manifolds on C and their linked profunctors; we write as fMf C (resp. σMf C) the full subcategory of finitely indexed (resp. σ-indexed) manifolds.

Many instances of Mf C will be studied in Chapter 4; σMf C and fMf C can also be of interest (see 4.1.5, 4.1.6).

(b) The left linking law (ii) is equivalent to the conjunction of the following conditions (for $i, j \in I$ and $h, k \in H$):

- $a^i_h e_{ih} = a^i_h$,

- $a^i_k e_{ih} \leqslant v^h_k a^i_h$.

In fact, from these we have: $v^h_k a^i_h = v^h_k a^i_h e_{ih} \leqslant a^i_k e_{ih}$.

(c) In a resolution of the linked profunctor a, each projector e_{ih} can be replaced with any $e'_{ih} \leqslant e_{ih}$ satisfying $a^i_h e'_{ih} = a^i_h$.

Thus, in the e-cohesive case, the linking condition (ii) can be more simply expressed using the supports $e_{ih} = \underline{e}(a^i_h)$, as we shall do in 3.6.8.

3.6.4 *Proposition* (Composing linked profunctors)

The composition of linked profunctors is well defined, and gives a category $\rho\text{Mf}\,\mathsf{C}$, *with identities* $1_U = (u^i_j)_{I,I}$.

C *embeds in* $\rho\text{Mf}\,\mathsf{C}$, *taking the object U to the manifold $(U, 1_U)$ (as in the inverse case, in* (3.54)).

Proof We use the same notation as in (3.74).

To prove that the composite c is well defined, we let (f_{hm}) be a resolution of $b = (b^h_m)$ and choose a family of projectors $e_{ihm} \in \text{Prj}\,U_i$ such that

$$f_{hm}\, a^i_h = a^i_h\, e_{ihm}, \qquad e_{ihm} \leqslant e_{ih} \qquad (i \in I,\ h \in H,\ m \in M). \tag{3.75}$$

Then each family $(b^h_m\, a^i_h)_{h \in H}$ of morphisms $U_i \to W_m$ is linked, with resolution $(e_{ihm})_{h \in H}$

$$\begin{aligned}
(b^h_m\, a^i_h)\, e_{ihm} &= b^h_m\, f_{hm}\, a^i_h = b^h_m\, a^i_h, \\
(b^k_m\, a^i_k)\, e_{ihm} &= b^k_m\, a^i_k . e_{ih}\, e_{ihm} \leqslant b^h_m\, v^h_k\, a^i_h \leqslant b^h_m a^i_h .
\end{aligned} \tag{3.76}$$

More generally, for $n \in M$

$$(b_n^k \, a_k^i) \, e_{ihm} = b_n^k \, a_k^i \, e_{ih} \, e_{ihm} = b_n^k \, v_k^h \, a_h^i \, e_{ihm}$$
$$\leqslant b_n^h \, a_h^i \, e_{ihm} = b_n^h \, f_{hm} \, a_h^i = w_n^m \, b_m^h \, a_h^i, \tag{3.77}$$

and (c_m^i) is a linked profunctor, with resolution $\hat{e}_{im} = \bigvee_h e_{ihm}$

$$c_m^j \, u_j^i = \bigvee_h (b_m^h \, a_h^j \, u_j^i) \leqslant \bigvee_h (b_m^h \, a_h^i) = c_m^i,$$
$$c_m^i \, \hat{e}_{im} = \bigvee_{h,k} (b_m^h \, a_h^i \, e_{ikm}) = \bigvee_h b_m^h \, a_h^i = c_m^i \qquad \text{(by (3.76))}, \tag{3.78}$$
$$c_n^i \, \hat{e}_{im} = \bigvee_{h,k} (b_m^h \, a_h^i \, e_{ikm}) \leqslant \bigvee_k (w_n^m . b_m^k \, a_k^i)$$
$$= w_n^m . \bigvee_k b_m^k \, a_k^i = w_n^m . c_m^i \qquad \text{(by (3.77))}.$$

The composition is associative, and $\rho\mathsf{Mf}\,\mathsf{C}$ is indeed a category. $\qquad\square$

3.6.5 Theorem and Definition (The prj-structure)

C *is always a ρ-cohesive prj-category.*

(a) The category $\rho\mathsf{Mf}\,\mathsf{C}$ is prj-cohesive, once we define its projectors as those of $\rho\mathsf{IMf}\,\mathsf{K}$, described in 3.5.4 and 3.5.5(d).

(b) If $a\colon U \to V$ is a morphism in $\rho\mathsf{Mf}\,\mathsf{C}$, $e = (u_j^i e_i)_{I,I} \in \mathsf{Prj}\,(U)$ and $f = (v_k^h f_h)_{H,H} \in \mathsf{Prj}\,(V)$, then

$$(fae)_h^i = f_h a_h^i e_i. \tag{3.79}$$

(c) If $a = (a_h^i)$ and $b = (b_h^i)$ are parallel morphisms in $\rho\mathsf{Mf}\,\mathsf{C}$

(i) $\quad a \leqslant b \quad \Leftrightarrow \quad$ *the profunctors a, b have resolutions $(e_{ih}), (f_{ih})$*
$\qquad\qquad\qquad\qquad$ *such that: $e_{ih} \leqslant f_{ih}$ and $a_h^i = b_h^i e_{ih}$ $(\forall\, i, h)$,*
$\qquad\qquad \Leftrightarrow \quad a_h^i \leqslant b_h^i$ $(\forall\, i, h)$,

(ii) $\quad a\,!\,b \quad \Leftrightarrow \quad$ *the profunctors a, b have resolutions $(e_{ih}), (f_{ih})$*
$\qquad\qquad\qquad\qquad$ *such that: $a_h^i f_{ik} \leqslant b_h^i$ and $b_h^i e_{ik} \leqslant a_h^i$ $(\forall\, i, h, k)$,*
$\qquad\qquad \Leftrightarrow \quad a_h^i\,!\,b_h^i$ $(\forall\, i, h)$ *and $(a_h^i \vee b_h^i)$ is a linked profunctor,*

(iii) $\quad a\,!\,b \quad \Rightarrow \quad a \wedge b = (a_h^i \wedge b_h^i)_{I,H}, \quad a \vee b = (a_h^i \vee b_h^i)_{I,H}.$

Proof (b) We begin by proving this point, with respect to the projectors e, f of $\rho\mathsf{IMf}\,\mathsf{K} \subset \rho\mathsf{Mf}\,\mathsf{C}$. In fact, property (3.79) extends (3.56), and follows again from the following inequalities (with equality for $j = i$ and $k = h$):

$$f_h^k a_k^j e_j^i = f_h v_h^k a_k^j u_j^i e_i \leqslant f_h a_h^i e_i.$$

(a) The axiom (PCH.1) holds in ρMf C, because ρIMf K is inverse. As to (PCH.2), given the linked profunctor $a\colon U \to V$ in ρMf C, with resolution (e_{ih}), and $f \in \mathsf{Prj}\,V$, we choose a family of projectors $e'_{ih} \in \mathsf{Prj}\,U_i$ such that

$$f_h a^i_h = a^i_h e'_{ih}, \qquad e'_{ih} \leqslant e_{ih} \qquad (i \in I,\ h \in H). \tag{3.80}$$

Furthermore, we let

$$e'_i = \bigvee{}_h\, e'_{ih}, \qquad \hat{e}_i = \bigvee{}_j\, u^j_i e'_j u^i_j, \tag{3.81}$$

so that, by (3.60), $\hat{e} = (u^i_j \hat{e}_i)$ is the projector of the manifold U spanned by the family (e'_i). We prove that $fa = a\hat{e}$

$$\begin{aligned}
a^i_h e'_i &= a^i_h \bigvee{}_k\, e'_{ih} = \bigvee{}_k\, a^i_h e_{ih} e'_{ih} = \bigvee{}_k\, v^k_h a^i_k . e'_{ih} \\
&= \bigvee{}_k\, v^k_h f_k a^i_k = \bigvee{}_k\, f_h v^k_h a^i_k = f_h a^i_h,
\end{aligned} \tag{3.82}$$

$$(fa)^i_h = f_h a^i_h = a^i_h e'_{ih} \leqslant a^i_h e'_i \leqslant a^i_h \hat{e}_i = (a\hat{e})^i_h, \tag{3.83}$$

$$\begin{aligned}
(a\hat{e})^i_h &= a^i_h \hat{e}_i = \bigvee{}_j\, a^i_h u^j_i e'_j u^i_j \leqslant \bigvee{}_j\, a^j_h e'_j u^i_j \\
&= \bigvee{}_j\, f_h a^j_h u^i_j = f_h a^i_h = (fa)^i_h \quad (\text{by } (3.82)).
\end{aligned} \tag{3.84}$$

(c) We prove (i). If $a \leqslant b$, then $a = b.e$ and $a^i_h = b^i_h.e_i \leqslant b^i_h$.

Assume now that $a^i_h \leqslant b^i_h$ for all i and h; let (e'_{ih}), (f_{ih}) be resolutions of a and b, respectively, and let us choose projectors e''_{ih} such that $a^i_h = b^i_h.e''_{ih}$. Then the family $e_{ih} = e'_{ih}.e''_{ih}.f_{ih}$ is a resolution of a (by 3.6.3) that satisfies, with (f_{ih}), our conditions.

Last, if the resolutions (e_{ih}) and (f_{ih}) satisfy these conditions, we write $e_i = \bigvee_h e_{ih}$, and $\hat{e}$ the projector of U spanned by the family (e_i), as in (3.81), so that $a = b\hat{e}$

$$a^i_h = b^i_h.e_{ih} \leqslant b^i_h.\hat{e}_i = (b\hat{e})^i_h, \tag{3.85}$$

$$\begin{aligned}
(b\hat{e})^i_h &= b^i_h.\hat{e}_i = b^i_h.\bigvee{}_j\, u^j_i e_j u^i_j = \bigvee{}_j\, b^i_h.u^j_i e_j u^i_j \\
&\leqslant \bigvee{}_{j,k}\, b^j_h.e_{jk} u^i_j \leqslant \bigvee{}_{j,k}\, b^j_h.f_{jk} e_{jk} u^i_j \\
&= \bigvee{}_{j,k}\, v^k_h b^j_k e_{jk} u^i_j = \bigvee{}_{j,k}\, v^k_h a^j_k u^i_j = a^i_h.
\end{aligned} \tag{3.86}$$

The proof of (ii) and (iii) is similar (see 3.5.5). $\qquad\square$

3.6.6 Proposition

For the ρ-cohesive prj-category C, *the inverse ρ-gluing completion of* K $=$ IC *coincides with the inverse core of the ρ-gluing completion of* C

$$\rho\text{IMf K} = \mathrm{I}(\rho\text{Mf C}). \tag{3.87}$$

Proof We have already seen in 3.6.3–3.6.4 that $\rho \mathsf{IMf}\,\mathsf{K}$ is a subcategory of $\mathsf{I}(\rho \mathsf{Mf}\,\mathsf{C})$. A symmetric profunctor $a = (a_h^i)_{I,H}$ on $\mathsf{K} = \mathsf{IC}$ is a linked profunctor on C, provided with a Morita inverse $(a_h^{i\,\sharp})_{H,I}$ in $\rho \mathsf{Mf}\,\mathsf{C}$ (see 3.4.6).

Conversely, take a linked profunctor

$$a = (a_h^i)_{I,H} \colon (U_i, u_j^i)_I \to (V_h, v_k^h)_H$$

with resolution (e_{ih}) and Morita inverse $b = (b_i^h)_{H,I}$. Then ba and ab are projectors of $\rho \mathsf{Mf}\,\mathsf{C}$, and all the composites $b_i^h\,a_h^i$ and $a_h^i\,b_i^h$ are also

$$b_i^h\,a_h^i \leqslant (ba)_i^i \leqslant 1.$$

Finally we verify that $(ba)_i^i\,e_{ih} = b_i^h\,a_h^i$ and $a_h^i = a_h^i\,b_i^h\,a_h^i$

$$(ba)_i^i\,e_{ih} = \mathsf{V}_k\,b_i^k\,a_k^i\,e_{ih} \leqslant \mathsf{V}_k\,b_i^k\,v_k^h\,a_h^i \leqslant b_i^h\,a_h^i \leqslant (ba)_i^i\,e_{ih},$$

$$a_h^i = (aba)_h^i\,e_{ih} = (\mathsf{V}_j\,a_h^j\,(ba)_j^i)\,e_{ih} = \mathsf{V}_j(a_h^j\,u_j^i\,(ba)_i^i\,e_{ih})$$
$$\leqslant a_h^i\,(ba)_i^i\,e_{ih} = a_h^i\,b_i^h\,a_h^i \leqslant (aba)_h^i \leqslant a_h^i.$$

$\square$

3.6.7 Theorem (The gluing completion of a prj-category)

(a) The prj-category $\rho \mathsf{Mf}\,\mathsf{C}$ is the ρ-gluing completion of the prj-cohesive category C.

In other words, $\rho \mathsf{Mf}\,\mathsf{C}$ is ρ-gluing, and for every ρ-gluing prj-category D and every ρ-cohesive prj-functor $F \colon \mathsf{C} \to \mathsf{D}$ there is a ρ-cohesive prj-functor $G \colon \rho \mathsf{Mf}\,\mathsf{C} \to \mathsf{D}$ that extends F, determined up to isomorphism

$$
\begin{array}{ccc}
\mathsf{C} & \xrightarrow{\ J\ } & \rho \mathsf{Mf}\,\mathsf{C} \\
{\scriptstyle F}\downarrow & \swarrow{\scriptstyle G} & \\
\mathsf{D} & &
\end{array}
\tag{3.88}
$$

(b) If F is faithful, so is G. If F is full and faithful, so is G.

(c) (Manifold realisation) For a ρ-manifold U_I on C, $G(U_I) = \mathrm{gl}\,\underline{F}(U)$ is the gluing in D of the manifold $\underline{F}(U)$, defined in (3.61).

If F is faithful, G is also, and a morphism $a \colon U_I \to V_H$ of $\rho \mathsf{Mf}\,\mathsf{C}$ is determined as a morphism $\hat{a} = G(a) \colon \mathrm{gl}\,\underline{F}(U) \to \mathrm{gl}\,\underline{F}(V)$ of D whose 'components'

$$c_h^i = v^{h\sharp}\,\hat{a}\,u^i \colon FU_i \to FV_h \qquad (i \in I,\ h \in H), \tag{3.89}$$

'belong' to C: there is a (unique) $a_h^i \colon U_i \to V_h$ in C such that $F(a_h^i) = c_h^i$.

(d) (Gluing density) *Let $F\colon \mathsf{C} \to \mathsf{D}$ be a ρ-cohesive prj-embedding with values in a ρ-gluing prj-category. Then F satisfies the previous universal property (and D is the ρ-gluing completion of C) if and only if C is gluing ρ-dense in D: every ρ-manifold on C has a gluing in D.*

Proof (a) It is an easy consequence of the inverse gluing completion theorem 3.5.8, applied to the ρ-cohesive category $\mathsf{K} = \mathrm{I}\mathsf{C}$, and the previous part of this section.

In fact, by Proposition 3.6.6 and Theorem 3.5.8, $\mathrm{I}(\rho\mathrm{Mf}\,\mathsf{C}) = \rho\mathrm{IMf}\,\mathsf{K}$ is ρ-gluing, as an inverse category; hence the prj-category $\rho\mathrm{Mf}\,\mathsf{C}$ is ρ-gluing (by Theorem 3.6.2).

Now, if $F\colon \mathsf{C} \to \mathsf{D}$ is a ρ-cohesive prj-functor with values in a ρ-gluing prj-category, F takes ρ-manifolds on C to ρ-manifolds on D, which have a gluing in D (also with respect to linked profunctors).

Points (b)–(d) are proved as in Theorem 3.5.8. $\qquad\square$

3.6.8 The e-cohesive case

Let C be a ρ-cohesive e-category. The terminology of prj-structures, for manifolds, profunctors and gluing, is still used, but the previous results take a simpler form. (Note that, for every u in $\mathsf{K} = \mathrm{I}\mathsf{C}$, the support of u in C is $\underline{e}(u) = u^{\sharp}u$, as in K.)

Given two ρ-manifolds $U = (U_i, u_j^i)_I$ and $V = (V_h, v_k^h)_H$ on C, a linked profunctor $a = (a_h^i)_{I,H}\colon U \to V$ (defined in 3.6.3) is characterised by:

(i) $a_h^j u_j^i \leqslant a_h^i, \qquad v_k^h a_h^i \leqslant a_k^i$ $\qquad\qquad$ *(profunctor laws)*,

(ii) $a_k^i\,\underline{e}(a_h^i) = v_k^h\,a_h^i$ $\qquad\qquad\qquad\qquad$ *(left linking law)*.

We have replaced the left linking law 3.6.3(ii) with the present one, which says that $(\underline{e}(a_h^i))$ is a resolution of a (the least one). Equivalently this can be written as:

(ii′) $a_k^i\,\underline{e}(a_h^i) \leqslant v_k^h\,a_h^i$.

The prj-category $\rho\mathrm{Mf}\,\mathsf{C}$ of ρ-manifolds and linked profunctors on C is now an e-category, with

$$(\underline{e}(a))_i = \bigvee{}_h\,\underline{e}(a_h^i), \qquad (\underline{e}(a))_j^i = u_j^i(\underline{e}(a))_i = (\underline{e}(a))_j\,u_j^i. \qquad (3.90)$$

Indeed, $\underline{e}(a)$ is a projector of the manifold U_I, according to 3.5.4(iv)

$$a_h^i(u_i^j\underline{e}(a_h^j)u_j^i) \leqslant a_h^j\,\underline{e}(a_h^j)u_j^i = a_h^j\,u_j^i \leqslant a_h^i,$$

$$(u_i^j\underline{e}(a_h^j)u_j^i) \leqslant \underline{e}(a_h^i), \qquad\qquad\qquad (3.91)$$

$$(u_i^j(\underline{e}(a))_j\,u_j^i) = (u_i^j\,\bigvee{}_h\,\underline{e}(a_h^j)u_j^i) \leqslant \bigvee{}_h\,\underline{e}(a_h^i) = (\underline{e}(a))_i.$$

We verify now the axioms of the e-structure.

For (ECH.1): $(a.\underline{e}(a))_h^i = a_h^i$, as follows from the argument below (for $b = 1_V$). On the other hand, $a = ae$ in $\rho\mathrm{Mf}\,\mathsf{C}$ implies $a_h^i = a_h^i e_i$ (by (3.79)), whence $e_i \geqslant \underline{e}(a_h^i)$ for every h, and $e \geqslant \underline{e}(a)$.

Last, for (ECH.2), given a second profunctor $b = (b_m^h) \colon V_H \to W_M$, we have

$$(a.\underline{e}(ba))_h^i = a_h^i\,\underline{e}(ba)_i = a_h^i \bigvee{}_k (\underline{e}(ba))_k^i = a_h^i \bigvee{}_{k,m} \underline{e}(b_m^k\,a_k^i)$$

$$= a_h^i \bigvee{}_{k,m} (\underline{e}(a_k^i)\underline{e}(b_m^k\,a_k^i)) = \bigvee{}_{k,m} (a_h^i\,\underline{e}(a_k^i)\underline{e}(b_m^k\,a_k^i))$$

$$= \bigvee{}_{k,m} (v_h^k\,a_k^i\,\underline{e}(b_m^k\,a_k^i)) = \bigvee{}_k (v_h^k\,(\bigvee{}_m\underline{e}(b_m^k))\,a_k^i) \qquad (3.92)$$

$$= \bigvee{}_k (v_h^k(\underline{e}(b))_k\,a_k^i) = \bigvee{}_k ((\underline{e}(b))_h\,v_h^k\,a_k^i)$$

$$= (\underline{e}(b))_h\,a_h^i = (\underline{e}(b)\,a)_h^i.$$

Finally, from 3.6.5, for two parallel linked profunctors a, b

$$a \leqslant b \quad \Leftrightarrow \quad a_h^i \leqslant b_h^i\,\underline{e}(a_h^i) \quad (\forall\,i, h), \qquad (3.93)$$

$$a\,!\,b \quad \Leftrightarrow \quad a_h^i\,\underline{e}(b_k^i) \leqslant b_h^i \text{ and } b_h^i\,\underline{e}(a_k^i) \leqslant a_h^i \quad (\forall\,i, h, k). \qquad (3.94)$$

3.6.9 Lifting gluing structures

Let $U \colon \mathsf{C} \to \mathsf{D}$ be a faithful functor with values in a ρ-gluing e-category.

We recall, from 3.3.6(e) and 3.3.6(f), that C *has a ρ-cohesive e-structure lifted from* D if the following conditions are fulfilled:

(i) for every a in $\mathsf{C}(X, Y)$ the support $\underline{e}(Ua)$ in D lifts to C: there is a (unique) $e \in \mathsf{C}(X, X)$ such that $Ue = \underline{e}(Ua)$ (and then $e = \underline{e}(a)$ in C),

(ii) for every linked ρ-set α of morphisms $X \to Y$ in C, the join of $U(\alpha)$ in D lifts to C: there is a (unique) $c \in \mathsf{C}(X, Y)$ such that $Uc = \bigvee U(\alpha)$ (and then $c = \bigvee \alpha$).

In these hypotheses, we say that C *has a ρ-gluing e-structure lifted from* D if the following trivially equivalent conditions hold:

(iii) every ρ-manifold $(X_i, u_j^i)_I$ on C has a gluing in C, preserved by U,

(iii') C is ρ-*gluing* and the functor U is also.

In a concrete case, to verify (iii), we take the 'underlying manifold' $(UX_i, Uu_j^i)_I$ on D, then its gluing $(Y, v^i \colon UX_i \to Y)_I$ in D, and prove that this lax cocone can be lifted to a lax cocone $(X, u^i \colon X_i \to X)$ in C, which is the gluing of the original manifold.

3.7 The cohesive completion

The last three sections of this chapter deal with the cohesive completion theorem, and the proof of the gluing completion theorem. They are of a technical interest.

In this section we prove that a cohesive category has a totally cohesive completion, together with a finitely cohesive and a σ-cohesive one. We still use a cardinal bound ρ, as a wild cart referring to finite, countable and arbitrary linked joins (see 3.2.3).

3.7.1 Theorem and Definition (The cohesive completion)

Every cohesive category C *has a universal cohesive embedding* $\eta\colon \mathsf{C} \to \rho c\mathsf{C}$ *into a* ρ-*cohesive category, that preserves the existing linked* ρ-*joins; it will be called the* ρ-*cohesive completion of* C.

The universal property of η *means that: for every cohesive functor* $F\colon$ $\mathsf{C} \to \mathsf{D}$ *that preserves the existing linked* ρ-*joins, with values in a* ρ-*cohesive category, there is a unique* ρ-*cohesive functor* $G\colon \rho c\mathsf{C} \to \mathsf{D}$ *that extends* F *(i.e.* $F = G\eta$*).*

Note. If C is small, $\eta\colon \mathsf{C} \to \rho c\mathsf{C}$ is a universal arrow from C to the inclusion-functor of the category of small cohesive categories in that of small ρ-cohesive categories. A set theory with a further level of 'hyperclasses', or a hierarchy of Grothendieck universes can give a better framework for such statements.

Proof See 3.7.5 and 3.7.6.   □

3.7.2 A description of the cohesive completion

The ρ-cohesive completion $\rho c\mathsf{C}$ of a cohesive category C can be built in the following way.

(a) First we form the category $\mathcal{P}_\rho\mathsf{C}$, with the same objects as in C and morphisms $\alpha\colon X \to Y$ given by the linked ρ-sets $\alpha \subset \mathsf{C}(X, Y)$. The composition extends that of C

$$\beta\alpha = \{ba \mid a \in \alpha,\ b \in \beta\}. \tag{3.95}$$

The category $\mathcal{P}_\rho\mathsf{C}$ is equipped with a preorder $\alpha \prec \beta$ defined by the following equivalent properties:

(i) $\alpha\,!\,\beta$ and every $a \in \alpha$ is a linked join $a = \bigvee_{b \in \beta} (a \wedge b)$,

(i') $\alpha\,!\,\beta$ and every $a \in \alpha$ is a linked join $\bigvee \{x \mid x \in {\downarrow}\beta,\ x \leqslant a\}$, where ${\downarrow}\beta$ is the down-closed subset of $\mathsf{C}(X, Y)$ generated by β.

In fact, (i) implies (i′) by the extension property 3.2.1(d) of compositive joins, that also holds for linked joins (as we remarked in 3.2.3). Conversely, (i′) implies (i) by the expansion property 3.2.1(c), as the conditions $x \in \downarrow\beta$, $x \leqslant a$ imply $x \leqslant a \wedge b$ for some $b \in \downarrow\beta$.

(b) This preorder determines the quotient category

$$\rho c\mathsf{C} = \mathcal{P}_\rho\mathsf{C}/\!\sim, \tag{3.96}$$

where $\alpha \sim \beta$ is the congruence associated to $\prec$, namely $\alpha \prec \beta$ and $\beta \prec \alpha$.

The order and the linking relation in $\rho c\mathsf{C}$ are given by

$$\begin{aligned}
[\alpha] \leqslant [\beta] \quad &\Leftrightarrow \quad \alpha \prec \beta, \\
[\alpha] \,!\, [\beta] \quad &\Leftrightarrow \quad \alpha \,!\, \beta \text{ as linked sets of } \mathsf{C},
\end{aligned} \tag{3.97}$$

independently of the choice of representatives.

Linked meets and linked ρ-joins are given in $\rho c\mathsf{C}$ by the following formulas

$$[\alpha] \wedge [\beta] = [\{a \wedge b \mid a \in \alpha,\, b \in \beta\}], \tag{3.98}$$

$$\bigvee \Sigma' = [\textstyle\bigcup \Sigma], \tag{3.99}$$

where $[\alpha] \,!\, [\beta]$, Σ is a linked ρ-set of ρ-sets of C (i.e. $\alpha \,!\, \alpha'$, for all $\alpha, \alpha' \in \Sigma$) and $\Sigma' = \{[\alpha] \mid \alpha \in \Sigma\}$.

In particular, in $\rho c\mathsf{C}$, for any linked ρ-set α of C

$$\bigvee \alpha' = [\alpha]. \tag{3.100}$$

(c) The universal embedding $\eta\colon \mathsf{C} \to \rho c\mathsf{C}$ takes the object X to itself and the morphism a to the equivalence class of $\{a\}$.

(d) The σ-cohesive completion of a *finitely cohesive* category C can be given a simpler description: for each morphism a in $\sigma c\mathsf{C}$ there is an increasing sequence of parallel morphisms $(a_n)_{n\in\mathbb{N}}$ of C such that $a = [\{a_n \mid n \in \mathbb{N}\}]$ (as we have seen in Exercise 3.2.8(b)).

3.7.3 *Proposition and Definition* (Density)

If C is a cohesive subcategory of a ρ-cohesive category D, with the same objects, the embedding $F\colon \mathsf{C} \to \mathsf{D}$ is the ρ-cohesive completion of C if and only if:

(i) the functor F preserves the existing linked ρ-joins,

(ii) for every morphism b in D there is a linked ρ-set α in C whose join in D is b.

The second condition will be expressed saying that C is ρ-dense in D, as a cohesive subcategory.

Proof The necessity of these conditions is obvious.

We assume that they hold and prove that the ρ-cohesive functor G: $\rho c\mathsf{C} \to \mathsf{D}$ that extends F is an isomorphism of cohesive categories. Since G is bijective on objects and surjective on morphisms, by (ii), it suffices to show that it reflects the order (hence it is injective) and the linking relation.

Let α and β be parallel, linked ρ-sets in C. If $G[\alpha] \leqslant G[\beta]$ in D, for every $a \in \alpha$ we have $a = Ga \leqslant G[\beta] = \bigvee_b b$, hence $a = \bigvee_b (a \wedge b)$ is a linked join in D. Since a and all $a \wedge b$ are in C, the linked join subsists in C, and therefore $[\alpha] \leqslant [\beta]$ in $\rho c\mathsf{C}$.

Last, if $G[\alpha] \,!\, G[\beta]$ in D, then $\bigvee \alpha = \bigvee \beta$ in D, whence $a \,!\, b$, in D for every $a \in \alpha$ and $b \in \beta$, and the same holds in the cohesive subcategory C; in other words, $[\alpha] \,!\, [\beta]$ in $\rho c\mathsf{C}$. $\qquad\square$

3.7.4 Theorem (The cohesive completion of an inverse category)

The ρ-cohesive completion $\rho c\mathsf{K}$ of an inverse category K (with respect to its canonical cohesive structure) is an inverse category, provided with the canonical cohesive structure.

The involution of $\rho c\mathsf{K}$ is given by

$$[\alpha] \mapsto [\alpha^\sharp], \qquad \alpha^\sharp = \{a^\sharp \mid a \in \alpha\}. \tag{3.101}$$

Its projectors are the classes $[\varepsilon]$, where ε is any ρ-set of parallel projectors of K.

Proof The mapping $\alpha \mapsto \alpha^\sharp = \{a^\sharp \mid a \in \alpha\}$ plainly defines an involution on $\mathcal{P}_\rho\mathsf{K}$, and then in $\rho c\mathsf{K}$. The *latter* is regular, because (using (3.43))

$$\alpha^\sharp\alpha = \{a^\sharp b \mid a, b \in \alpha\} \sim \{a^\sharp a \mid a \in \alpha\},$$

$$\alpha\alpha^\sharp\alpha \sim \{a.b^\sharp b \mid a, b \in \alpha\} \sim \{aa^\sharp a \mid a \in \alpha\} = \alpha.$$

An endomorphism $[\alpha]$ is a projector (with regard to the regular involution, see 2.2.2) if and only if $\alpha \sim \alpha^\sharp\alpha \sim \{a^\sharp a \mid a \in \alpha\}$, if and only if α is equivalent to a ρ-set of projectors of K. Therefore the projectors of $\rho c\mathsf{K}$ commute and the latter is an inverse category (by Theorem 2.3.7).

We have now to prove that the 'completion' cohesive structure of $\rho c\mathsf{K}$ coincides with the 'inverse' one, determined by supports and cosupports.

If $[\alpha] \leqslant [\beta]$ in the completion structure, the projector $[\varepsilon] = [\alpha^\sharp\alpha] = [\{a^\sharp a \mid a \in \alpha\}]$ yields

$$[\beta].[\varepsilon] = [\{b.a^\sharp a \mid a \in \alpha, \, b \in \beta\}] = [\{a \wedge b \mid a \in \alpha, \, b \in \beta\}] = [\alpha] \wedge [\beta] = [\alpha],$$

and $[\alpha] \leqslant [\beta]$ in the inverse structure.

Conversely, if $[\alpha] = [\beta].e$ for some projector e of $\rho c\mathsf{K}$, the relation $e \leqslant 1$ in $\rho c\mathsf{K}$ implies $[\alpha] \leqslant [\beta]$ in the completion structure.

Last, the relation $[\alpha]\,!\,[\beta]$ in the completion structure is equivalent to each of the following conditions:

- $\alpha\,!\,\beta$ as ρ-sets of K,
- $a\,!\,b$, for all $a \in \alpha$ and $b \in \beta$,
- all the endomorphisms $b^\sharp a$ and $ba^\sharp$ are projectors (for $a \in \alpha$, $b \in \beta$),
- $[\beta^\sharp \alpha]$ and $[\beta \alpha^\sharp]$ are projectors of $\rho c\mathsf{K}$,
- $[\alpha]\,!\,[\beta]$ in the inverse structure. $\qquad\square$

3.7.5 A category of linked sets

We now begin the proof of the cohesive completion theorem 3.7.1, by a preliminary part.

Let C be a category provided with a proximity relation $!$ (and no order): we embed C in a category $\mathcal{P}_\rho\mathsf{C}$ with order and proximity relations, satisfying (CH.1–3) and that part of (CH.5ρ) which deals with joins and their compositive property.

The objects are the same. A morphism $\alpha \in \mathcal{P}_\rho\mathsf{C}(X,Y)$ is given by any ρ-set $\alpha \subset \mathsf{C}(X,Y)$, linked in C (including the empty subset 0_{XY}). The composition of $\alpha\colon X \to Y$ with $\beta\colon Y \to Z$ is obviously:

$$\beta\alpha = \{ba \mid a \in \alpha,\, b \in \beta\}, \tag{3.102}$$

that is a linked ρ-set of C-morphisms from X to Z.

$\mathcal{P}_\rho\mathsf{C}$ is obviously a category, with the identity of X given by the subset $\{1_X\}$; we equip $\mathcal{P}_\rho\mathsf{C}$ with the inclusion relation $\alpha \subset \alpha'$ (for parallel maps) and the linking relation:

$$\alpha\,!\,\alpha' \quad \text{if} \quad a\,!\,a' \text{ in } \mathsf{C}, \text{ for all } a \in \alpha \text{ and } a' \in \alpha'. \tag{3.103}$$

Now the axioms (CH.1–3) are trivially satisfied. Furthermore, let $\Sigma \subset \mathcal{P}_\rho\mathsf{C}(X,Y)$ be a linked ρ-set of $\mathcal{P}_\rho\mathsf{C}$ and let $\beta = \bigcup\Sigma \subset \mathsf{C}(X,Y)$: this is again a ρ-set of parallel morphisms of C, clearly linked. The morphism β is the join of the set Σ with respect to the order of $\mathcal{P}_\rho\mathsf{C}$. Moreover, the join is compositive: if $\gamma\colon X' \to X$ and $\delta\colon Y \to Y'$ are in $\mathcal{P}_\rho\mathsf{C}$, we have

$$\delta\beta\gamma = \{dbc \mid c \in \gamma,\, b \in \bigcup\Sigma,\, d \in \delta\}$$
$$= \bigcup_{\alpha\in\Sigma}\{dbc \mid c \in \gamma,\, b \in \alpha,\, d \in \delta\} = \bigvee_{\alpha\in\Sigma}\delta\alpha\gamma. \tag{3.104}$$

It can be noted that $\mathcal{P}_\rho\mathsf{C}$ has arbitrary non-empty meets; however these are not compositive, even in the binary case, and will play no role in the following steps.

3.7.6 The proof

We complete the proof of Theorem 3.7.1. C is now a cohesive category and $\mathcal{P}_\rho C$ is the category of its linked ρ-sets, constructed as above (in 3.7.5) on $(C, !)$.

Consider the binary relation $\alpha \prec \beta$, on parallel morphisms of $\mathcal{P}_\rho C$ defined, as in 3.7.2, by the equivalent properties:

(i) $\alpha ! \beta$ and every $a \in \alpha$ is a linked join $a = \bigvee_{b \in \beta} (a \wedge b)$,

(i') $\alpha ! \beta$ and every $a \in \alpha$ is a linked join $\bigvee \{x \mid x \in {\downarrow}\beta, \, x \leqslant a\}$.

To prove that it is indeed a preorder of categories, let $\alpha \prec \beta \prec \gamma$. If $a \in \alpha$ and $c \in \gamma$, it follows that $(a \wedge b) ! c$ (for all $b \in \beta$) and $a = \bigvee_b (a \wedge b) ! c$.

Moreover, applying the properties 3.2.1(a), (d), we have linked joins:

$$a = \bigvee \{b \mid b \in {\downarrow}\beta, \, b \leqslant a\} = \bigvee \{c \mid c \in {\downarrow}\gamma, \, c \leqslant b \leqslant a \text{ for some } b \in {\downarrow}\beta\}$$

$$= \bigvee \{c \mid c \in {\downarrow}\gamma, \, c \leqslant a\}.$$

The preorder is consistent with composition, because linked joins and meets are.

Let $\sim$ be the congruence associated to the preorder $\prec$, and let us consider the quotient category

$$\rho c C = (\mathcal{P}_\rho C)/{\sim}, \tag{3.105}$$

equipped with the order $\leqslant$ induced by the preorder $\prec$, namely $[\alpha] \leqslant [\beta]$ if $\alpha \prec \beta$ (independently of the choice of representatives).

The linking relation is defined by $[\alpha] ! [\beta]$ if $\alpha ! \beta$ as linked sets of C.

To show that this is independent of the choice of representatives, we suppose that $\alpha \prec \alpha'$, $\beta \prec \beta'$ and $\alpha' ! \beta'$ and prove that $\alpha ! \beta$. If $a \in \alpha$ and $b \in \beta$, we have linked joins

$$a = \bigvee_{a'} (a \wedge a'), \qquad b = \bigvee_{b'} (b \wedge b'),$$

where $a' ! b'$ implies $(a \wedge a') ! (b \wedge b')$, and therefore $a ! b$, applying Exercise 3.2.3(b).

As to (CH.4, 5ρ), linked meets and linked ρ-joins are calculated in $\rho c C$ by the following formulas:

$$[\alpha] \wedge [\beta] = [\{a \wedge b \mid a \in \alpha, \, b \in \beta\}] \quad (\text{for } [\alpha] ! [\beta]), \tag{3.106}$$

$$\bigvee \Sigma' = [\bigcup \Sigma], \tag{3.107}$$

where Σ is any linked ρ-set of ρ-sets of C ($\alpha ! \alpha'$, for all $\alpha, \alpha' \in \Sigma$) and $\Sigma' = \{[\alpha] \mid \alpha \in \Sigma\}$.

Last, we define the functor $\eta : C \to \rho c C$ taking the object X to itself and the morphism a to the equivalence class of $\{a\}$. Clearly, η reflects the order and linking relations, is cohesive and preserves the existing linked ρ-joins of C.

To verify the universal property, we let $G([\alpha]) = \bigvee_{a \in \alpha} Fa$ and check that G is a ρ-cohesive functor; its uniqueness is obvious. $\qquad\qquad$ □

3.8 Adequate prj-cohesive categories

We examine conditions ensuring that the cohesive completion of a prj-cohesive category is again prj-cohesive.

C is always a prj-cohesive category.

3.8.1 Resolutions of sets of maps

It is easy to show that a set $\alpha \subset C(X, Y)$ of parallel morphisms in our prj-cohesive category is linked if and only if there is a family of projectors $e_{ab} \in \mathsf{Prj}\,(X)$ (for $a, b \in \alpha$) such that

$$a = a.e_{ab}, \qquad a.e_{ba} = b.e_{ab} \qquad \text{(for all } a,\, b \in \alpha\text{).} \tag{3.108}$$

More particularly, a *resolution* of the set α will be a family $(e_a)_{a \in \alpha}$ of projectors of X such that

$$a = a.e_a, \qquad a.e_b = b.e_a \qquad \text{(for all } a,\, b \in \alpha\text{),} \tag{3.109}$$

or equivalently

$$a = a.e_a, \qquad a.e_b \leqslant b \qquad \text{(for all } a,\, b \in \alpha\text{).} \tag{3.110}$$

A set admitting a resolution is clearly linked, but these two facts are equivalent in most of the cases we are interested in. For instance, *if* C is an e-category, a set α of parallel morphisms is linked if and only if, equivalently:

$$a.\underline{e}(b) = b.\underline{e}(a) \qquad \text{(for all } a,\, b \in \alpha\text{),} \tag{3.111}$$

$$a.\underline{e}(b) \leqslant b \qquad \text{(for all } a,\, b \in \alpha\text{).} \tag{3.112}$$

These properties mean that the family of supports $e_a = \underline{e}(a)$ ($a \in \alpha$) is a resolution of α, the least one; it will be called the *e-resolution* of α. Other cases of existence of resolutions are examined in 3.8.3, 3.8.4.

Every prj-cohesive functor preserves resolutions of sets of parallel maps.

3.8.2 Transfer of resolutions

A resolution (e_a) of the set α can be *transferred* by composition in the following way. Given two morphisms x, y

$$X' \xrightarrow{\;x\;} X \xrightarrow{\;a\;} Y \xrightarrow{\;y\;} Y' \qquad\qquad (a \in \alpha), \qquad (3.113)$$

we can choose, for each $a \in \alpha$, a projector $e'_a \in \mathsf{Prj}\,(X')$ such that $e_a.x = x.e'_a$. Then, a trivial checking shows that

$$(e'_a) \text{ is a resolution of the set } y\alpha x = \{yax \mid a \in \alpha\} \subset \mathsf{C}(X', Y'). \quad (3.114)$$

3.8.3 Exercises and complements (Existence of resolutions)

C is always a prj-cohesive category.

(a) Every set ε of projectors of a given object has a canonical resolution, namely $(e)_{e \in \varepsilon}$.

(b) More generally, every set α that has an upper bound c has a resolution. Therefore, if C is ρ-cohesive each linked ρ-set has a resolution.

(c) If C has non-empty ρ-meets of parallel projectors (in $\mathsf{Prj}\,(X)$ or equivalently in $\mathsf{C}(X, X)$), that are *compositive in* C, every linked ρ-set α has a resolution (e_a).

Then, if α is not empty, it has a compositive meet $\wedge \alpha$, which can be expressed as follows, after fixing any morphism $b \in \alpha$

$$\wedge \alpha = b.\wedge_a e_a. \qquad (3.115)$$

(d) Each finite linked set of morphisms of a prj-category has a resolution.

3.8.4 Adequate prj-cohesive categories

We say that the prj-cohesive category C is ρ-*adequate* if it satisfies:

(PCH.3ρ)   each linked ρ-set of C has a resolution,

(PCH.4ρ)   C has ρ-joins of parallel projectors, compositive in C.

Let us recall that joins of projectors of the object X, in $\mathsf{Prj}\,(X)$ or in $\mathsf{C}(X, X)$, are the same (by 3.3.2). We have seen that the first condition is always satisfied in the finite case (in Exercise 3.8.3(d)), or when C is e-cohesive (in 3.8.1).

A prj-category which is ρ-cohesive is also ρ-adequate (by 3.8.3(b)).

A ρ-*adequate functor* will be a prj-cohesive functor between ρ-adequate prj-cohesive categories, which preserves ρ-joins of projectors.

3.8.5 Proposition

(a) If C *is a ρ-adequate prj-category, a linked ρ-set of parallel morphisms has a linked join (cf. 3.2.3) if and only if it has an upper bound.*

Every existing ρ-join of morphisms is linked, in the sense of 3.2.3 (i.e. preserved by composition and by intersection with a compatible morphism).

(b) A ρ-adequate functor preserves all the existing ρ-joins.

Proof (a) First, every ρ-set α with an upper bound c has a compositive join. In fact, if $a = c.e_a$, for $a \in \alpha$, let $\hat{e} = \bigvee e_a$ (a compositive join) and $\hat{b} = c.\hat{e}$. Then

$$\hat{b} = c.\bigvee e_a = \bigvee_a ce_a = \bigvee_a a = \bigvee \alpha$$

is a compositive join.

It is also a linked join, by Theorem 3.3.3(b), and we have also proved that every existing ρ-join in C is a linked join.

(b) If $F \colon \mathsf{C} \to \mathsf{D}$ is a ρ-adequate functor, $\hat{b} = \bigvee \alpha$ is a ρ-join with $a = \hat{b}.e_a$ ($a \in \alpha$) and $\hat{e} = \bigvee e_a$ as above

$$F(\bigvee_{a \in \alpha} a) = F(\hat{b}) = F(\hat{b}.\hat{e}) = F(\hat{b}).F(\hat{e}) = F(\hat{b}).F(\bigvee e_a)$$
$$= F(\hat{b}).(\bigvee Fe_a) = \bigvee(F\hat{b}.Fe_a) = \bigvee F(\hat{b}.e_a) = \bigvee_{a \in \alpha} Fa.$$

$\square$

3.8.6 Corollary

(a) For a prj-category C *the following conditions are equivalent:*

(i) C *is ρ-cohesive,*

(ii) in C *every linked ρ-set of parallel morphisms has a join, preserved by composition,*

(iii) C *is ρ-adequate and every linked ρ-set of parallel morphisms has an upper bound,*

(iv) C *has compositive ρ-joins of parallel projectors, and every linked ρ-set of parallel morphisms has an upper bound.*

(b) A functor between ρ-cohesive prj-categories is ρ-cohesive if and only if it is ρ-adequate.

(c) In particular, the previous points hold for e-categories.

Proof A straightforward consequence of 3.8.3(b) and the previous proposition. □

3.8.7 Theorem (The completion of an adequate prj-category)

Let C be a ρ-adequate prj-category and ρcC its ρ-cohesive completion (described in 3.7.2).

(a) The projectors of C coincide with the endo-maps $[\alpha] \leqslant 1$ of ρcC.

(b) The category ρcC is prj-cohesive, with respect to these projectors.

(c) For parallel morphisms $[\alpha]$ and $[\beta]$ of ρcC:

(i) $[\alpha] \,!\, [\beta] \iff$ there is a resolution $(e_x)_{x \in \alpha \cup \beta}$ of the set $\alpha \cup \beta$ in C,

(d) For parallel morphisms $[\alpha]$ and $[\beta]$ of ρcC, the following conditions are equivalent:

(ii) $[\alpha] \leqslant [\beta]$,

(iii) there is a resolution $(e_x)_{x \in \alpha \cup \beta}$ of $\alpha \cup \beta$ with $a = a.(\bigvee_{b \in \beta} e_b)$ for all $a \in \alpha$,

(iv) there is a resolution $(e_x)_{x \in \alpha \cup \beta}$ of $\alpha \cup \beta$ with $\bigvee_{a \in \alpha} e_a \leqslant \bigvee_{b \in \beta} e_b$,

(v) for every $a \in \alpha$, $a \,!\, \beta$ and $a = \bigvee_{b \in \beta} a \wedge b$.

(e) The embedding $C \to \rho cC$ is ρ-adequate, and can also be considered as the universal ρ-adequate functor from C to a ρ-cohesive prj-category.

Proof (a) If ε is a ρ-set of parallel projectors of C, $e = \bigvee \varepsilon$ is a linked join in C, by Proposition 3.8.5; since linked joins are preserved by the embedding in ρcC, we have $e = [\varepsilon]$ in ρcC.

It follows that the projectors of C coincide with the endo-maps $[\alpha] \leqslant 1$ of ρcC: indeed the relation $e \leqslant 1$ in C is preserved by the embedding, while if $[\alpha] \leqslant 1$ in ρcC, each morphism $a \in \alpha$ is a projector ($a \leqslant [\alpha] \leqslant 1$ in ρcC, hence $a \leqslant 1$ in C) and $[\alpha] = \bigvee \alpha \leqslant 1$ in C.

(b)–(d). Now we have to prove that the ρ-cohesive category ρcC is prj-cohesive, with respect to the projectors of C.

Because of 3.3.4, this reduces to checking the conditions 3.3.2(i)–(ii) for the order and the linking relation in ρcC; in the same time we shall also verify the present characterisations of these relations.

We first consider the linking relation. If $\alpha \,!\, \beta$ in $\mathcal{P}_\rho C$, then $\alpha \cup \beta$ is a linked ρ-set, with resolution $(e_x)_{x \in \alpha \cup \beta}$.

Given such a resolution, the subsets

$$\varepsilon = \{e_a \mid a \in \alpha\}, \qquad \eta = \{e_b \mid b \in \beta\},$$

and their joins $e = \bigvee \varepsilon = [\varepsilon]$, $f = \bigvee \eta = [\eta]$ yield a resolution of $[\alpha]$ and $[\beta]$ in $\rho c\mathsf{C}$, in agreement with 3.3.2(ii)

$$\alpha.e = \{a.e \mid a \in \alpha\} = \alpha, \qquad \beta.f = \beta, \tag{3.116}$$

$$[\alpha].f = [\alpha].[\eta] = [\{a.e_b \mid a \in \alpha, b \in \beta\}]$$
$$= [\{b.e_a \mid a \in \alpha, b \in \beta\}] = [\beta].[\varepsilon] = [\beta].e. \tag{3.117}$$

Conversely, we suppose that there is a resolution of $[\alpha]$ and $[\beta]$ in $\rho c\mathsf{C}$:

$$[\alpha] = [\alpha].e, \qquad [\beta] = [\beta].f, \qquad [\alpha].f = [\beta].e.$$

By (PCH.3ρ) there are resolutions (e_a) of α and (f_b) of β; thus

$$a = [\alpha].e_a = [\alpha].ee_a = ae \qquad \text{(for } a \in \alpha\text{)},$$

and we may assume that $e_a \leqslant e$, for all $a \in \alpha$; similarly $f_b \leqslant f$, for $b \in \beta$. Then, in $\rho c\mathsf{C}$

$$a.f_b = [\alpha].e_a.f_b = [\alpha]f.e_a f_b = [\beta]e.e_a f_b = [\beta].e_a f_b = b.e_a, \tag{3.118}$$

whence $a \, ! \, b$, in C (for all a and b) and $[\alpha] \, ! \, [\beta]$ in $\rho c\mathsf{C}$.

Secondly, we check the ordering.

(ii) $\Rightarrow$ (iii). If $[\alpha] \leqslant [\beta]$ in $\rho c\mathsf{C}$, every resolution (e_x) of $\alpha \cup \beta$ (with $e = \bigvee e_a$ and $f = \bigvee e_b$) gives: $a = \bigvee_b (a \wedge b) = \bigvee_b ae_b = af$.

(iii) $\Rightarrow$ (iv). Given such a resolution, we replace each e_a with $e_a.f$; this gives a new resolution of $\alpha \cup \beta$ such that $e \leqslant f$.

(iv) $\Rightarrow$ (ii). If this property holds we have, by (3.116) and (3.117), $[\alpha] = [\alpha].e = [\alpha].f = [\beta].e$, as required by 3.3.2(i). Now, if $[\alpha] = [\beta].h$ for some projector h, the relation $h \leqslant 1$ in $\rho c\mathsf{C}$ implies $[\alpha] \leqslant [\beta]$.

(v) $\Leftrightarrow$ (ii). Property (v) is equivalent to 3.7.2(i), by Proposition 3.8.5.

(e) Finally the embedding $\mathsf{C} \to \rho c\mathsf{C}$ preserves the projectors by the above remarks, and their ρ-joins (as all the existing linked ρ-joins) by definition.

The new universal property is a particular case of the original one, in Theorem 3.7.1. $\square$

3.8.8 Theorem (The completion of an adequate e-category)

If C *is a ρ-adequate e-cohesive category, then* ρcC *is e-cohesive, with supports*

$$\underline{e}[\alpha] = [\{\underline{e}(a) \mid a \in \alpha\}] = \bigvee_{a \in \alpha} \underline{e}(a), \tag{3.119}$$

for every linked ρ-set α of C.

The embedding $\eta \colon$ C $\to$ ρcC *is a ρ-adequate e-functor. It is the universal ρ-adequate e-functor from* C *to a ρ-cohesive prj-category.*

Proof For a linked ρ-set α of C, let us consider the ρ-set of projectors $\varepsilon = \{\underline{e}(a) \mid a \in \alpha\}$; it is an endo-map in $\mathcal{P}_\rho$C.

Clearly $\alpha.\varepsilon = \{a.\underline{e}(a') \mid a, a' \in \alpha\} \sim \alpha$; on the other hand, if $[\alpha].e = [\alpha]$ then (as in the proof of Theorem 3.8.7) $ae = a$ for all $a \in \alpha$, i.e. $\underline{e}(a) \leqslant e$, for all a, and $[\varepsilon] \leqslant e$. Hence, in ρcC, $[\alpha]$ has support $[\varepsilon] = \bigvee \underline{e}(a)$. This also proves that the embedding η preserves supports.

As to (ECH.2), if β is a linked ρ-set of C, composable with α, then (for all $a, a' \in \alpha$ and $b \in \beta$)

$$a.\underline{e}(ba') \leqslant a.\underline{e}(ba')\underline{e}(a) = a.\underline{e}(ba'.\underline{e}(a)) = a.\underline{e}(ba.\underline{e}(a')) \leqslant a.\underline{e}(ba),$$

$$\underline{e}[\beta].[\alpha] = [\{\underline{e}(b).a \mid a \in \alpha, \, b \in \beta\}] = [\{a.\underline{e}(ba) \mid a \in \alpha, \, b \in \beta\}]$$

$$= [\{a.\underline{e}(ba') \mid a, a' \in \alpha, \, b \in \beta\}] = \alpha.\underline{e}[\beta\alpha]. \qquad \square$$

3.9 Proof of the gluing completion theorems

We end this chapter completing the main results of Section 3.5, with the notation used there.

K is always a ρ-cohesive inverse category. M $= \rho$IMf K is the category of ρ-manifolds on K and symmetric profunctors, built in 3.5.3.

Here we prove Theorem 3.5.4 (on the inverse structure of M), Theorem 3.5.5 (on the cohesive structure of M), and our main theoretical result, Theorem 3.5.8 on the inverse gluing completion of K.

As the proof of Theorem 3.6.7, on the gluing completion of prj-cohesive categories, relies on Theorem 3.5.8, we are actually proving both gluing theorems.

3.9.1 *Proof of Theorem 3.5.4*

(a) First, the category $\mathsf{M} = \rho\mathrm{IMf}\,\mathsf{K}$ has a natural regular involution, that takes the symmetric profunctor $a = (a_h^i)\colon U_I \to V_H$ to:

$$a^\sharp = (a_i^h)\colon V_H \to U_I, \qquad (a_i^h = a_h^{i\,\sharp}),$$

$$(a_h^i).(a_i^h).(a_h^i) = (\vee_{k,j}\, a_h^j a_j^k a_k^i) = (a_h^i)_{I,H}. \tag{3.120}$$

We recall that $a_i^h = a_h^{i\,\sharp}\colon V_h \to U_i$ is the extended component of a. The last equality above follows from $a_h^j a_j^k a_k^i \leqslant a_h^j u_j^i \leqslant a_h^i$ for all k and j, with equality for $j = i$ and $k = h$.

(b) We now prove the equivalence of properties (i)–(iv), in 3.5.4. A projector in M

$$e = (e_j^i)\colon U \to U$$

is meant with respect to the previous involution: $e = ee = e^\sharp$, with $e_i^j = e_j^{i\,\sharp}$ for all i, j.

(i) $\Rightarrow$ (ii). Indeed $e = e^\sharp e = (\vee_h\, e_j^h e_h^i)$ and $e_j^i = \vee_h\, e_j^h e_h^i \leqslant u_j^i$.

(ii) $\Rightarrow$ (iv). The endo-map $e_i = e_i^i \leqslant u_i^i = 1$ is a projector of U_i and:

$$e_j^i = e_j^i e_i^j e_j^i \leqslant e_j^i e_i \leqslant u_j^i e_i = u_j^i e_i^i \leqslant e_j^i,$$

so that $e_j^i = u_j^i e_i$ and $u_i^j e_j u_j^i = e_i^j u_j^i \leqslant e_i^i = e_i$ (for all i, j).

(iv) $\Rightarrow$ (iii). It is easy to show that $u_j^i e_i = e_j u_j^i$. The family $e = (e_j^i) = (u_j^i e_i)_{I,I}$ is an endomorphism of U, since (for $i, j, h \in I$)

$$u_h^j e_j^i = u_h^j(u_j^i e_i) \leqslant u_h^i e_i = e_h^i, \qquad e_h^j e_j^i = (e_h u_h^j)(u_j^i e_i) \leqslant u_h^i.$$

(iii) $\Rightarrow$ (i). By the following inequalities

$$(e^\sharp e)_j^i = \vee_h\, e_j^h e_h^i = \vee_h(e_j u_j^h u_h^i e_i) \leqslant e_j u_j^i e_i = e_j^i,$$

$$(e^\sharp e)_j^i = \vee_h\, e_j^h e_h^i \geqslant e_j^j e_j^i = e_j u_j^i e_i = e_j^i.$$

(c) To prove that M is inverse (by the Characterisation Theorem 2.3.7) it is now sufficient to show that the product of two projectors of U, $e = (e_j^i)$ and $f = (f_j^i)$, is a projector, verifying the previous condition (iii)

$$(ef)_j^i = \vee_h\, e_j^h f_h^i = \vee_h\, e_j u_j^h u_h^i f_i = e_j u_j^i f_i,$$

$$(ef)_i^i = e_i f_i \in \mathsf{Prj}\,(U_i),$$

$$(ef)_j^i = e_j u_j^i f_i = u_j^i(ef)_i^i = (ef)_j^j u_j^i. \qquad \square$$

3.9.2 Proof of Theorem 3.5.5

(a) Property (3.56) is a consequence of the following inequality

$$f_h^k a_k^j e_j^i = f_h v_h^k a_k^j u_j^i e_i \leqslant f_h a_h^i e_i,$$

with equality for $j = i$ and $k = h$.

(b) Part One. We check the characterisation (3.57) of the order of M. If $a_h^i \leqslant b_h^i$, for all $i \in I$, $h \in H$

$$(ab^\sharp a)_h^i = \bigvee_{k,j} a_h^j b_j^k a_k^i \geqslant \bigvee_{k,j} a_h^j a_j^k a_k^i = (aa^\sharp a)_h^i = a_h^i,$$

$$(ab^\sharp a)_h^i = \bigvee_{k,j} a_h^j b_j^k a_k^i \leqslant \bigvee_{k,j} a_h^j b_j^k b_k^i \leqslant \bigvee_{k,j} a_h^j u_j^i \leqslant a_h^i,$$

hence $ab^\sharp a = a$ and $a \leqslant b$, in M.

(b) Part Two. To prove the characterisation (3.58) of the linking relation in M, we assume that $a \, ! \, b$, in the inverse category M; then $b^\sharp a$ and $ba^\sharp$ are projectors. Applying 3.5.4(ii) we have (for $i, j \in I$ and $h \in H$):

$$b_j^h a_h^i \leqslant (b^\sharp a)_j^i \leqslant u_j^i, \qquad b_k^i a_i^h \leqslant (ba^\sharp)_k^h \leqslant v_k^h. \tag{3.121}$$

If the last conditions hold, we have $b_i^h a_h^i \leqslant u_i^i = 1$ and $b_h^i a_i^h \leqslant v_h^h = 1$, i.e. $a_h^i \, ! \, b_h^i$ in K (for all i, h); moreover $(a_h^i \vee b_h^i)_{I,H}$ is a symmetric profunctor

$$v_k^h(a_h^i \vee b_h^i) = (v_k^h a_h^i) \vee (v_k^h b_h^i) \leqslant a_k^i \vee b_k^i,$$

$$(a_h^j \vee b_h^j)^\sharp (a_h^i \vee b_h^i) = (a_j^h a_h^i) \vee (a_j^h b_h^i) \vee (b_j^h a_h^i) \vee (b_j^h b_h^i) \leqslant u_j^i.$$

Last, if $x = (a_h^i \vee b_h^i)_{I,H}$ is a symmetric profunctor, then $a, b \leqslant x$ (by (3.57)), and $a \, ! \, b$.

(b) Part Three. If $a \, ! \, b$, one shows as before that $y = (a_h^i \wedge b_h^i)_{I,H}$ is a symmetric profunctor. By (3.57), $x = a \vee b$ and $y = a \wedge b$.

(c) To prove that the inverse category $\mathsf{M} = \rho \mathrm{IMf} \, \mathsf{K}$ is ρ-cohesive, we start from a linked ρ-set α of parallel maps $(U_i, u_j^i)_I \to (V_h, v_k^h)_H$ and write as $\alpha_h^i : U_i \to V_h$ the ρ-set of its (i, h)-components.

Now α_h^i is a linked ρ-subset of $\mathsf{K}(U_i, V_h)$, by (3.58), and has a join $b_h^i : U_i \to V_h$ in K. It is now easy to check that $b = (b_h^i) : U \to V$ is the linked join of α.

This also proves that the embedding $\mathsf{K} \to \mathsf{M}$ is a ρ-cohesive functor.

(d) Follows easily from the characterisation of projectors, in Theorem 3.5.4.

(e) Obvious. $\qquad\qquad\qquad\qquad\qquad\qquad\qquad\qquad\qquad\qquad\qquad\qquad\square$

3.9.3 Proof of Theorem 3.5.8 (Inverse gluing completion)

(a) We already know that the inverse category $\mathsf{M} = \rho\mathsf{IMf}\,\mathsf{K}$ is ρ-cohesive, and the embedding $\mathsf{K} \to \mathsf{M}$ is a ρ-cohesive functor, by 3.5.5(c).

To prove that M is ρ-gluing we have to show that each ρ-manifold $U = (U_r, Z_s^r)_R$ of M has a gluing in M. Loosely speaking, we put together all the given atlases, noting that a ρ-indexed union of ρ-sets is still a ρ-set.

Explicitly, the manifold U_R is given by a ρ-indexed family of ρ-manifolds U_r on K, *which we index on the same ρ-set I* (this is possible, because charts can be repeated)

$$U_r = (U_{ri}, u_{rj}^{ri})_I \qquad (r \in R). \tag{3.122}$$

The transition morphisms $Z_s^r \colon U_r \to U_s$, written as

$$Z_s^r \colon (U_{ri}, u_{rj}^{ri})_I \to (U_{si}, u_{sj}^{si})_I,$$
$$Z_s^r = (z_{sj}^{ri} \colon U_{ri} \to U_{sj})_{i,j\in I} \qquad (r, s \in R), \tag{3.123}$$

satisfy the following conditions (for $r, s, t \in R$ and $i, j, h \in I$):

$$Z_r^r = 1, \qquad \text{i.e.} \quad z_{rj}^{ri} = u_{rj}^{ri},$$
$$Z_t^s Z_s^r \leqslant Z_t^r, \qquad \text{i.e.} \quad z_{th}^{sj} z_{sj}^{ri} \leqslant z_{th}^{ri}, \tag{3.124}$$
$$(Z_s^r)^\sharp = Z_r^s, \qquad \text{i.e.} \quad (z_{sj}^{ri})^\sharp = z_{ri}^{sj}.$$

Now we form a global diagram in K, indexed by the ρ-set $R \times I$

$$Z = (Z_{ri}, z_{sj}^{ri})_{R\times I}, \qquad Z_{ri} = U_{ri}, \tag{3.125}$$

which is a symmetric manifold Z on K, because of (3.124). We have re-named the object U_{ri} as Z_{ri} when it plays the role of a chart of Z, instead of U_r.

There is a family of morphisms $(Z^r)_{r\in R}$ in M (symmetric profunctors):

$$Z^r \colon U_r \to Z, \qquad Z^r = (z_{sj}^{ri} \colon U_{ri} \to Z_{sj})_{i\in I, (s,j)\in R\times I}, \tag{3.126}$$

and we note that the morphism $Z^{s\sharp} \colon Z \to U_s$ has the following components

$$Z^{s\sharp} = (z_{sj}^{ri} \colon Z_{ri} \to U_{sj})_{(r,i)\in R\times I, j\in I}.$$

The maps (Z^r) say that Z is the gluing in M of the manifold U_R. Indeed, they satisfy the characterisation 3.5.2 for the gluing:

$$(Z^{s\sharp}.Z^r)_j^i = \bigvee_{t,h} z_{sj}^{th} z_{th}^{ri} = z_{sj}^{ri} = (Z_s^r)_j^i,$$
$$(Z^t.Z^{t\sharp})_{sj}^{ri} = \bigvee_h z_{sj}^{th} z_{th}^{ri}, \tag{3.127}$$
$$(\bigvee_t Z^t.Z^{t\sharp})_{sj}^{ri} = \bigvee_{t,h} z_{sj}^{th} z_{th}^{ri} = z_{sj}^{ri} = (1_Z)_{sj}^{ri}.$$

(b) We show now that the embedding $J \colon \mathsf{K} \to \mathsf{M}$ satisfies the required universal property, with respect to a ρ-cohesive functor $F \colon \mathsf{K} \to \mathsf{L}$ with values in a ρ-gluing inverse category.

The functor $F \colon \mathsf{K} \to \mathsf{L}$ gives a commutative solid square

$$
\begin{array}{ccc}
\mathsf{K} & \xrightarrow{\ \ F\ \ } & \mathsf{L} \\
{\scriptstyle J}\downarrow & & {\scriptstyle J'}\downarrow\big\uparrow\,{\scriptstyle \mathrm{gl}} \\
\rho\mathrm{IMf}\,\mathsf{K} & \xrightarrow[\ \underline{F}\ ]{} & \rho\mathrm{IMf}\,\mathsf{L}
\end{array}
\tag{3.128}
$$

where the (obvious) extension $\underline{F}$ is defined in Theorem 3.5.5.

But L is a ρ-gluing inverse category, and the embedding J' is an equivalence (by Lemma 3.5.7).

Its pseudo inverse $\mathrm{gl} \colon \rho\mathrm{IMf}\,\mathsf{L} \to \mathsf{L}$ is also an equivalence of ρ-gluing inverse categories, and therefore a ρ-cohesive functor (by 3.4.5(c)). Moreover, we can choose this pseudo inverse so that $\mathrm{gl}.J' = \mathrm{id}\,\mathsf{L}$, as remarked in 3.5.7(b).

In this way, the ρ-cohesive functor $G = \mathrm{gl}.\underline{F} \colon \rho\mathrm{IMf}\,\mathsf{K} \to \mathsf{L}$ forms a commutative triangle (3.65)

$$
GJ = \mathrm{gl}.\underline{F}J = \mathrm{gl}.J'F = F.
\tag{3.129}
$$

Conversely, if the functor $G \colon \mathsf{M} \to \mathsf{L}$ is ρ-cohesive and $GJ = F$, let us recall that G preserves all ρ-gluings (by Proposition 3.5.2). As we have seen, in Lemma 3.5.6(c), for every manifold U_I on K, the M-object U is the gluing of the manifold $\underline{J}(U_I)$. Therefore $G(U)$ is *a* gluing of the manifold $\underline{F}(U)$, isomorphic to the chosen gluing $\mathrm{gl}.\underline{F}(U)$ in L.

With some care, these isomorphisms can be organised, forming a natural family.

(c) As $G = \mathrm{gl}.\underline{F}$, and $\mathrm{gl} \colon \rho\mathrm{IMf}\,\mathsf{L} \to \mathsf{L}$ is an equivalence, it is sufficient to prove both claims for $\underline{F}$.

If F is faithful and $a, b \colon U_I \to V_H$ are parallel arrows in $\rho\mathrm{IMf}\,\mathsf{K}$, $\underline{F}(a) = \underline{F}(b)$ means that

$$
F(a_h^i) = F(b_h^i) \colon U_i \to V_h \quad \text{(for all } i \in I, \, h \in H),
$$

and then $a = b$. If F is full and faithful, and

$$
c = (c_h^i)_{I,H} \colon (FU_i, Fu_j^i)_I \to (FV_h, Fv_k^h)_H,
$$

each component $c_h^i \colon FU_i \to FV_h$ has a unique lifting $a_h^i \colon U_i \to V_h$; this family forms a symmetric profunctor $a \colon U \to V$ (as F reflects the order relation and the involution), and $\underline{F}(a) = c$.

(d) The functor G is faithful, by the previous point, and the morphisms $a\colon U_I \to V_H$ are determined by their image $\hat{a} = G(a)\colon G(U) \to G(V)$ in L.

(e) The functor F determines as above (up to isomorphism), a ρ-cohesive functor $G\colon \mathsf{M} \to \mathsf{L}$ such that $GJ = F$.

G is full and faithful, by (c), and is essentially surjective on the objects by the density hypothesis. $\qquad\square$

4

From topological manifolds to fibre bundles

In this chapter we examine various 'local structures', viewed as manifolds on the e-cohesive category of the corresponding models.

The first two sections include C^r-manifolds, manifolds with boundary, foliated manifolds, fibre and vector bundles, covering maps and simplicial complexes.

Then we deal with new topics, which were not present in [G4]. Links with Directed Algebraic Topology are investigated in Sections 4.3–4.5, presenting — in particular — a notion of 'locally cartesian ordered manifold'. We end with embedded manifolds, G-bundles and fundamental groupoids.

The term *map* is often used for 'continuous mapping', total or partial. Neighbourhood can be shortened to nbd.

Literature. We are investigating, at an elementary level, structures of diverse fields of mathematics, which cannot be developed here. A basic knowledge of these fields should be sufficient to follow the present approach.

Further information can be found in specific books, like

- for differentiable manifolds: Auslander–MacKenzie [AM], Kobayashi–Nomizu [KN],

- for foliations: Tamura [Ta], Candel–Conlon [CaC],

- for simplicial complexes and covering maps: a textbook on algebraic topology, like Spanier [Sp], Hilton–Wylie [HiW], or Hatcher [Hat],

- for fibre bundles: Steenrod [St], Husemoller [Hus],

- for directed spaces: the author's [G8].

4.1 Differentiable manifolds and simplicial complexes

We begin by dealing with basic local structures, C^r-manifolds and simplicial complexes.

Technically, we start from a totally cohesive e-category C that contains the elementary objects we want to glue (like $\mathcal{C}^r$), and form its gluing completion $\mathrm{Mf}\,\mathsf{C}$. The morphisms are defined as linked profunctors, and described by their realisation in some gluing e-category D (like $\mathcal{C}$), using Theorem 3.6.7.

C^r-manifolds are only assumed to be paracompact or Hausdorff when this is explicitly said.

4.1.1 Topological and differentiable manifolds

We begin by extending the analysis of inverse gluing completions, in 3.5.9, to the $\mathcal{S}$-concrete categories of which they form the inverse core.

(a) The e-cohesive categories $\mathcal{S}$ and $\mathcal{C}$ are already gluing, because their inverse cores are: we just apply Theorem 3.6.2.

(b) The gluing completion of the totally cohesive e-category $\mathcal{C}^r$ 'is' the category $\mathrm{Mf}\,\mathcal{C}^r$ *of* C^r-*manifolds and partial* C^r-*mappings, defined on open subsets.*

Its inverse core $\mathrm{IMf}\,\mathcal{C}^r$ was described in 3.5.9(c), as the category of C^r-manifolds and partial C^r-diffeomorphisms. Now, the inclusion $F\colon \mathcal{C}^r \to \mathcal{C}$ extends to a gluing functor

$$G\colon \mathrm{Mf}\,\mathcal{C}^r \to \mathcal{C},$$

the *topological realisation* of C^r-manifolds.

Applying the gluing completion theorem 3.6.7(c), a symmetric profunctor $a\colon U_I \to V_H$ in $\mathrm{Mf}\,\mathcal{C}^r$ amounts to its gluing $\hat{a}\colon \mathrm{gl}\,U \to \mathrm{gl}\,V$ in $\mathcal{C}$, a partial map between open subspaces whose components

$$a_h^i = v^{h\sharp}\hat{a}u^i\colon U_i \to V_h$$

are partial C^r-mappings.

(c) For a fixed $n \geqslant 0$, the full subcategory $\mathcal{C}^r(n) \subset \mathcal{C}^r$ of the open subspaces of $\mathbb{R}^n$ gives the full subcategory

$$\mathrm{Mf}\,\mathcal{C}^r(n) \subset \mathrm{Mf}\,\mathcal{C}^r,$$

of C^r-*manifolds of dimension* n (including the empty manifold).

For any r, the category $\mathcal{C}^r(0)$ is the full subcategory of $\mathcal{S}$ with two objects, $\emptyset$ and $\{*\}$. As any set is a gluing of singletons, $\mathrm{Mf}\,\mathcal{C}^r(0)$ is equivalent to $\mathcal{S}$.

*(d) The category $\mathrm{Mf}\,\mathcal{C}^r$ can also be obtained (up to equivalence of categories) as the gluing completion of the full subcategory $\mathcal{C}_0^r$ of $\mathcal{C}^r$ whose objects are the euclidean spaces $\mathbb{R}^n$, since each open euclidean space is a union (and a gluing in $\mathcal{C}^r$) of open balls, which are diffeomorphic to $\mathbb{R}^n$.

Let us note that the category $\mathcal{C}^r$ is the projector completion of the subcategory $\mathcal{C}_0^r$ (cf. 2.2.6).

*(e) Oriented C^r-manifolds can be dealt with in this way, starting from the totally cohesive e-category of open euclidean spaces and orientation-preserving partial C^r-mappings between them, defined on open subsets.

This is easy for $r \geqslant 1$, when orientation-preserving is defined by the positivity of the Jacobian determinant of first-order partial derivatives; for topological manifolds it is much more involved (and based on relative homology). We shall only give a few hints in the next subsection.

*(f) Banach manifolds [La] can also be dealt with in this way, starting from the totally cohesive e-category of open subspaces of Banach spaces and partial C^r-mappings between them, defined on open subsets.

4.1.2 Examples and complements (The spheres as manifolds)

(a) The circle $\mathbb{S}^1$ can be described by an intrinsic smooth atlas with two charts

$$(4.1)$$

where U_1 and U_2 are open intervals of the line $\mathbb{R}$.

The atlas is determined by the transition morphism $u_2^1 \colon U_1 \twoheadrightarrow U_2$, a partial homeomorphism which takes the open subinterval A to the open subinterval A', and B to B'. Analytically, u_2^1 can be realised by two translations, $A \to A'$ and $B \to B'$, preserving the orientation of U_1 and U_2 in the figure above. (This coherence will be relevant in Section 4.4.)

This atlas can be extended to all dimensions, replacing the open intervals U_i with open balls of $\mathbb{R}^n$, but the analytic expression of the transition map becomes complicated.

(b) The circle $\mathbb{S}^1$ is also produced by an atlas which can be easily extended to higher dimension.

We start from the pierced euclidean line $U = \mathbb{R} \setminus \{0\}$ and the total diffeomorphism

$$f \colon U \to U, \qquad f(x) = 1/x, \qquad (4.2)$$

which is involutive: $ff = \mathrm{id}\, U$.

Letting $u = u^\sharp \colon \mathbb{R} \rightarrowtail \mathbb{R}$ be the corresponding partial diffeomorphism, we have an intrinsic C^∞-atlas of $\mathbb{S}^1$, formed of two (equal) charts

$$U_1 = U_2 = \mathbb{R},$$
$$u_2^1 = u \colon U_1 \rightarrowtail U_2, \qquad u_1^2 = u^\sharp = u \colon U_2 \rightarrowtail U_1, \tag{4.3}$$

$$\tag{4.4}$$

The function $f(x) = 1/x$ is locally antitone (on the connected components of its domain). One may prefer to use the locally monotone diffeomorphism $f(x) = -1/x$, to get an orientable atlas.

(c) Extending the previous point to any $n \geqslant 0$, we start from the pierced space $U = \mathbb{R}^n \setminus \{0\}$ and the total diffeomorphism

$$f \colon U \rightarrowtail U, \qquad f(x) = x/||x||^2 = x.(x_1^2 + \ldots + x_n^2)^{-1}, \tag{4.5}$$

which takes each point $x \in U$ to the point $f(x)$ of the same half-line starting at 0, with $||f(x)|| = 1/||x||$. (This shows that $ff = \operatorname{id} U$.)

Again, we let $u = u^\sharp \colon \mathbb{R}^n \rightarrowtail \mathbb{R}^n$ be the corresponding partial diffeomorphism, and we have an intrinsic C^∞-atlas of $\mathbb{S}^n$, formed of two charts

$$U_1 = U_2 = \mathbb{R}^n,$$
$$u_2^1 = u \colon U_1 \rightarrowtail U_2, \qquad u_1^2 = u^\sharp = u \colon U_2 \rightarrowtail U_1. \tag{4.6}$$

The diffeomorphism f reverses orientation. One can modify it, e.g. composing f with a sign-change in the first coordinate.

4.1.3 Manifolds with boundary

The reader likely knows something about this topic; we only review the basic definitions.

An open disc of the euclidean plane is a manifold, but a compact disc is a manifold with boundary: its boundary, a circle, is an ordinary manifold. An n-dimensional manifold with boundary X has a boundary which is a manifold of dimension $n - 1$; the boundary is empty if and only if X is actually a manifold.

The term 'boundary' has here an intrinsic meaning, and should not be confused with the topological boundary of a subset of a space, a relative notion; the elementary exercises below show the difference.

Manifolds with boundary are locally homeomorphic to the open subspaces of the euclidean 'half-spaces'

$$H^n = \{(t_1, t_2, ...t_n) \in \mathbb{R}^n \mid t_1 \geqslant 0\} \qquad (n \geqslant 0). \tag{4.7}$$

These subspaces will be called *half-open euclidean spaces*; they include the open euclidean spaces (up to homeomorphism). We recall that a mapping $U \to \mathbb{R}^m$ defined on an open subspace of some H^n is of class C^r if it has a C^r-extension $V \to \mathbb{R}^m$ defined on an open subspace of $\mathbb{R}^n$.

Extending what we have reviewed at the beginning of the general Introduction, a manifold with boundary is classically defined as a topological space X equipped with an atlas $(u^i)_{i \in I}$, where each chart

$$u^i \colon U_i \to X_i, \tag{4.8}$$

is a homeomorphism between a half-open euclidean space U_i and an open subspace X_i of X. We assume that $X = \bigcup X_i$ and that every transition map (between half-open euclidean spaces)

$$u^i_j = u_j u^i \colon U_i \twoheadrightarrow U_j \qquad (i, j \in I), \tag{4.9}$$

is of class C^r (in the sense recalled above). Again $u_j = (u^j)^{-1} \colon X_j \to U_j$ (and the composition in (4.9) is an abuse of notation, as in Section 0.1).

Now, the category $C^r \mathcal{H}$ of half-open euclidean spaces and partial C^r-mappings defined on an open subspace is a subcategory of C, and has a totally cohesive e-structure lifted from C (being closed in C under supports, binary linked meets and arbitrary linked joins).

Its gluing completion Mf $C^r \mathcal{H}$ is the category of C^r-manifolds with boundary, and partial C^r-mappings defined on open subspaces.

Exercises and complements. (a) If U is open in H^n, the *intrinsic interior* $\underline{\mathrm{Int}}\, U$ and the *intrinsic boundary* $\underline{\partial}\, U$ of U are defined as

$$\underline{\mathrm{Int}}\, U = \mathrm{Int}_{\mathbb{R}^n} U, \qquad \underline{\partial}\, U = U \setminus \underline{\mathrm{Int}}\, U = U \cap \partial_{\mathbb{R}^n} U \tag{4.10}$$

so that $x \in \underline{\mathrm{Int}}\, U$ if and only if x belongs to some open euclidean set contained in U.

A convenient half-disc U, open in H^n, can show the difference between $\underline{\partial}\, U$, $\partial_{H^n} U$ and $\partial_{\mathbb{R}^n} U$.

*(b) Intrinsic interior points and intrinsic boundary points are preserved by partial homeomorphisms (not by partial continuous mappings), and there are totally cohesive functors

$$\underline{\mathrm{Int}} \colon \mathrm{I}(C^r \mathcal{H}) \to \mathrm{I}(C^r), \qquad \underline{\partial} \colon \mathrm{I}(C^r \mathcal{H}) \to \mathrm{I}(C^r). \tag{4.11}$$

4.1.4 Foliated manifolds

A *trivial foliation* will be a cartesian product $U \times V$, where U and V are open euclidean spaces; the subsets $V_x = \{x\} \times V$ are its *leaves* (for $x \in U$).

Here we must distinguish the object $\mathbb{R}^2 \times \mathbb{R}$ (which has 1-dimensional leaves) from $\mathbb{R} \times \mathbb{R}^2$ (which has 2-dimensional leaves). More precisely, the objects we are considering are pairs (U, V) of open euclidean spaces, with an underlying space $|(U, V)| = U \times V$; they will be written as pairs when useful.

A *partial C^r-map* $f \colon U \times V \dashrightarrow U' \times V'$ of trivial foliations is a partial C^r-mapping, defined on an open subset of $U \times V$, which preserves the leaves: if (x, y_1) and (x, y_2) are in $V_x \cap \operatorname{Def} f$, their f-images are in the same leaf of $U' \times V'$ (i.e. have the same projection on U').

All this forms a category $C^r\mathcal{F}$ *of trivial C^r-foliations and partial C^r-maps*. It has a faithful functor

$$|-| \colon C^r\mathcal{F} \to C^r, \qquad |(U, V)| = U \times V, \quad |f| = f, \qquad (4.12)$$

which lifts to $C^r\mathcal{F}$ the totally cohesive e-structure of C^r.

The gluing completion $\operatorname{Mf} C^r\mathcal{F}$ *is* the category of C^r-foliations and partial C^r-maps of foliations, and inherits from (4.12) a forgetful gluing functor

$$|-| \colon \operatorname{Mf} C^r\mathcal{F} \to \operatorname{Mf} C^r. \qquad (4.13)$$

This produces a topological realisation in $\mathcal{C}$.

Exercises and complements. (a) Let $p \colon U \times V \to U$ and $p' \colon U' \times V' \to U'$ be cartesian projections. A morphism $f \colon U \times V \dashrightarrow U' \times V'$ in $C^r\mathcal{F}$ can be characterised as a C^r-morphism satisfying:

(i) there exists a partial mapping $\overline{f} \colon U \dashrightarrow U'$ in $\mathcal{S}$ such that

$$
\begin{array}{ccc}
U \times V & \overset{f}{\dashrightarrow} & U' \times V' \\
{\scriptstyle p}\downarrow & \geqslant & \downarrow{\scriptstyle p'} \\
U & \underset{\overline{f}}{\dashrightarrow} & U'
\end{array}
\qquad p'f \leqslant \overline{f}p \ \text{ in } \mathcal{S}. \qquad (4.14)
$$

Then $\overline{f} \colon U \dashrightarrow U'$ is uniquely determined. It is defined on the open set $p(\operatorname{Def} f) \subset U$, and belongs to C^r. (The other inequality generally fails, since $\operatorname{Def} f$ need not be a union of leaves.)

(b) There is thus, after (4.12), a second functor

$$P \colon C^r\mathcal{F} \to C^r, \qquad P(U, V) = U, \quad P(f) = \overline{f}, \qquad (4.15)$$

which is not faithful. Prove that it is a totally cohesive e-functor, so that it extends to manifolds.

4.1.5 Countably-indexed and finitely-indexed manifolds

Let us say something about paracompactness, which is important for differentiable manifolds.

A topological space X is said to be paracompact if every open cover has a countable subcover. In the exercises below we recall that all euclidean spaces are second countable and paracompact, while a locally euclidean space is paracompact if and only if it is second countable.

Therefore, the countable gluing completion $\sigma \mathrm{Mf}\, \mathcal{C}^r$ 'is' the category of paracompact (possibly non-Hausdorff) C^r-manifolds and partial C^r-maps.

Similarly, the finite gluing completion $\mathrm{fMf}\, \mathcal{C}^r$ contains all compact manifolds, and also all open euclidean spaces.

Exercises and complements. (a) If X is second countable, then it is paracompact.

(b) Every euclidean space is second countable, and paracompact.

(c) A locally euclidean space is paracompact if and only if it is second countable.

*(d) The 'long line' is a classical example of a connected topological space which is locally homeomorphic to the euclidean line, and not paracompact ([KN], p. 166). (If we do not require connectedness, any uncountable sum of real lines would do.)

4.1.6 Simplicial complexes

A *simplicial complex* is a set X equipped with a family of finite subsets, the *linked parts*, such that the empty subset and all singletons are linked, and every subset of a linked part is linked [Sp].

A *morphism of simplicial complexes* is a mapping that preserves the linked parts. A *simplicial subcomplex* of X is a subset A, equipped with the linked parts of X contained in A.

(Requiring that the empty subset be linked is not 'entirely' redundant: omitting this condition, the empty set would have two different simplicial structures, which is not convenient. Equivalently, one can modify the axioms so that they only concern the non-empty subsets of our set.)

The linked parts are meant to express a notion of 'proximity'. Simplicial complexes should not be confused with simplicial sets, i.e. presheaves on finite positive ordinals (see 1.6.6). In fact, they are more close to 'symmetric simplicial sets', i.e. presheaves on finite positive cardinals [G6].

Let us write as $\mathcal{D}_0 \subset \mathcal{S}$ the full subcategory of finite cardinals $m > 0$ and partial mappings between them. $\mathcal{D}_0$ is a totally cohesive e-subcategory of $\mathcal{S}$.

The gluing completion Mf $\mathcal{D}_0$ can be interpreted as the category of *simplicial complexes and partial simplicial maps, defined on a simplicial subcomplex.* Finite simplicial complexes are obtained by the *finite* gluing completion fMf $\mathcal{D}_0$.

The category $\mathcal{D}_0$ embeds in the category $\mathcal{D}$ of topological spaces and continuous partial mappings defined on a closed subspace, by means of the *standard-simplex* functor:

$$\Delta \colon \mathcal{D}_0 \to \mathcal{D},$$
$$m \mapsto \Delta^{m-1} = \{(t_0, t_1, \ldots t_{m-1}) \in \mathbb{R}^m \mid t_i \geqslant 0, \ \Sigma t_i = 1\}, \tag{4.16}$$

that takes the cardinal $m = \{0, 1, \ldots m - 1\}$ to the convex hull of the 'unit points' e_i^m of $\mathbb{R}^m$, and the partial mapping $f \colon m \rightarrowtail n$ to its affine extension

$$\Delta(f) \colon \Delta^{m-1} \rightarrowtail \Delta^{n-1},$$

identifying $i \in m$ with e_i^m.

Let us note that $\Delta(f)$ is defined on the convex hull of Def f, a closed affine subspace of Δ^{m-1}, so that Δ does take values in $\mathcal{D}$. (In particular, the cardinal 1 is sent to $\Delta^0 = \mathbb{R}^0$, the singleton.)

Now, $\mathcal{D}$ has a finitely gluing e-structure, but the e-functor $\Delta \colon \mathcal{D}_0 \to \mathcal{D}$ is not finitely cohesive. For instance, the join of the projectors

$$e, e' \colon \{0, 1, 2\} \rightarrowtail \{0, 1, 2\}, \qquad \text{Def } e = \{0, 1\}, \ \text{Def } e' = \{1, 2\},$$

is the identity, but the projector $\Delta(e) \vee \Delta(e') \colon \Delta^2 \rightarrowtail \Delta^2$ is only defined on the union of two edges of the triangle Δ^2.

4.2 Fibre bundles, vector bundles and covering maps

We deal now with fibre bundles as 'manifolds of trivial bundles', forming a category Mf $\mathcal{B}$. The topological realisation takes place in a gluing category $\mathcal{F}$ of surjective open maps $p \colon X \to B$, that plays the role of $\mathcal{C}$ for differentiable manifolds.

Similarly, the vector bundles will form the gluing completion Mf $\mathcal{V}$ of a cohesive category of trivial vector bundles.

The homotopy lifting property of fibre bundles is automatically extended from trivial fibre bundles (see 4.2.3). Linking this section with the previous one, the tangent bundle functor of a differentiable manifold is dealt with in 4.2.5.

4.2.1 A gluing category

We begin by constructing a category $\mathcal{F}$ of *surjections and partial maps*.

A *surjection* will be any surjective open continuous mapping $p\colon X \to B$ (everywhere defined) between topological spaces, viewed as a 'generalised fibration'. We already remarked, in 1.4.9(b), that B has the finest topology making p continuous.

A *partial map* $(f, \overline{f})\colon p \rightsquigarrow p'$ of surjections is given by a commutative square in $\mathcal{C}$

$$
\begin{array}{ccc}
X & \overset{f}{\dashrightarrow} & X' \\
{\scriptstyle p}\downarrow & & \downarrow{\scriptstyle p'} \\
B & \underset{\overline{f}}{\dashrightarrow} & B'
\end{array}
\qquad (4.17)
$$

Thus f and $\overline{f}$ are partial continuous mappings, defined on open subsets of X and B respectively, with

$$
\operatorname{Def} f = p^{-1}(\operatorname{Def} \overline{f}), \qquad \operatorname{Def} \overline{f} = p(\operatorname{Def} f), \qquad (4.18)
$$

and $\overline{f}$ is determined by f. Note that $W = \operatorname{Def} f = p^{-1}(p(W))$ is an open subset of X saturated for p (cf. 1.4.9) — a union of equivalence classes of the equivalence relation R_p associated to the mapping p.

We shall express property (4.18) saying that $(\operatorname{Def} f, \operatorname{Def} \overline{f})$ is a *saturated pair* for p, of open subsets of X and B. (Compare this with the weaker condition of (4.14).)

Composition and identities of these partial maps are obvious, forming the category $\mathcal{F}$. This category is concrete on $\mathcal{C}$ (and therefore on $\mathcal{S}$) by the forgetful functor

$$
U\colon \mathcal{F} \to \mathcal{C},
$$
$$
(p\colon X \to Y) \mapsto X, \qquad (f, \overline{f}) \mapsto f. \qquad (4.19)
$$

As proved in the exercises below, $\mathcal{F}$ has a gluing e-cohesive structure, lifted from $\mathcal{C}$ by U.

*The full subcategory $\mathcal{SF}$ of *Serre fibrations* (defined by the homotopy lifting property for singular simplices, or singular cubes) has similar properties.*

4.2.2 Exercises and complements

(a) The partial map $(f, \overline{f})\colon p \twoheadrightarrow p'$ is equivalent to a partial map $f\colon X \twoheadrightarrow X'$ in $\mathcal{C}$ such that

(i) Def f is saturated for p, and for every $x, y \in \text{Def } f$, $p(x) = p(y)$ implies $p'f(x) = p'f(y)$.

(b) A *projector* $(e, \overline{e})\colon p \twoheadrightarrow p$ of $\mathcal{F}$ is an endomorphism satisfying the equivalent conditions

$$
\begin{array}{ccc}
X & \overset{e}{\dashrightarrow} & X \\
p \downarrow & & \downarrow p' \\
B & \underset{\overline{e}}{\dashrightarrow} & B
\end{array}
\tag{4.20}
$$

(ii) $U(e, \overline{e}) = e\colon X \twoheadrightarrow X$ is idempotent in $\mathcal{C}$,

(iii) e and $\overline{e}$ are idempotent in $\mathcal{C}$,

(iv) $(e, \overline{e})$ is idempotent in $\mathcal{F}$.

The idempotent $(e, \overline{e})$ is equivalent to a partial identity $e\colon X \twoheadrightarrow X$ defined on an open subset W of X, saturated for p. Then $\overline{e}$ is defined on $p(W)$, open in B.

(c) With these projectors, the category $\mathcal{F}$ becomes an e-category: the support $(e, \overline{e})\colon p \twoheadrightarrow p$ of the partial map $(f, \overline{f})\colon p \twoheadrightarrow p'$ is the pair of partial identities on Def f and Def $\overline{f}$. Then

$$
(f, \overline{f}) \leqslant (g, \overline{g}) \text{ in } \mathcal{F} \quad \Leftrightarrow \quad f \leqslant g \text{ in } \mathcal{C}.
\tag{4.21}
$$

The inverse category $\mathrm{I}\mathcal{F}$ has the same objects. A morphism $(u, \overline{u})\colon p \twoheadrightarrow p'$ is a morphism of $\mathcal{F}$ formed of partial homeomorphisms $u\colon X \twoheadrightarrow X'$ and $\overline{u}\colon B \twoheadrightarrow B'$ between saturated pairs of p and p'

$$
\begin{array}{ccc}
X & \overset{u}{\dashrightarrow} & X' \\
p \downarrow & & \downarrow p' \\
B & \underset{\overline{u}}{\dashrightarrow} & B'
\end{array}
\tag{4.22}
$$

$$
\text{Def } u = p^{-1}(\text{Def } \overline{u}), \qquad \text{Val } u = p^{-1}(\text{Val } \overline{u}),
$$

$$
\text{Def } \overline{u} = p(\text{Def } u), \qquad \text{Val } \overline{u} = p(\text{Val } u).
$$

(d) $\mathcal{F}$ is totally cohesive. The join of a linked family $(f_i, \overline{f}_i)\colon p \to p'$ is computed pairwise, in $\mathcal{C}$

$$
\bigvee_i (f_i, \overline{f}_i) = (\bigvee_i f_i, \bigvee_i \overline{f}_i).
\tag{4.23}
$$

(e) A manifold M on $\mathcal{F}$ consists of a family of commutative diagrams

$$
\begin{array}{ccc}
X_i & \xrightarrow{u^i_j} & X_j \\
{\scriptstyle p_i}\downarrow & & \downarrow{\scriptstyle p_j} \\
B_i & \xrightarrow[\overline{u}^i_j]{} & B_j
\end{array}
\qquad M = (p_i \colon X_i \to B_i, (u^i_j, \overline{u}^i_j))_I, \qquad (4.24)
$$

where $(X_i, (u^i_j))_I$ and $(B_i, (\overline{u}^i_j))_I$ are manifolds on $\mathcal{C}$. Equivalently:

(v) $(X_i, (u^i_j))_I$ is a manifold on $\mathcal{C}$,

(vi) each partial map $u^i_j \colon X_i \dashrightarrow X_j$ defines an $\mathcal{F}$-morphism $p_i \dashrightarrow p_j$ (as characterised by (i)).

(f) $\mathcal{F}$ is gluing. *Hints.* The gluing $p \colon X \to B$ of the manifold M in $\mathcal{F}$ can be obtained gluing in $\mathcal{C}$ the manifolds $(X_i, (u^i_j))_I$ and $(B_i, (\overline{u}^i_j))_I$,

$$
u^i \colon X_i \dashrightarrow X, \qquad \overline{u}^i \colon B_i \dashrightarrow B, \qquad (4.25)
$$

and constructing $p \colon X \to B$ so that all the following squares commute in $\mathcal{C}$

$$
\begin{array}{ccc}
X_i & \xrightarrow{u^i} & X \\
{\scriptstyle p_i}\downarrow & & \downarrow{\scriptstyle p} \\
B_i & \xrightarrow[\overline{u}^i]{} & B
\end{array}
\qquad\qquad (4.26)
$$

4.2.3 Fibre bundles

The 'elementary spaces' that we want to patch together are now the *trivial fibre bundles*, i.e. the cartesian projections $p \colon B \times F \to B$, where B and F are topological spaces and $B \times F$ has the product topology.

Let $\mathcal{B}$ be the full subcategory of $\mathcal{F}$ determined by these objects, with the lifted e-category structure: this is still totally cohesive (as a full e-subcategory) but not gluing.

A morphism $(f, \overline{f}) \colon p \dashrightarrow p'$ in $\mathcal{B}$

$$
\begin{array}{ccc}
B \times F & \xrightarrow{f} & B' \times F' \\
{\scriptstyle p}\downarrow & & \downarrow{\scriptstyle p'} \\
B & \xrightarrow[\overline{f}]{} & B'
\end{array}
\qquad\qquad (4.27)
$$

is determined by a partial map $f \colon B \times F \dashrightarrow B' \times F'$ satisfying 4.2.2(i), with $\operatorname{Def} f = p^{-1}(\operatorname{Def} \overline{f}) = (\operatorname{Def} \overline{f}) \times F$.

Expressing f by its components

$$f_1 = p'f = f\bar{p}\colon B \times F \to B', \qquad f_2\colon B \times F \to F',$$

we have:

$$f(b, y) = (\bar{f}(b), f_2(b, y)), \qquad \mathrm{Def}\, f = \mathrm{Def}\, f_2 = (\mathrm{Def}\, \bar{f}) \times F. \qquad (4.28)$$

Finally, the morphism $(f, \bar{f})$ of $\mathcal{B}$ is equivalent to giving two partial maps in $\mathcal{C}$ such that:

(i) $\bar{f}\colon B \to B', \qquad f_2\colon B \times F \to F', \qquad \mathrm{Def}\, f_2 = (\mathrm{Def}\, \bar{f}) \times F.$

The trivial fibre bundle $p\colon B \times F \to B$ will also be denoted as $B \times F$.

The inverse category $I\mathcal{B}$ has the same objects. A morphism, as in 4.2.2(c), is a pair $(u, \bar{u})\colon p \to p'$ of $\mathcal{B}$ composed by partial homeomorphisms between saturated pairs of p and p'. This gives two morphisms of $\mathcal{C}$

$$\bar{u}\colon B \to B', \qquad u_2\colon B \times F \to F', \qquad \mathrm{Def}\, u_2 = (\mathrm{Def}\, \bar{u}) \times F, \qquad (4.29)$$

where $\bar{u}$ is in $I\mathcal{C}$ (a partial homeomorphism between open subsets), and for every $b \in \mathrm{Def}\,\bar{u}$, the map $u_2(b, -)\colon F \to F'$ is a homeomorphism. Thus, provided that the morphism u is not empty, the fibres F and F' are homeomorphic.

The gluing completion $\mathrm{Mf}\,\mathcal{B}$ has for objects the 'manifolds'

$$M = ((B_i \times F_i), (u_j^i))_I$$

on $\mathcal{B}$, for morphisms their linked profunctors: it is *the category of fibre bundles and partial maps*.

The inclusion $\mathcal{B} \to \mathcal{F}$ extends to the *topological realisation* functor

$$\mathrm{Mf}\,\mathcal{B} \to \mathcal{F},$$

that takes the previous object M to its gluing, in 4.2.2(f).

*The realisation $\mathrm{Mf}\,\mathcal{B} \to \mathcal{SF}$, with values in the category of Serre fibrations and partial maps (cf. 4.2.1), *automatically proves the homotopy lifting property of fibre bundles*; more precisely, it extends this property from local — where it is obvious — to global.*

By the previous characterisation of the morphisms of $I\mathcal{B}$, the topological type of the fibre $F_b = p^{-1}(\{b\})$ at the point b of the base $B = \mathrm{gl}\,(B_i, \bar{u}_j^i)_I$ is locally constant, and constant on every connected component of B.

4.2.4 Vector bundles

We deal now with fibre bundles whose fibres are real vector spaces of finite dimension.

Let us recall that any linear mapping $\mathbb{R}^m \to \mathbb{R}^n$ is continuous (for the euclidean topologies), and any linear isomorphism is a homeomorphism.

A finite dimensional vector space F on $\mathbb{R}$ will always be equipped with its *linear topology*, transferred by any linear isomorphism $F \to \mathbb{R}^n$. Any linear mapping $F \to F'$ between finite dimensional real vector spaces is continuous.

A *trivial vector bundle* is a trivial fibre bundle $p\colon B \times F \to B$, where B is a topological space and F is a finite dimensional, real vector space.

With these objects, we want to form a category $\mathcal{V}$, concrete on $\mathcal{B}$ and $\mathcal{F}$

$$\mathcal{V} \to \mathcal{B} \to \mathcal{F}, \qquad (4.30)$$

having 'fibrewise linear' partial mappings $f\colon B \times F \nrightarrow B' \times F'$.

More precisely, following 4.2.3(i), a morphism $(f, \overline{f})\colon p \nrightarrow p'$ in $\mathcal{V}$

$$
\begin{array}{ccc}
B \times F & \overset{f}{\dashrightarrow} & B' \times F' \\
{\scriptstyle p}\big\downarrow & & \big\downarrow{\scriptstyle p'} \\
B & \underset{\overline{f}}{\dashrightarrow} & B'
\end{array}
\qquad (4.31)
$$

is determined by two partial maps $\overline{f}$, f_2 in $\mathcal{C}$ such that:

(i) $\overline{f}\colon B \nrightarrow B'$, $\qquad f_2\colon B \times F \nrightarrow F'$, $\qquad \operatorname{Def} f_2 = (\operatorname{Def} \overline{f}) \times F$,

(ii) for every $b \in \operatorname{Def} \overline{f}$, the mapping $f_2(b, -)\colon F \to F'$ is $\mathbb{R}$-linear.

Given $(f, \overline{f})$, f_2 is its second component (on the same definition set). Given $\overline{f}$ and f_2 we let:

$$f(b, y) = (\overline{f}(b), f_2(b, y)), \qquad \operatorname{Def} f = \operatorname{Def} f_2 = (\operatorname{Def} \overline{f}) \times F. \qquad (4.32)$$

$\mathcal{V}$ is totally e-cohesive, with a structure lifted from $\mathcal{B}$ and $\mathcal{F}$, by the forgetful functors (4.30). A morphism

$$u\colon B \times F \nrightarrow B' \times F'$$

of the inverse core $\mathrm{I}\mathcal{V}$ is in $\mathrm{I}\mathcal{B}$ (by 4.2.3); moreover, for every b in $\operatorname{Def} \overline{u}$, $u_2(b, -)\colon F \to F'$ is a linear isomorphism.

The gluing completion $\operatorname{Mf} \mathcal{V}$ is the *category of vector bundles and partial morphisms* (defined, on saturated pairs). The algebro-topological type of the fibre $F_b = p^{-1}(\{b\})$ is locally constant on the basis.

Also here we have the topological realisation in $\mathcal{F}$ (or in $\mathcal{SF}$).

Tensor calculus automatically extends from finite dimensional vector spaces to vector bundles (cf. [AbM]). For instance, the n-th tensor power

$$(-)^{\otimes n}: \mathcal{V} \to \mathcal{V}, \qquad U \times F \mapsto U \times F^{\otimes n}, \qquad (4.33)$$

is a totally cohesive functor, and extends to $\mathrm{Mf}\,\mathcal{V}$.

4.2.5 Tangent bundles

Consider again the category $\mathcal{C}^r$ of trivial C^r-manifolds, with $r \geqslant 1$. The (trivial) tangent bundle functor, with the abuse of notation described in 4.2.3, is:

$$T: \mathcal{C}^r \to \mathcal{V}, \qquad U \mapsto U \times \mathbb{R}^{\dim U}, \quad f \mapsto Tf, \qquad (4.34)$$

$$Tf(x, h) = (fx, D_h f(x)), \qquad \text{for } x \in \mathrm{Def}\, f \text{ and } h \in \mathbb{R}^{\dim U}, \qquad (4.35)$$

where $D_h f(x)$ is the derivative of f at x, along the vector h.

Since $\mathcal{C}^r$, $\mathcal{V}$ and T are totally cohesive, T extends to a gluing functor, the *tangent bundle functor* of C^r-manifolds

$$T: \mathrm{Mf}\,\mathcal{C}^r \to \mathrm{Mf}\,\mathcal{V} \qquad (r \geqslant 1). \qquad (4.36)$$

4.2.6 Covering maps

A *covering map* $p: X \to B$ is a surjective continuous mapping between topological spaces such that every $b \in B$ has an open nbd V whose inverse image $p^{-1}(V)$ is a union of disjoint open sets in X (the *sheets* of p over V), each of which is mapped homeomorphically onto V by p

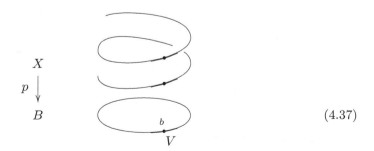

$$(4.37)$$

Then p is a surjective local homeomorphism, and an open map (see Exercises 4.2.7(a), (b)). The nbd V is said to be *evenly covered* by p.

Realising the standard circle $\mathbb{S}^1$ in the complex plane, we have classical examples, like the *exponential map* represented in the figure above

$$p\colon \mathbb{R} \to \mathbb{S}^1, \qquad p(t) = e^{2\pi i t} = \cos 2\pi t + i.\sin 2\pi t$$

$$\text{(the } \textit{universal covering of the circle}\text{).} \tag{4.38}$$

For a positive integer n, we also have the endo-map

$$p\colon \mathbb{S}^1 \to \mathbb{S}^1, \qquad p(z) = z^n$$

$$\text{(the } \textit{n-fold covering of the circle}\text{).} \tag{4.39}$$

The spaces X and B are called the *total space* and the *base space* of the covering. For every point $b \in B$, the *fibre* $F_b = p^{-1}\{b\}$ is a discrete subspace of X.

Over an *evenly covered nbd* V of b, the projection $p_V \colon p^{-1}(V) \to V$ is equivalent to a cartesian projection $p\colon V \times F_b \to V$, under a homeomorphism h that commutes with these projections (see Exercise 4.2.7(c))

$$
\begin{array}{ccc}
p^{-1}(V) & \xrightarrow{\ h\ } & V \times F_b \\
{\scriptstyle p_V}\downarrow & & \downarrow{\scriptstyle p} \\
V & = & V
\end{array}
\tag{4.40}
$$

A covering map *is thus the same as a fibre bundle with discrete fibres*. The cardinal of F_b is locally constant at b, hence constant on every connected component of B. For instance, the exponential map (4.38) has countable fibres, while the endo-mapping (4.39) is an *n-fold covering*: the cardinal of each fibre is n.

The full subcategory $c\mathcal{F}$ of $\mathcal{F}$ determined by *covering maps* is totally cohesive and gluing. Viewing it as a category of manifolds (on trivial covering maps) can be useful, to simplify its interactions with other categories of manifolds, reducing them to the trivial case.

Of course, a *trivial covering map* is a trivial fibre bundle with discrete fibre, i.e. a cartesian projection $B \times F \to B$, where B and F are topological spaces, F is discrete and $B \times F$ has the product topology.

Let $c\mathcal{B}$ be the full subcategory of $\mathcal{B}$ (and $\mathcal{F}$) determined by such objects, with the induced e-cohesive structure: this is totally cohesive but not gluing. The inclusion $c\mathcal{B} \to c\mathcal{F}$ extends to an equivalence Mf $c\mathcal{B} \to c\mathcal{F}$ with the e-cohesive category of covering maps and partial maps defined on open subspaces.

4.2.7 Exercises and complements (Local homeomorphisms)

(a) A mapping $f\colon X \to Y$ between topological spaces is said to be a *local homeomorphism* if

(i) every $x \in X$ has an open nbd U such that f restricts to a homeomorphism $U \to f(U)$ with values in an open subspace of Y.

This is equivalent to:

(ii) the mapping f is continuous, open and locally injective: every $x \in X$ has a nbd on which f is injective.

(b) A covering map $p\colon X \to B$ is a surjective local homeomorphism.

(c) Define the homeomorphism h, in diagram (4.40).

4.3 Preordered spaces and directed spaces

Directed Algebraic Topology is a recent subject, which arose in the 1990's; it is studied in many articles, and in a book of the present author [G8]. The existent applications deal mainly with the analysis of concurrent processes, but other domains where privileged directions appear should be eligible: rewrite systems, traffic networks, space-time models, etc.

We review here some settings where Directed Algebraic Topology makes sense: the elementary category pTop of preordered topological spaces, and the richer category dTop of 'd-spaces', or spaces with distinguished paths. Theorem 4.3.6 proves that the category dC of d-spaces and partial d-maps defined on open subsets is a gluing e-category.

The next section will apply this approach to topological and differentiable manifolds.

As in [G8], the prefix ↑ denotes a directed enrichment of a reversible structure.

4.3.1 Directed spaces

Directed Algebraic Topology studies 'directed spaces', of different kinds.

While Topology and Algebraic Topology deal with reversible worlds, where a path can always be travelled backwards, the study of non-reversible phenomena requires wider worlds, where a 'directed space' *can* have non-reversible paths (but reversible spaces are still present).

The classical topics, consisting of ordinary homotopies, fundamental groups and fundamental n-groupoids, are now extended with (possibly) non-reversible versions: *directed homotopies, fundamental monoids* and *fundamental n-categories*.

Similarly, the homology theories of the directed settings take values in 'directed' algebraic structures, like *preordered* abelian groups or abelian *monoids*.

In the first part of this section we give a few hints at the starting points of this subject, taken from the book [G8].

Notation and complements. We use the following notation, for paths in an ordinary topological space X.

(a) A continuous mapping $a\colon \mathbb{I} \to X$ defined on the standard interval $\mathbb{I} = [0, 1]$ is called a *path* in X, from $x_0 = a(0)$ to $x_1 = a(1)$; the latter are the *endpoints* of the path.

Concatenation of two consecutive paths $a, b\colon \mathbb{I} \to X$ (with $a(1) = b(0)$) will be written as

$$(a * b)(t) = \begin{cases} a(2t) & \text{if } 0 \leqslant t \leqslant 1/2, \\ a(2t - 1) & \text{if } 1/2 \leqslant t \leqslant 1. \end{cases} \tag{4.41}$$

The constant path at the point x is written as e_x. The *reversed path* $t \mapsto a(1 - t)$ is written as $a^\sharp$. A loop is a path whose endpoints coincide.

(b) (*The fundamental groupoid*) The fundamental groupoid $\Pi_1(X)$ of the space X is studied in an excellent book by R. Brown [Bw]. We briefly review this well known construction.

An object of $\Pi_1(X)$ is any point $x \in X$; an arrow $[a]\colon x \to y$ is a class of paths in X, from x to y, up to homotopy with fixed end-points. In other words, two paths $a_0, a_1\colon \mathbb{I} \to X$ from x to y are identified if there exists a continuous mapping $h\colon \mathbb{I} \times \mathbb{I} \to X$ which 'transforms one into the other', without moving the endpoints

$$h(t, i) = a_i(t), \quad h(0, t) = x, \quad h(1, t) = y \qquad (t \in \mathbb{I}, \ i = 0, 1). \tag{4.42}$$

Consecutive arrows $[a]\colon x \to y$ and $[b]\colon y \to z$ are composed by concatenation of paths:

$$[a].[b] = [a * b]\colon x \to z. \tag{4.43}$$

This operation is well defined and gives a groupoid, with identities $1_x = [e_x]$ represented by constant loops and inverses $[a]^{-1} = [a^\sharp]$ represented by reversed paths.

This groupoid contains all the classical fundamental groups $\pi_1(X, x) = \Pi_1(X)(x, x)$. Two of them are linked by an 'inner isomorphism' of $\Pi_1(X)$, when the base-points are connected by a path a

$$\pi_1(X, x) \to \pi_1(X, y),$$
$$[\omega] \mapsto [a].[\omega].[a]^{-1} \qquad (\text{for } [a]\colon x \to y). \tag{4.44}$$

One of the main tools to compute fundamental groups is the van Kampen Theorem. Its extension to groupoids, in [Bw], is more powerful: it applies to an open cover (X_1, X_2) of a space X, without assuming that X_1, X_2 and $X_1 \cap X_2$ be path-connected.

4.3.2 Preordered topological spaces

The simplest topological setting where one can study directed paths and directed homotopies is likely the category pTop of *preordered topological spaces* and monotone continuous mappings.

We are not requiring any particular relationship between topology and preorder. (But interesting links will appear below: 'path-preordered spaces' in 4.3.7 and 'locally generated preorders' in 4.5.3.)

The *standard directed interval* $\uparrow\!\mathbb{I} = \uparrow[0,1]$ has the euclidean topology and the natural order. A (directed) *path* in a preordered space X is — by definition — a map $a\colon \uparrow\!\mathbb{I} \to X$ in pTop, continuous and monotone.

The product $X = \Pi X_i$ of preordered spaces has the product topology and the product preorder: $(x_i) \prec_X (x_i')$ if and only if, for each index i, $x_i \prec x_i'$ in X_i.

More generally, the category pTop has all limits (resp. colimits), constructed as in Top and equipped with the initial (resp. final) preorder for their structural maps.

The standard embedding of Top in pTop will be given by the *indiscrete preorder*, so that all (ordinary) paths in a topological space X are still directed in the associated preordered space. This embedding is right adjoint to the forgetful functor pTop $\to$ Top.

4.3.3 Some examples

This elementary setting already allows a few hints at notions and applications dealt with in [G8], which we mention here without proof.

Consider the following order relation in the plane

$$(x,y) \leqslant (x',y') \quad \Leftrightarrow \quad |y' - y| \leqslant x' - x. \tag{4.45}$$

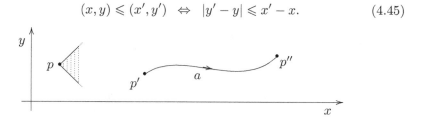

The picture shows the 'cone of the future' at a point p (i.e. the set of points $\geqslant p$) and a *directed path* from p' to p'', i.e. a continuous mapping $a \colon [0,1] \to \mathbb{R}^2$ which is monotone with respect to the natural order of the standard interval and the previous order of the plane: if $t \leqslant t'$ in $[0,1]$, then $a(t) \leqslant a(t')$ in the plane.

Take now the following (compact) subspaces X, Y of the plane, with the induced order (the cross-marked open rectangles are taken out). A directed path in X or Y satisfies the same conditions as above

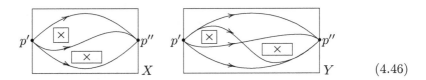

$$(4.46)$$

Then — as displayed in the figures above and easy to guess — there are, respectively, 3 or 4 homotopy classes of directed paths from the point p' to the point p'', in the fundamental categories $\uparrow\Pi_1(X)$ and $\uparrow\Pi_1(Y)$; in both cases there are none from p'' to p', and every loop is constant.

First, we can view each of these 'directed spaces' as a stream with two islands, and the induced order as an upper bound for the relative velocity feasible in the stream. Secondly, we can interpret the horizontal coordinate as (a measure of) time, the vertical coordinate as position in a 1-dimensional physical medium, and the order as the possibility of going from (x, y) to (x', y') with velocity $\leqslant 1$ (with respect to a 'rest frame', linked to the medium). The two forbidden rectangles are now linear obstacles in the medium, with a bounded duration in time. Thirdly, our figures can be viewed as execution paths of concurrent automata subject to some conflict of resources, as in [FGR], fig. 14.

In all these cases, the fundamental category distinguishes between obstructions (islands, temporary obstacles, conflict of resources) which occur *essentially* together (in the earlier diagram on the left) or one after the other, *in a discernible way* (on the right). On the other hand, the underlying topological spaces are homeomorphic, and topology, or algebraic topology, cannot distinguish these two situations.

We also note that all the fundamental monoids $\uparrow\pi_1(X, x_0)$ are trivial: as a striking difference from the classical case, the fundamental monoids can carry a very minor part of the information of the fundamental category $\uparrow\Pi_1(X)$.

4.3.4 Spaces with distinguished paths and partial maps

The main setting studied in [G8] is in fact a richer one, the category dTop of *spaces with distinguished paths*, where an object can have *vortices* (i.e. non-reversible loops). This framework, introduced in [G7], is also used in many articles, by different authors.

A *d-space* X, or *space with distinguished paths*, is a topological space equipped with a set dX of (continuous) maps $a \colon \mathbb{I} \to X$, called *distinguished paths*, or *directed paths*, or *d-paths*, satisfying three axioms:

(i) (*Constant paths*) every constant map $\mathbb{I} \to X$ is directed,

(ii) (*Partial reparametrisation*) dX is closed under composition with every monotone map $\mathbb{I} \to \mathbb{I}$,

(iii) (*Concatenation*) dX is closed under path-concatenation: if the d-paths a, b are consecutive in X ($a(1) = b(0)$), then their concatenation $a * b$ is also a d-path.

A *map of d-spaces*, or *d-map*, is a continuous mapping $f \colon X \to Y$ between d-spaces which preserves the directed paths: if $a \in dX$, then $fa \in dY$. They form the category dTop. Products and equalisers, sums and coequalisers in dTop are easily constructed as the corresponding limits or colimits in Top, equipped with the adequate d-structure.

A preordered space X will always be viewed as a d-space, where a directed path is any map $a \colon {\uparrow}\mathbb{I} \to X$ in pTop. We have two full embeddings

$$\mathsf{Top} \subset \mathsf{pTop} \subset \mathsf{dTop}. \tag{4.47}$$

Reversing d-paths yields the *opposite* d-space $RX = X^{\mathrm{op}}$, where $a \in d(X^{\mathrm{op}})$ if and only if $a^{\sharp}$ is in dX. This defines the *reversor* endo-functor

$$R \colon \mathsf{dTop} \to \mathsf{dTop}, \qquad RX = X^{\mathrm{op}}. \tag{4.48}$$

The d-space X is said to be *reversive* if it is isomorphic to X^{op}. More particularly, it is *reversible* if $X = X^{\mathrm{op}}$, i.e. if its distinguished paths are closed under reversion.

Technically it is useful to consider also Moore paths $a \colon [t_0, t_1] \to X$, defined on any non-trivial compact interval; obviously, we say that a is directed if the standard path $a\varphi \colon \mathbb{I} \to X$ is, where φ is the affine homeomorphism

$$\varphi \colon \mathbb{I} \to [t_0, t_1], \qquad t \mapsto t_0 + t/(t_1 - t_0), \tag{4.49}$$

or, equivalently, any isotone homeomorphism $\mathbb{I} \to [t_0, t_1]$.

Other 'topological' settings for directed algebraic topology are studied in [G8]. In particular, cubical sets yield a richer directed structure, if a more complex one: it lacks concatenation of paths, and is studied via a realisation functor with values in dTop.

Extending the relationship Top $\subset$ $\mathcal{C}$, the category dTop of d-spaces is a subcategory of the category $d\mathcal{C}$ *of d-spaces and partial d-maps*, where a morphism $f\colon X \rightarrowtail Y$ is a partial map, defined on an open subspace, which preserves d-paths: if $a\colon \mathbb{I} \to X$ is a directed path with image in Def f, then $fa\colon \mathbb{I} \to Y$ is a directed path of Y.

Forgetting the d-structure we have a faithful functor

$$d\mathcal{C} \to \mathcal{C}. \tag{4.50}$$

We shall prove, in Theorem 4.3.6, that $d\mathcal{C}$ has a gluing e-structure, lifted from $\mathcal{C}$.

4.3.5 Standard models

The euclidean spaces $\mathbb{R}^n$, I^n, $\mathbb{S}^n$ will have their *natural* (reversible) d-structure, where all paths are distinguished. $\mathbb{I}$ is the natural interval.

The *directed real line*, or *d-line* $\uparrow\mathbb{R}$, is the euclidean line with the natural order; its directed paths are the increasing maps $\mathbb{I} \to \mathbb{R}$. Its cartesian power in dTop (and pTop) is the *n-dimensional real d-space* $\uparrow\mathbb{R}^n$, with the cartesian order ($x \leqslant x'$ if and only if $x_i \leqslant x'_i$ for all i). The *standard d-interval* $\uparrow\mathbb{I} = \uparrow[0, 1]$ has the subspace structure of the d-line; the *standard d-cube* $\uparrow\mathbb{I}^n$ is its *n*-th power, and a subspace of $\uparrow\mathbb{R}^n$.

These d-spaces are reversive (isomorphic to their opposite); in particular, the canonical reflecting isomorphism

$$r\colon \uparrow\mathbb{I} \to R(\uparrow\mathbb{I}), \qquad t \mapsto 1 - t, \tag{4.51}$$

reflects paths and homotopies into the opposite d-space.

The *standard directed circle* $\uparrow\mathbb{S}^1$ is the standard circle with the *anticlockwise structure*, where the directed paths $a\colon \mathbb{I} \to \mathbb{S}^1$ move this way, in the oriented plane $\mathbb{R}^2$: $a(t) = (\cos \vartheta(t), \sin \vartheta(t))$, with a monotone continuous argument $\vartheta\colon \mathbb{I} \to \mathbb{R}$

$$\uparrow\mathbb{S}^1 \tag{4.52}$$

$\uparrow\mathbb{S}^1$ can be obtained as the coequaliser in dTop of the following pair of maps

$$\partial^0, \partial^1\colon \{*\} \rightrightarrows \uparrow\mathbb{I}, \qquad \partial^0(*) = 0, \quad \partial^1(*) = 1. \tag{4.53}$$

Indeed, as in 2.5.2(a), the coequaliser is the quotient $\uparrow\mathbb{I}/\partial\mathbb{I}$, which identifies the endpoints, with the topology of $\mathbb{S}^1$; the quotient d-structure, generated by the projected paths, is the desired one – because of the axioms on concatenation and reparametrisation of d-paths.

The *directed n-dimensional sphere* is defined, for $n > 0$, as the quotient of the directed cube $\uparrow\mathbb{I}^n$ modulo its (ordinary) boundary $\partial\mathbb{I}^n$

$$\uparrow\mathbb{S}^n = (\uparrow\mathbb{I}^n)/(\partial\mathbb{I}^n) \qquad (n > 0), \tag{4.54}$$

while $\uparrow\mathbb{S}^0 = \{-1, 1\}$ has the discrete topology and the unique d-structure (which is discrete, and corresponds to the discrete order, see 4.3.7).

The directed n-dimensional torus is the n-th cartesian power of the directed circle

$$\uparrow\mathbb{T}^n = \uparrow\mathbb{S}^1 \times ... \times \uparrow\mathbb{S}^1. \tag{4.55}$$

The directed circle can also be described as an orbit space (see 2.5.6)

$$\uparrow\mathbb{S}^1 = \uparrow\mathbb{R}/\mathbb{Z}, \tag{4.56}$$

with respect to the action of the group of integers on the directed line $\uparrow\mathbb{R}$ (by translations). The distinguished paths of $\uparrow\mathbb{S}^1$ are simply the projections of the increasing paths $\vartheta \colon \mathbb{I} \to \mathbb{R}$ in the line.

Remarks and complements. (a) The shortcoming of pTop in expressing directed spaces is evident here: the coequaliser of diagram (4.53) in pTop is $\mathbb{S}^1$ with the chaotic preorder, as already seen in Exercise 1.7.8(f). In the category of ordered topological spaces we even get the singleton.

*(b) The d-spaces $\uparrow\mathbb{S}^n$ can be obtained as suspensions in dTop of the pointed d-space $\uparrow\mathbb{S}^0$ ([G8], Section 1.7.5), as in the classical case recalled in 2.6.2. (Unpointed suspension gives different interesting structures: see 4.5.4(f).)

4.3.6 Theorem

The category $d\mathcal{C}$ of d-spaces and partial d-maps (introduced in 4.3.4) is a gluing e-category, with structure lifted by the forgetful functor $d\mathcal{C} \to \mathcal{C}$.

Proof We verify conditions 3.6.9(i)–(iii) for lifting a gluing e-structure.

(a) Given a partial d-map $f \colon X \to Y$ in $d\mathcal{C}$, its support $\underline{e}(f) \colon X \to X$ in $\mathcal{C}$ is the partial identity on the open subspace Def f; it obviously preserves any directed path with image in Def f.

(b) Given a family of partial d-maps $f_i \colon X \to Y$ and their linked join $f = \bigvee f_i \colon X \to Y$ in $\mathcal{C}$, we have to prove that f is also a partial d-map.

If $a: \mathbb{I} \to X$ is a directed path with image in Def f, the subsets $U_i = a^{-1}(\text{Def } f_i)$ form an open cover of $\mathbb{I}$, a compact metric space. Choosing a Lebesgue number $\varepsilon > 0$ for this cover, every subinterval of $\mathbb{I}$ of a smaller length is contained in some U_i, and $a(U_i) \subset \text{Def } f_i$. Letting $n > 1/\varepsilon$, we decompose a in an n-ary 'regular' concatenation of paths

$$a = a_1 * a_2 * \ldots * a_n,$$

$$a_k: \mathbb{I} \to X, \qquad a_k(t) = a((t + k - 1)/n), \qquad (4.57)$$

$$a_k(\mathbb{I}) = a[(k-1)/n, \, k/n] \qquad (k = 1, \ldots, n).$$

Each path a_k has image contained in some Def f_i. All paths $f a_k$ are directed in Y, and their concatenation $f a$ is also.

(c) Given a manifold $(X_i, u_j^i)_I$ on $d\mathcal{C}$, we take the topological space X, which is the gluing of the underlying manifold in $\mathcal{C}$, with the open embeddings $u^i: X_i \to X$ interpreted as inclusions of open subspaces. We let $u_i = u^{i\sharp}: X \rightarrowtail X_i$, so that $u_j^i = u_j u^i: X_i \rightarrowtail X_j$ is the restriction of $u^j u_j u^i u_i: X \rightarrowtail X$; the latter, being the partial identity on $X_i \cap X_j$, preserves and reflects directed paths.

We define a d-structure on X saying that a path $a: \mathbb{I} \to X$ is directed if there is a subdivision of $\mathbb{I}$

$$0 = t_0 < t_1 < \ldots < t_n = 1, \qquad (4.58)$$

such that each restricted Moore path $a_k: [t_{k-1}, t_k] \to X$ has image in some X_i, and is directed there. (This does not depend on the choice of X_i, by the previous remark.)

The axioms on constant paths and binary concatenations, in 4.3.4, are obviously fulfilled.

For the third axiom, let $a: \mathbb{I} \to X$ be directed, and choose a subdivision $(t_k)_{0 \leqslant k \leqslant n}$ of the standard interval fulfilling the previous condition. Let $\varphi: \mathbb{I} \to \mathbb{I}$ be a monotone map, with $\varphi(\mathbb{I}) = [t', t'']$. We fix the integers $k' < k''$ so that

$$t_{k'-1} \leqslant t' < t_{k'}, \qquad t_{k''-1} < t'' \leqslant t_{k''}.$$

Let $m = k'' - k' \geqslant 1$ (and note that $t_{k'} = t_{k''-1}$ for $m = 1$). The $m + 2$ points

$$t', \quad t_{k'}, \quad t_{k'+1}, \quad \ldots, \quad t_{k''-1}, \quad t'',$$

form a subdivision of the interval $\varphi(\mathbb{I})$, and we can choose a subdivision $s_0, \ldots, s_{m+1}$ of $\mathbb{I}$ whose φ-image is the former subdivision

$$0 = s_0 < s_1 < \ldots < s_m < s_{m+1} = 1,$$

$$\varphi(s_0) = t', \quad \varphi(s_1) = t_{k'}, \quad \ldots, \quad \varphi(s_m) = t_{k''-1}, \quad \varphi(s_{m+1}) = t''.$$

(The choice is not determined, generally, as φ need not be injective.)

Now the map $\varphi \colon \mathbb{I} \to \mathbb{I}$ restricted to each interval $[s_{h-1}, s_h]$ of the new subdivision has image contained in some interval $[t_{k-1}, t_k]$ of the old one; therefore the path $a\varphi \colon \mathbb{I} \to X$ restricted to each interval $[s_{h-1}, s_h]$ is a reparametrisation of a restriction $a\|[t_{k-1}, t_k]$, which takes values in some X_i and is directed there.

With this structure, the d-space X is the gluing of the given manifold, in $d\mathcal{C}$. $\qquad\qquad\square$

4.3.7 Comparing preordered spaces and d-spaces

The interplay between pTop and dTop is described (in [G8], 1.4.5) by two adjoint functors

$$\mathsf{p} \colon \mathsf{dTop} \rightleftarrows \mathsf{pTop} \colon \mathsf{d}, \qquad\qquad \mathsf{p} \dashv \mathsf{d}. \qquad\qquad (4.59)$$

The functor d equips a preordered space Y with the (monotone) maps $\uparrow\mathbb{I} \to Y$ as distinguished paths, while p provides a d-space X with the *path preorder* $x \preceq x'$ (x' is *reachable* from x), meaning that there exists a distinguished path from x to x'.

Both functors act 'identically' on the morphisms: for instance, if $f \colon Y \to Y'$ is in pTop, the morphism $f = \mathsf{d}(f) \colon \mathsf{d}Y \to \mathsf{d}Y'$ is the same continuous mapping between the associated d-spaces. (More precisely, the identity functor of Top underlies both functors.)

The adjunction is expressed by a natural bijection, which acts as an identity on the morphisms (more precisely, on the underlying morphisms of Top)

$$\varphi_{XY} \colon \mathsf{pTop}(\mathsf{p}X, Y) \to \mathsf{dTop}(X, \mathsf{d}Y),$$
$$(f \colon \mathsf{p}X \to Y) \mapsto (f \colon X \to \mathsf{d}Y). \qquad\qquad (4.60)$$

In fact, if $f \colon X \to \mathsf{d}Y$ is a d-map, the relation $x \preceq x'$ in X implies the existence of a d-path in X from x to x', whose f-image is a monotone path in Y, and implies that $f(x) \prec_Y f(x')$.

Conversely, we suppose that $f \colon \mathsf{p}X \to Y$ is a map of preordered spaces: if $x \preceq x'$ then $fx \prec_Y fx'$ in Y. We have to prove that f is also a d-map $X \to \mathsf{d}Y$. Let a be a d-path in X; if $t \leqslant t'$ in $\mathbb{I}$, the restriction of a to the interval $[t, t']$ proves that $a(t) \preceq a(t')$, and therefore $fa(t) \prec_Y fa(t')$; therefore fa is a monotone path in Y, and a d-path in $\mathsf{d}Y$.

The functor $\mathsf{d} \colon \mathsf{pTop} \to \mathsf{dTop}$ preserves limits (as any right adjoint) but does not preserve colimits: the coequaliser of the endpoints $\{*\} \rightrightarrows \uparrow\mathbb{I}$ gives the circle with chaotic preorder in pTop, and the directed circle $\uparrow\mathbb{S}^1$ in dTop.

A d-space will be said to be *of preorder type* (resp. *of order type*) if it can be obtained, as above, from a preordered (resp. ordered) space. Thus the d-spaces $\uparrow\mathbb{R}^n$, $\uparrow\mathbb{I}^n$ are of order type; $\mathbb{R}^n$, $\mathbb{I}^n$ and $\mathbb{S}^n$ are of *chaotic preorder type*; the product $\mathbb{R} \times \uparrow\mathbb{R}$ is of preorder type. The d-space $\uparrow\mathbb{S}^1$ is not of preorder type.

A preordered space will be said to be *path-preordered* if it can be obtained, as above, from a d-space, or equivalently if its preorder coincides with the path preorder $\preceq$. Every convex subspace of $\uparrow\mathbb{R}^n$ inherits a path order; a disconnected subspace inherits an order relation which is not path ordered, generally.

Comments and complements. (a) The reader will note that the subspace $X = [0, 1] \cup [2, 3]$ inherits from $\uparrow\mathbb{R}$ an ordering where the relation $1 < 2$ seems to have no topological interest. The associated path-preordered space $\mathsf{pd}X$, where we only keep the restricted order on each connected component, seems to be more interesting. Similar facts will reappear many times, below.

(b) The adjunction $\mathsf{p} \dashv \mathsf{d}$ restricts to an isomorphism between:

- the full subcategory of dTop of d-spaces of preorder type,

- the full subcategory of pTop of path-preordered spaces.

4.3.8 *Complements

All this becomes clearer at the light of the theory of idempotent adjunctions, reviewed in 1.8.6: in fact the adjunction $\mathsf{p} \dashv \mathsf{d}$ is idempotent, in a strict sense.

The components of the unit and counit

$$\eta_X : X \to \mathsf{dp}X, \qquad \varepsilon_Y : \mathsf{pd}Y \to Y \qquad (X \text{ in } \mathsf{dTop}, Y \text{ in } \mathsf{pTop}), \qquad (4.61)$$

'are' identities of topological spaces. Moreover, the first says that every d-path in X is also a d-path in $\mathsf{dp}X$, the second says that $y \preceq y'$ in $\mathsf{pd}Y$ implies $y \prec y'$ in Y.

The morphisms of pTop

$$\mathsf{p}\eta_X : \mathsf{p}X \to \mathsf{pdp}X, \qquad \varepsilon_{\mathsf{p}X} : \mathsf{pdp}X \to \mathsf{p}X, \qquad (4.62)$$

prove that the preordered spaces $\mathsf{p}X$ and $\mathsf{pdp}X$ are (precisely) the same; similarly, the d-space $\mathsf{d}Y$ coincides with $\mathsf{dpd}Y$. The four natural transformations $\mathsf{p}\eta$, $\varepsilon\mathsf{p}$, $\eta\mathsf{d}$, $\mathsf{d}\varepsilon$ are identities, and the adjunction $\mathsf{p} \dashv \mathsf{d}$ is strictly idempotent.

As in 1.8.6, the adjunction restricts to an isomorphism

$$\mathsf{p} \colon \mathrm{Alg}(\mathsf{dTop}) \rightleftarrows \mathrm{Coalg}(\mathsf{pTop}) \colon \mathsf{d}, \qquad (4.63)$$

between:

- the reflective full subcategory of algebraic objects in dTop, namely the d-spaces X such that the 'topological identity' $\eta_X \colon X \to \mathsf{dp}X$ is invertible (and therefore an identity), i.e. *the d-spaces of preorder type*,

- the coreflective full subcategory of coalgebraic objects in pTop, namely the preordered spaces Y such that $\varepsilon_Y \colon \mathsf{pd}Y \to Y$ is the identity, i.e. *the path-preordered spaces*.

4.4 Locally cartesian ordered manifolds

In this section we present a (new) notion of 'locally cartesian ordered manifold', where every point has a neighbourhood isomorphic to $\uparrow\mathbb{R}^n$, the cartesian ordered euclidean n-space, and the fibres of the tangent bundle are ordered vector spaces.

This can include the directed circle $\uparrow\mathbb{S}^1$, the directed torus and a directed structure on the Klein bottle. It cannot involve the higher directed spheres $\uparrow\mathbb{S}^n$, because of the presence of 'pointlike vortices' in their directed structure.

More general notions of 'directed manifolds' could be developed, to take this into account; but this should rely on a research of the most useful approach, or approaches.

4.4.1 Reviewing the directed spheres

(a) (*The directed circle*) Our notion of 'locally cartesian ordered manifold' of class C^r will be based on the ordered spaces $\uparrow\mathbb{R}^n$ together with 'locally monotone' maps, more general than the monotone ones.

To see why we need them, we consider (informally) the directed circle $\uparrow\mathbb{S}^1$, as produced by an intrinsic atlas with two charts

$$
\begin{array}{ccc}
U_1 & u_2^1 & U_2 \\[-2pt]
\rule{3.5cm}{0.4pt} & \xrightarrow{\ \bullet\ } & \rule{3.5cm}{0.4pt} \\[-6pt]
A \qquad\quad B & & B' \qquad\quad A'
\end{array}
\qquad (4.64)
$$

as in (4.1), but U_1 and U_2 are now open intervals of the *ordered* line $\uparrow\mathbb{R}$.

The atlas is determined by the transition morphism $u_2^1 \colon U_1 \rightsquigarrow U_2$, a partial homeomorphism which takes the open subinterval B (resp. A) to the open subinterval B' (resp. A'). The directed structure of $\uparrow\mathbb{S}^1$ is the quotient of the (obvious) directed structure of the ordered space $\uparrow U_1 + \uparrow U_2$, modulo the equivalence relation generated by u_2^1: we identify each $x \in A \cup B$ with $u_2^1(x) \in A' \cup B'$.

Now, the partial mapping u_2^1 is isotone on each connected component of Def u_2^1, but is not monotone on the latter (as an ordered subspace of U_1): A precedes B in U_1, while B' precedes A' in U_2.

True, we can also realise $\uparrow\mathbb{S}^1$ by an atlas with three charts (or more), where all transitions maps u_j^i are defined on subintervals and monotone, as suggested by the following picture

$$
\begin{array}{llllll}
\underline{\quad U_1 \quad} & & \underline{\quad U_2 \quad} & & \underline{\quad U_3 \quad} & \\
A \qquad\quad B & & B' \qquad\quad C & & C' \qquad\quad A' &
\end{array}
\qquad (4.65)
$$

But the problem is still there: the category $p\mathcal{C}$ of preordered topological spaces and monotone partial maps defined on open subsets, with its obvious e-structure lifted from $\mathcal{C}$, is not even finitely cohesive: the *locally monotone* partial map $u_2^1: U_1 \rightarrowtail U_2$ of (4.64) is the join of two monotone partial mappings, which have no monotone join.

(b) (*Higher directed spheres*) A similar presentation, as a space 'locally isomorphic' to $\uparrow\mathbb{R}^n$, is impossible for the higher directed spheres, because of the presence of 'point-like vortices'.

We say that the d-space X has a *point-like vortex* at x if every neighbourhood of x in X contains some non-reversible loop ([G8], Section 1.4.7). This cannot happen in $\uparrow\mathbb{S}^1$, but it is easy to realise a directed disc with a point-like vortex at the centre

$$(4.66)$$

All higher directed spheres $\uparrow\mathbb{S}^n$, for $n \geqslant 2$, have a point-like vortex at the class $[0]$ (of the boundary points), as showed by the following sequence of 'arbitrarily small' non-reversible loops in $\uparrow\mathbb{S}^2 = (\uparrow\mathbb{I}^2)/(\partial\mathbb{I}^2)$, the quotient of the directed square that collapses the boundary

(4.67)

Since every point $x \neq [0]$ of $\uparrow\mathbb{S}^n$ has a neighbourhood isomorphic to $\uparrow\mathbb{R}^n$, this also shows that $\uparrow\mathbb{S}^n$ is not locally isomorphic to any fixed 'model'. (But of course one could use diverse models of dimension n: for instance, all the d-spaces underlying the preordered real vector spaces of that dimension.)

4.4.2 Locally monotone continuous mappings

We extend the category pTop to the category lmTop *of preordered topological spaces and locally monotone continuous mappings.*

A morphism $f \colon X \to Y$, also called an *lm-map*, is a continuous mapping such that every $x \in X$ has a neighbourhood U on which f is monotone.

Equivalently, for every $x \in X$ and every nbd V of $f(x)$ in Y there exists a nbd U of x in X such that f restricts to a monotone mapping $U \to V$. Equivalently again, there exists an open cover (U_i) of X such that f is continuous and monotone on each U_i.

The morphism $f \colon X \to Y$ is invertible in lmTop if and only if it is a homeomorphism and its inverse f^{-1} is locally monotone. This is the same as a homeomorphism which is *locally isotone*: every $x \in X$ has a nbd U on which f restricts to an isotone mapping $U \to f(U)$ between ordered subsets of X and Y.

The category lmTop has *all sums, finite products* and *equalisers,* constructed as in Top, and equipped with the finest or coarsest preorder consistent with the structural maps. This is obvious for sums and equalisers, and examined below for finite products.

Exercises and complements. (a) A topological space X can admit different preorders, say $\prec$ and $\prec'$, which give 'the same lm-structure', in the sense that the map $\mathrm{id}\, X$ is locally isotone from $(X, \prec)$ to $(X, \prec')$.

One may prefer to 'identify' the objects $(X, \prec)$ and $(X, \prec')$, under the previous condition; but this would simply produce an equivalent category.

(b) The category lmTop has finite products, the same as in pTop.

(c) The forgetful functor $U \colon$ lmTop $\to$ Top has both a left and a right adjoint, $D \dashv U \dashv C$ where DS (resp. CS) is the space S equipped with the discrete order (resp. the indiscrete preorder).

(d) If X is a path-preordered space (see 4.3.7), any locally monotone map $f: X \to Y$ (in lmTop) is monotone.

4.4.3 Manifolds and local cartesian orders

We now introduce the category $lm\mathcal{C}$ of *preordered topological spaces and partial lm-maps defined on an open subspace*. By this we mean that the partial mapping $f: X \to Y$ is defined on an open subset Def f of X, continuous and locally monotone on it. Below, the term *partial lm-map* will have this meaning, leaving understood that the definition-set is open in the domain.

It is a totally cohesive e-category, with structure lifted by the forgetful functor $U: lm\mathcal{C} \to \mathcal{C}$. The support $\underline{e}(f)$ of a partial lm-map $f: X \to Y$ is the partial identity on the open subspace Def $f \subset X$. The topological join f of a linked family of partial lm-maps $f_i: X \to Y$ is a partial lm-map: if $x \in$ Def $f = \bigcup$ Def f_i, some f_i is defined on x and monotone on some subset U, open in Def f_i and therefore in X; but f coincides with f_i on U, and is monotone there.

We also have the totally cohesive e-subcategory $lm\mathcal{C}^r \subset lm\mathcal{C}$ of cartesian ordered open euclidean spaces *and partial lm-maps of class* C^r (defined on an open subspace). An object is an open ordered subspace of some $\uparrow\mathbb{R}^n$ (with cartesian order); a morphism $f: U \to V$ is a partial map defined on an open subset of U, on which f is locally monotone and of class C^r.

When useful, we extend the objects of $lm\mathcal{C}^r$: an object is any ordered space equipped with an isotone C^r-diffeomorphism with an object of the previous type.

Its gluing completion Mf $(lm\mathcal{C}^r)$ is, by definition, the category of *locally cartesian ordered C^r-manifolds and partial lm-maps of class* C^r.

The gluing completion of the forgetful functor $lm\mathcal{C}^r \to d\mathcal{C}$

$$\text{Mf } lm\mathcal{C}^r \to d\mathcal{C} \qquad (4.68)$$

will be called the *d-space realisation of a locally cartesian ordered manifold*. (We have seen in Theorem 4.3.6 that $d\mathcal{C}$ is a gluing e-category.)

4.4.4 The locally cartesian ordered circle

(a) We have already seen, in (4.64) and (4.65), two intrinsic atlases on $lm\mathcal{C}^\infty$, whose d-space realisation in $d\mathcal{C}$ is the directed circle.

All this can be easily extended to the n-dimensional torus $\uparrow\mathbb{T}^n$, a cartesian power of ↑$\mathbb{S}^1$. (We have also seen that the higher directed spheres cannot be given this structure.)

(b) Alternatively, we can use a procedure depending on group actions (reviewed in 2.5.6).

The circle $\mathbb{S}^1$ is the orbit space $\mathbb{R}/\mathbb{Z}$ of the action of the group $\mathbb{Z}$ on the ordered euclidean line, by translations: $x \mapsto x + k$ (see 2.5.6(b)). The group $\mathbb{Z}$ acts isotonically on $\mathbb{R}$: any operator gives an isomorphism of ordered topological spaces. Making use of a stronger condition 4.4.5(ii), the next theorem shows that ↑$\mathbb{S}^1$ is the gluing of a locally cartesian ordered manifold.

This stronger condition is essentially based on the fact that, if in (a) we use 'small open arcs' (lesser than half-circles), the intersection of two of them is always connected (possibly empty).

The same procedure will be used to define a locally cartesian ordered structure on the torus and the Klein bottle; the last point is the goal of this second part of the section.

4.4.5 Theorem

We have an ordered topological space X and a group G which acts (isotonically) on it: each operator $g \in G$ has a continuous monotone action $X \to X$, $x \mapsto gx$ (an isomorphism of ordered topological spaces). We let $p: X \to B$ be the projection onto the orbit space $B = X/G$.

(a) We suppose that this is a covering space *action, in the sense that:*

(i) each $x \in X$ has an open nbd U_x such that $U_x \cap gU_x = \emptyset$, for all $g \neq 1$.

Then the projection p is a covering map (cf. 4.2.6).

(b) Suppose now that X is an open euclidean space, with cartesian order.

The following condition, stronger than (i), ensures that the orbit space B is the realisation of a locally cartesian ordered C^∞-manifold, that is an object of $\mathrm{Mf}\,(lm\mathcal{C}^\infty)$:

(ii) for each $x \in X$ we can choose an open nbd U_x, so that:

- $U_{gx} = gU_x$ (for $g \in G$),

- if $U_x \cap U_y \neq \emptyset$, the set $U = U_x \cup U_y$ satisfies $U \cap gU = \emptyset$, for all $g \neq 1$.

Proof (a) First, let us recall that the projection $p: X \to B = X/G$ onto the orbit space is an open mapping (see (2.61)).

For every $a \in B$ we choose a point x in the fibre $F_a = p^{-1}\{a\}$ and an open nbd U_x of x in X such that $U_x \cap gU_x = \emptyset$, for all $g \neq 1_G$.

Now $V_a = p(U_x)$ is an open nbd of a in B, which is evenly covered by p (as defined in 4.2.6): we have a disjoint union of open subspaces of X:

$$p^{-1}(V_a) = \bigcup_x gU_x \qquad (x \in F_a), \qquad (4.69)$$

and the projection p restricts to a family of homeomorphisms $gU_x \to V_a$.

(b) In the new hypotheses, the space X is an open euclidean space, with cartesian order. We want to prove that the orbit space $B = X/G$ is the realisation of a manifold $(V_a, (v_b^a))$ on lmC^∞, indexed by B itself.

We use a choice (U_x) of open nbds of the points of X that satisfies (ii), and we let $V_{px} = pU_x$, its homeomorphic image in B. For the sake of simplicity, we view V_{px} as a cartesian ordered open euclidean space (which is strictly true of U_x).

For $a, b \in B$, we let $v_b^a \colon V_a \dashrightarrow V_b$ be the partial identity on $V_a \cap V_b$, and we want to show that this partial diffeomorphism is isotone (where defined). If $V_a \cap V_b = \emptyset$, then v_b^a is empty and trivially isotone. Otherwise

$$p^{-1}(V_a \cup V_b) = p^{-1}(V_a) \cup p^{-1}(V_b)$$

is not empty, and we can choose some $x \in F_a$ and $y \in F_b$ such that $U_x \cap U_y \neq \emptyset$.

Now the open set $U = U_x \cup U_y$ has $U \cap gU = \emptyset$, for all $g \neq 1_G$. Therefore all the subsets $gU \subset X$ induce the same order relation on $pU = V_a \cup V_b$, and $V_a \cap V_b$ inherits the same ordering from V_a and V_b. $\qquad \square$

4.4.6 The locally ordered n-torus

We take the cartesian ordered euclidean space $X = \mathbb{R}^n$ with the right action of the group $G = \mathbb{Z}^n$, by translations: $x \mapsto x + h$ $(h \in \mathbb{Z}^n)$. The orbit space is the n-dimensional torus, a cartesian power of the circle

$$\mathbb{T}^n = (\mathbb{S}^1)^n = \mathbb{R}^n/\mathbb{Z}^n, \qquad (4.70)$$

and the circle itself for $n = 1$.

As a cartesian power of the circle, we already know that it is a locally cartesian ordered C^∞-manifold $\uparrow\mathbb{T}^n$, of dimension n. But this can also be deduced from the previous theorem.

We use the l_∞-norm and l_∞-metric on X, which is convenient because

$$d_\infty(x, x + h) = ||h||_\infty = \max |h_i|. \qquad (4.71)$$

To verify condition 4.4.5(ii) we choose:

$$U_x = \{x' \in \mathbb{R}^n \mid d_\infty(x, x') < 1/4\},$$
$$x' \in U_x \iff |x_i - x_i'| < 1/4, \text{ for } i = 1, ..., n. \qquad (4.72)$$

Suppose that $U_x \cap U_y \neq \emptyset$, so that $d_\infty(x, y) < 1/2$ and the l_∞-distance of two points $z, z' \in U = U_x \cup U_y$ is (strictly) smaller than $1/4 + 1/2 + 1/4 = 1$. Suppose also that $U \cap (U + h) \neq \emptyset$ for some $h \in \mathbb{Z}^n$: there is thus some $z \in U$ with $z + h \in U$ and $||h||_\infty = d_\infty(z, z + h) < 1$. Therefore $h = 0$.

Theorem 4.4.5(b) gives the result we are looking for.

4.4.7 Reviewing the Klein bottle

The *Klein bottle* $\mathbb{K}$ is a non-orientable compact surface. It has a classical model in $\mathbb{R}^4$, that can be effectively drawn in the plane, as one can see in many books and sites (e.g. in [Hat], p. 19).

We shall use the well-known presentation of $\mathbb{K}$ as an orbit space

$$p: \mathbb{R}^2 \to \mathbb{K} = \mathbb{R}^2/G, \tag{4.73}$$

of the euclidean plane under the action of a (non-commutative) group of transformations G of $\mathbb{R}^2$. G is generated by two operators $a, b: \mathbb{R}^2 \to \mathbb{R}^2$

$$a(x, y) = (x + 1, -y), \qquad b(x, y) = (x, y + 1), \tag{4.74}$$

under the relation $bab = a$. Each operator $g \in G$ can be written as $g = a^h b^k$ $(h, k \in \mathbb{Z})$.

In fact, in a product of operators a, a^{-1}, b, b^{-1}, all terms b and b^{-1} can be taken to the right:

$$ba = ab^{-1}, \qquad b^{-1}a = ab, \qquad ba^{-1} = a^{-1}b^{-1}, \qquad b^{-1}a^{-1} = a^{-1}b.$$

Now, the operator $g = a^h b^k$ gives:

$$g(x, y) = (x + h, (-1)^h(y + k)), \tag{4.75}$$

so that, for the l_∞-metric on $\mathbb{R}^2$:

$$d_\infty((x, y), g(x, y)) = \begin{cases} \max(|h|, |k|), & \text{for } h \text{ even,} \\ \max(|h|, |2y + k|), & \text{for } h \text{ odd.} \end{cases} \tag{4.76}$$

We remark that:

$$d_\infty((x, y), g(x, y)) < 1 \implies g = 1. \tag{4.77}$$

In fact, in this hypothesis, the formulas (4.76) imply that $h = 0$; but then the first formula applies, and also $k = 0$.

4.4.8 The locally ordered Klein bottle

We want to make this surface into a locally cartesian ordered manifold $\uparrow\mathbb{K}$. We take on $\mathbb{R}^2$ the order relation of (4.45)

$$(x, y) \leqslant (x', y') \quad \Leftrightarrow \quad |y' - y| \leqslant x' - x. \tag{4.78}$$

For this ordering, the operators a and b are isotone (as easily proved, analytically), and all the operators of G are.

This is made evident by the following figure

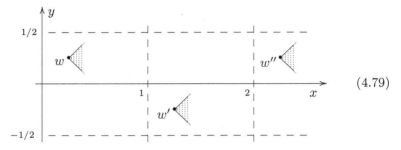

$$\tag{4.79}$$

which shows the points w, $w' = a(w)$ and $w'' = a(w')$ and their future-cone, preserved by the operator a. The operator b is a vertical translation, obviously isotone.

The plane, ordered by the relation (4.78), is cartesian ordered, being isomorphic to the naturally ordered $\mathbb{R}^2$, by a rotation of $45°$ around the origin.

We prove now that $\mathbb{K}$ inherits a locally cartesian ordered structure from the plane, applying Theorem 4.4.5(b). As in (4.72) (for the torus), we choose, for every point $w \in \mathbb{R}^2$, the neighbourhood

$$U_w = \{w' \in \mathbb{R}^2 \mid d_\infty(w, w') < 1/4\}. \tag{4.80}$$

Suppose that $U_w \cap U_{w'} \neq \emptyset$, so that $d_\infty(w, w') < 1/2$ and the l_∞-distance of two points $z, z' \in U = U_w \cup U_{w'}$ is (strictly) smaller than $1/4 + 1/2 + 1/4 = 1$. Suppose also that $U \cap gU \neq \emptyset$ for some $g = a^h b^k \in G$: there is thus some $z \in U$ with $gz \in U$ and $d_\infty(z, gz) < 1$. By (4.77), $g = 1$.

Remarks and complements. (a) The compact square $S = [-1/2, 1/2]^2$ meets all the orbits of G. The space $\mathbb{K}$ is homeomorphic to the quotient of S modulo the equivalence relation generated by G.

This is another well-known presentation of the Klein bottle, as a quotient of the compact square modulo an equivalence relation which is trivial outside of the boundary of the square.

(b) The action of G preserves the vector field

$$V_x = (x, v) \in T_x \mathbb{R}^2 = \{x\} \times \mathbb{R}^2, \qquad v = (1, 0) \in \mathbb{R}^2,$$
$$(Ta)(x, v) = (a(x), v), \qquad (Tb)(x, y) = (b(x), v). \tag{4.81}$$

Therefore $\mathbb{K}$ inherits a never-null vector field: it can be 'combed'. (This is easily visualised on the usual model of $\mathbb{K}$ in $\mathbb{R}^4$, drawn in $\mathbb{R}^2$.)

(c) The 2-dimensional sphere, or any even-dimensional one, cannot be combed: an interesting fact, whose proof by singular homology is beautiful and not really difficult [HiW, Sp, Hat].*

(d) The fundamental group of $\mathbb{K}$ is isomorphic to G. The homology group $H_1(\mathbb{K})$ is isomorphic to the abelianised group G^{ab}, i.e. to $\mathbb{Z} \oplus \mathbb{Z}/2$.

4.4.9 *Ordered tangent bundle

Consider again, as in 4.2.5, the trivial tangent bundle functor, with $r \geqslant 1$

$$T: \mathcal{C}^r \to \mathcal{V}, \qquad U \mapsto U \times \mathbb{R}^{\dim U}, \qquad f \mapsto Tf,$$
$$Tf(x, h) = (fx, D_h f(x)), \quad \text{for } x \in \operatorname{Def} f \text{ and } h \in \mathbb{R}^{\dim U}. \tag{4.82}$$

We can enrich it, replacing the vector space $\mathbb{R}^{\dim U}$ with the cartesian ordered vector space $\uparrow\mathbb{R}^{\dim U}$

$$T: lm\mathcal{C}^r \to or\mathcal{V}, \qquad U \mapsto U \times \uparrow\mathbb{R}^{\dim U}, \qquad f \mapsto Tf, \tag{4.83}$$

where Tf is computed as above.

An object of $or\mathcal{V}$ is a trivial fibre bundle $p: B \times F \to B$, where B is a topological space and F is a cartesian ordered finite dimensional, real vector space, isotonically isomorphic to $\uparrow\mathbb{R}^n$. A morphism is a 'fibrewise linear monotone' morphism $f: B \times F \rightarrowtail B' \times F'$; in other words, in 4.2.4(ii), each mapping $f_2(b, -): F \to F'$ is linear monotone.

We should only verify this property of $Tf: TU \rightarrowtail TV$.

The morphism $f: U \rightarrowtail V$ in $lm\mathcal{C}^r$ is a partial map defined on an open subset of U of $\mathbb{R}^n$, on which f is locally monotone and of class C^r. For a vector $(x, h) \in U \times \mathbb{R}^n$ with $h \geqslant 0$, we can find a non-degenerate interval $J = [0, \varepsilon]$ such that the path $a(t) = x + th$ stays in $\operatorname{Def} f$, for $t \in J$, and moreover $fa: J \to V$ is monotone. Then the derivative of fa at 0 is a weakly positive vector of $\uparrow\mathbb{R}^{\dim U}$, and the tangent vector $(fx, D_h f(x))$ is also weakly positive.

As in 4.2.5, T extends to the *ordered tangent bundle functor* of locally cartesian ordered C^r-manifolds

$$T: \mathrm{Mf}\, lm\mathcal{C}^r \to \mathrm{Mf}\,(or\mathcal{V}) \qquad (r \geqslant 1),$$

with values in a category Mf (*or* $\mathcal{V}$) of *cartesian ordered vector bundles and partial maps.*

4.5 Locally preordered spaces

Locally cartesian ordered manifolds can be generalised by a convenient notion of space with a 'local preorder'. A good candidate, called 'circulation', was introduced by S. Krishnan [Kr].

4.5.1 Locally preordered spaces

A *locally preordered topological space*, or *lp-space*, will be a topological space X equipped with a *local preorder*, or a 'circulation' (as in [Kr]). This means that every open set U is equipped with a preorder relation $\prec_U$ such that:

(i) if $U = \bigcup_i U_i$ is a union of open subsets of X, the preorder $\prec_U$ is the join of the preorders $\prec_{U_i}$.

More explicitly, condition (i) means that:

(ii) if $U = \bigcup_i U_i$ is a union of open subsets of X, then $x \prec_U x'$ if and only if there exists a finite sequence in X

$$x = x_0, \ x_1, \ ..., \quad x_n = x', \tag{4.84}$$

where each term precedes the consecutive one in some U_i (according to its preorder $\prec_{U_i}$).

The sequence (4.84) can be viewed as a *$\mathcal{U}$-monotone path* in U, where $\mathcal{U}$ is an open cover of U.

Condition (i) can also be expressed by the conjunction of the following ones:

(iii') if $U \subset V$ and $x \prec_U x'$, then $x \prec_V x'$,

(iii'') if $U = \bigcup_i U_i$ is a union of open subsets of X and $x \prec_U x'$, there exists a sequence (4.84) where each term precedes the consecutive one in some U_i (according to $\prec_{U_i}$).

An *lp-map* $f: X \to Y$ is a continuous mapping between lp-spaces which preserves the local preorder, in the sense that, for every U open in X and V open in Y, if $f(U) \subset V$ and $x \prec_U x'$, then $f(x) \prec_V f(x')$. This defines the category lpTop.

If X is an lp-space, an open subspace $X' \subset X$ has an obvious lp-structure, formed by the family $(\prec_U)$ of preorders where U is open in X'. We say then that X' is an *open lp-subspace* of X.

Examples and complements. (a) The *standard locally preordered circle* $\text{lp}\mathbb{S}^1$ will be the standard circle with a local preorder defined by an orientation of the circle. Namely:

- if U is a non-total open arc of the circle, $x \prec_U x'$ means that we can go from x to x' following the orientation,

- if U is a disconnected open subset, each connected component is ordered as above, and no other relation in U is added,

- if $U = \mathbb{S}^1$, $x \prec_U x'$ is the chaotic preorder.

This structure plainly corresponds to the d-space $\uparrow\!\mathbb{S}^1$ defined in 4.3.5. (The correspondence will be made precise in 4.5.7.)

(b) In [G8], Subsection 1.9.3, the term 'local preorder' was used for a more general structure, called a 'precirculation' in [Kr], and defined by condition (iii'). In fact, a circulation works better here.

4.5.2 Comparing preordered spaces and lp-spaces

As in 4.3.7, we can describe the interplay between pTop and lpTop by an adjunction

$$F\colon \text{lpTop} \rightleftarrows \text{pTop}\colon G, \qquad F \dashv G. \tag{4.85}$$

The functor F takes the lp-space X to the space itself, preordered by $\prec_X$, and acts 'identically' on morphisms, at the topological level (by the same abuse of notation as in 4.3.7).

The functor G equips the preordered space Y with the greatest local preorder $(\prec_V)$ smaller than $\prec$. Explicitly, for every open subset V in Y, we let $y \prec_V y'$ if

(i) for all open covers (V_i) of V there exists a finite sequence

$$y = y_0 \prec y_1 \prec \ ... \ \prec y_n = y', \tag{4.86}$$

consistent with the cover (V_i), in the sense that each pair of consecutive terms belongs to some V_i.

Also G acts 'identically' on morphisms: if $f\colon Y \twoheadrightarrow Y'$ is a map of preordered spaces, then $Gf\colon GY \to GY'$ is the same continuous mapping, viewed as an lp-map (by Exercise (a)).

The adjunction is expressed by a natural bijection, which is an identity at the underlying topological level (Exercise (b))

$$\varphi_{XY}\colon \text{pTop}(FX, Y) \to \text{lpTop}(X, GY),$$
$$(f\colon FX \to Y) \mapsto (f\colon X \to GY). \tag{4.87}$$

The ordered spaces $\uparrow\mathbb{R}^n$ and $\uparrow\mathbb{I}^n$ have associated lp-spaces, written as

$$\mathrm{lp}\mathbb{R}^n = G(\uparrow\mathbb{R}^n), \qquad \mathrm{lp}\mathbb{I}^n = G(\uparrow\mathbb{I}^n). \tag{4.88}$$

As in 4.3.8, the adjunction $F \dashv G$ is idempotent (see 1.8.6): the topological identities

$$F\eta_X \colon FX \to FGFX, \qquad \varepsilon_{FX} \colon FGFX \to FX,$$

are both monotone, and prove that the preordered spaces FX and $FGFX$ are the same.

Therefore the adjunction induces an isomorphism between the category of algebras on lpTop (for the monad GF) and the category of coalgebras on pTop (for the comonad FG).

The coalgebras on pTop are examined below, as preordered spaces with a 'locally generated preorder'.

The algebras on lpTop are the lp-spaces X such that $\eta_X \colon X \to GFX$ is the identity: this means that each relation $\prec_U$ is locally generated by the restriction of $\prec_X$ to U. For instance, the lp-space $X = \mathrm{lp}\mathbb{S}^1$ (in 4.5.1) is not, because $\prec_X$ is the chaotic preorder and GFX has the chaotic preorder on each open arc.

Exercises. (a) Every map $f \colon Y \to Y'$ of preordered spaces is also an lp-map $GY \to GY'$.

(b) The mapping φ in (4.87) is well-defined and bijective.

4.5.3 Locally generated preorders

The counit $\varepsilon_Y \colon FGY \to Y$ of the previous adjunction, in (4.87), is the identity of the underlying space, and a monotone map: $y \prec_Y y'$ in FGY implies $y \prec y'$ in Y.

We say that the preordered space Y *has a locally generated preorder*, or that it is an *lg-space*, if $\varepsilon_Y \colon FGY \to Y$ is an identity of preordered spaces. Equivalently, the relation $y \prec y'$ in Y satisfies condition 4.5.2(i), with $V = Y$.

We write as lgTop the category of lg-spaces and monotone maps, a full subcategory of pTop. It is also a full subcategory of lmTop (defined in 4.4.2), as proved in Exercise 4.5.4(e).

Moreover, the faithful functor $G \colon \mathsf{pTop} \to \mathsf{lpTop}$ of (4.85) restricts to a functor

$$\underline{G} \colon \mathsf{lgTop} \to \mathsf{lpTop}, \qquad Y \mapsto G(Y), \tag{4.89}$$

which is injective on objects and therefore an embedding. In fact, if Y

and Y' are lg-spaces and $GY = GY'$, the identities ε_Y and $\varepsilon_{Y'}$ prove that $Y = FGY = FGY' = Y'$.

4.5.4 Exercises and complements

(a) The cartesian ordered space $\uparrow\mathbb{R}^n$ is an lg-space. The same holds of any convex subspace X, for the restricted order.

(b) Let X be a path-preordered space, as defined in 4.3.7): if $x \prec x'$ in X there is a monotone path $a\colon \uparrow\mathbb{I} \to X$ from x to x'. Then X is an lg-space.

(c) The ordered subspace $X = [0,1] \cup [2,3]$ of the euclidean line is not an lg-space.

*(d) Find an lg-space which is not path-preordered. *Hints:* one can use the topological subspace $Y \subset \mathbb{R}^2$ often presented as a connected space which is not path-connected, and based on the sine function. Y is given the preorder where $(x,y) \prec (x',y')$ if $x \leqslant x'$ (as in the product $\uparrow\mathbb{R} \times \mathbb{R}$, where $\mathbb{R}$ has the chaotic preorder).

(e) Let $f\colon X \to Y$ be a locally monotone map of preordered spaces (see 4.4.2). If X is an lg-space, f is monotone.

*(f) A connected compact preordered space need not be an lg-space. *Hints:* use the standard circle C with the preorder $(x,y) \prec (x',y')$ if $y \leqslant y'$.

4.5.5 Preorders and partial maps

We have the category $lg\mathcal{C}$ of lg-spaces and *partial lg-maps*: by this we mean a partial map $f\colon X \nrightarrow Y$ defined on an open subspace $U = \mathrm{Def}\, f$, which is monotone for the preorder of $FG(U)$. (The difference with the restricted preorder can be seen in 4.5.4(c).)

We also have the category $lp\mathcal{C}$ of lp-spaces and monotone partial maps, defined on open subspaces (with their canonical structure of lp-spaces). The embedding $\underline{G}\colon \mathsf{lgTop} \to \mathsf{lpTop}$ of (4.89) can be extended to an embedding

$$\underline{G}\colon lg\mathcal{C} \to lp\mathcal{C}. \tag{4.90}$$

$\underline{G}$ is already defined on objects: an lg-space X is taken to the lp-space $G(X)$ defined in 4.5.2, where we let $x \prec_U x'$ if U is open in Y and

(i) for every open cover (U_i) of U there exists a sequence

$$x = x_0 \prec x_1 \prec \ldots \prec x_n = x'$$

consistent with the cover (U_i): each pair of consecutive terms belongs to some U_i.

A partial lg-map $f: X \nrightarrow Y$ is a total map $FG(U) \to Y$, and gives a total map $GU \to GY$, which is a monotone partial map $GX \nrightarrow GY$.

The embedding $\underline{G}: lgC \to lpC$ is consistent with the forgetful functors $lgC \to C$ and $lpC \to C$ (which forget all preorder relations).

4.5.6 Theorem (Locally preordered spaces as manifolds)

(a) The category lpC has a gluing e-structure, lifted by the forgetful functor lpC → C.

(b) The embedding $\underline{G}: lgC \to lpC$ is full, and gives to lgC a totally cohesive e-structure, lifted from lpC and from C.

(c) The embedding $\underline{G}: lgC \to lpC$ is the manifold completion of lgC.

Proof (a) We verify conditions 3.6.9(i)–(iii) for lifting a gluing e-structure. (The argument is similar to that of Theorem 4.3.6.)

(i) Given a partial lp-map $f: X \nrightarrow Y$ in lpC, its support $\underline{e}(f): X \nrightarrow X$ in C is the partial identity on the open subspace Def f, and obviously preserves the lp-structure of the latter.

(ii) Given a family of partial lp-maps $f_i: X \nrightarrow Y$ and their linked join $f = \bigvee f_i: X \nrightarrow Y$ in C, we have to prove that also f is a partial lp-map. For U open in Def f and V open in Y, suppose that $f(U) \subset V$ and $x \prec_U x'$.

Then U is covered by the open sets $U_i = U \cap \text{Def } f_i$, and there exists a sequence $x = x_0, x_1, ..., x_n = x'$ in X where each term precedes the consecutive one in some U_i (according to its preorder $\prec_{U_i}$). But f coincides with f_i on U_i, and we get a sequence $fx \prec_V fx_1 \prec_V ... \prec_V fx'$ in Y, showing that $fx \prec_V fx'$.

(iii) Given a manifold $(X_i, u_j^i)_I$ on lpC, we take the topological space X which is the gluing of the underlying manifold in C, with the open embeddings $u^i: X_i \to X$ interpreted as inclusions of open subspaces.

We let $u_i = u^{i\sharp}: X \nrightarrow X_i$, so that $u_j^i = u_j u^i: X_i \nrightarrow X_j$ is the restriction of $u^j u_j u^i u_i: X \nrightarrow X$; the latter, being the partial identity on $X_i \cap X_j$, preserves and reflects the lp-structure.

We take on X the lp-structure generated by its open cover (X_i), the finest that makes all the open embeddings $u^i: X_i \to X$ into lp-maps. Concretely, for each open subset U of X, the relation $x \prec_U x'$ means that there exists an open cover (U_j) of U where $U_j \subset X_{j(i)}$ and there exists a sequence $x = x_0, x_1, ..., x_n = x'$ in X where each term precedes the consecutive one in some U_j (according to its preorder as an open subset of $X_{j(i)}$).

With this structure, X is the gluing of the given manifold, in lpC.

(b) To prove that the embedding $\underline{G}\colon lg\mathcal{C} \to lp\mathcal{C}$ is full, we take two locally generated preordered spaces X, Y and a partial lp-map $f\colon X \dashrightarrow Y$ defined on an open subspace $U = \mathrm{Def}\, f$. Its restriction $f'\colon (U, \prec_U) \to (Y, \prec_Y)$ belongs to pTop.

Using the idempotent adjunction $F\colon \mathsf{lpTop} \rightleftarrows \mathsf{pTop}\colon G$ (in (4.85)), the lp-map $G(f')\colon G(U) \to G(Y)$ corresponds to a monotone map $FG(U) \to Y$, which is still f', topologically.

(c) We have to prove that $lg\mathcal{C}$ is gluing-dense in $lp\mathcal{C}$ (Theorem 3.5.8(e)).

Let X be an lp-space. We let I be the set of open sets of X, and $U_i = F(i) = (i, \prec_i)$ the corresponding lg-space. For $i \in I$, the inclusion $u^i\colon U_i \to F(X)$ is monotone, and the partial inverse $u_i = u^{i\sharp}\colon F(X) \dashrightarrow U_i$ is an lg-map (it is monotone on $FG(\mathrm{Def}\, u_i) = FG(U_i) = U_i$).

For $i, j \in I$, the partial map $u^i_j = u_j u^i\colon U_i \dashrightarrow U_j$ is also an lg-map. We have thus a manifold $U = (U_i, u^i_j)_I$ on $lg\mathcal{C}$. Its gluing in $lp\mathcal{C}$ is the topological space X with the finest lp-structure that makes all the embeddings $u^i\colon X_i \to X$ into lp-maps. This is the original structure of X. $\qquad\square$

4.5.7 *Exercises and complements

The comparison of the categories dTop, lpTop and pTop can be completed, forming a diagram of strictly idempotent adjunctions

$$\mathsf{dTop} \underset{G'}{\overset{F'}{\rightleftarrows}} \mathsf{pTop} \underset{G}{\overset{F}{\rightleftarrows}} \mathsf{pTop} \tag{4.91}$$

$$\mathsf{p} = FF'\colon \mathsf{dTop} \to \mathsf{pTop}, \qquad \mathsf{d} = G'G\colon \mathsf{pTop} \to \mathsf{dTop},$$

where the adjunction $F \dashv G$ has been studied in 4.5.2, and $\mathsf{p} \dashv \mathsf{d}$ in 4.3.7.

(a) Describe the new adjunction $F' \dashv G'$.

(b) Review the lp-space $lp\mathbb{S}^1$ described in 4.5.1(a), at the light of the new adjunction.

4.6 Embedded manifolds and inverse quantaloids

Manifolds on inverse quantaloids are a simple framework, where all profunctors compose. A fixed topological space S produces such a category $\mathcal{K}[S]$, formed of the subspaces of S and the partial identities between open subspaces of domain and codomain.

This framework can be used for differentiable manifolds, embedded in euclidean spaces.

4.6.1 Quantaloids and cohesion

As recalled in 3.1.5, a quantaloid is an ordered category C where all sets of parallel maps have a join, preserved by composition. In other words, all hom-sets $C(X, Y)$ are complete lattices, and composition distributes over arbitrary joins (it is not assumed to distribute over meets).

Combining quantaloids with the cohesive structures of Chapter 3, we have the following situations.

(a) A *prj-quantaloid* C will be a prj-category with arbitrary joins, preserved by composition.

Equivalently, a prj-quantaloid is a quantaloid which is a prj-category for its ordering (as in 3.3.4):

(i) C is pre-inverse, i.e. each endomorphism $\leqslant 1$ is idempotent (and called a projector, see 2.4.3),

(ii) $a \leqslant b$, if and only if there is a projector e such that $a = be$.

(b) An *e-quantaloid* C will be an e-category with arbitrary joins, preserved by composition.

Equivalently, an e-quantaloid is a quantaloid which is an e-category for its ordering, or a quantaloid fulfilling conditions (i), (ii) and the axioms (ECH.1, 2) of Definition 3.3.5.

(c) An *inverse quantaloid* C will be an inverse category with arbitrary joins (with respect to its canonical order), preserved by composition.

4.6.2 Exercises

(a) In a prj-quantaloid every set of parallel morphisms has an upper bound, and the linking relation $a \, ! \, b$ is indiscrete: always fulfilled, for parallel morphisms.

(b) A prj-quantaloid (resp. e-quantaloid, inverse quantaloid) is the same as a totally cohesive prj-category (resp. e-category, inverse category) with indiscrete linking relation.

(c) The inverse core of a prj-quantaloid C is an inverse quantaloid.

4.6.3 Categories of partial identities

Our main examples of the previous structures are categories of partial identities inside a fixed object S, a sort of local universe.

(a) For a topological space S, we write as $\mathcal{C}[S]$ the e-subcategory of $\mathcal{C}$

formed of all the subspaces of S and the partial identities $a\colon X \rightarrowtail Y$ on a subset W_a open in X, and contained in Y. Composition with $b\colon Y \rightarrowtail Z$ is by intersection

$$ba\colon X \rightarrowtail Z, \qquad W_{ba} = W_a \cap W_b. \tag{4.92}$$

(The subset W_b is open in Y; therefore $W_a \cap W_b$ is open in W_a, which is open in X.)

$\mathcal{C}[S]$ is an e-quantaloid: any family $a_i\colon X \rightarrowtail Y$ of parallel morphisms has a join

$$a = \bigvee\nolimits_{i \in I} a_i, \qquad W_a = \bigcup W_{a_i}, \tag{4.93}$$

the same as in $\mathcal{C}$.

(b) The inverse core $\mathcal{K}[S] = \mathrm{I}\mathcal{C}[S]$ is an inverse quantaloid. It is formed of the subspaces of S and the partial identities $a\colon X \rightarrowtail Y$ on a subset W_a open in X *and* in Y. Composition is defined by binary intersection of subsets of S, the canonical involution $a \mapsto a^{\sharp}$ is obvious and all endomorphisms are projectors.

$\mathsf{Prj}\,(X)$ is the frame $\mathcal{O}(X)$ of open subsets of X.

(c) We are also interested in the full subcategory $\mathcal{K}\mathcal{O}[S] \subset \mathcal{K}[S]$ formed of the open subspaces of S, and the partial identities on open subsets.

(d) A set S can be viewed as a discrete space, where every subset is open.

Now the categories $\mathcal{C}[S]$, $\mathcal{K}[S]$ and $\mathcal{K}\mathcal{O}[S]$ coincide. Their inverse core is an inverse quantaloid $\mathcal{I}[S] \subset \mathcal{I}$, consisting of the subsets of S and their partial identities.

4.6.4 Introducing embedded manifolds

We deal now with C^r-manifolds of dimension n, which can be C^r-embedded in $\mathbb{R}^{2n}$; in this way the transition maps between open subspaces of $\mathbb{R}^n$ can be seen as partial identities between subspaces of $\mathbb{R}^{2n}$, living in a quantaloid.

This topic is closely related to the Whitney embedding theorem, according to which a Hausdorff paracompact C^{∞}-manifold of dimension n can always be smoothly embedded in $\mathbb{R}^{2n}$ (cf. [Ad]).

As an elementary example, useful to fix notation, the standard circle $\mathbb{S}^1$, embedded in $\mathbb{R}^2$, is covered by two open arcs X_1, X_2, with homeomorphisms

$u^i \colon U_i \to X_i$ defined on open intervals of $\mathbb{R}$

$$(4.94)$$

The transition map $u^i_j \colon U_i \rightarrowtail U_j$ corresponds to a partial identity $w^i_j \colon$ $X_i \rightarrowtail X_j$ on a subset $W^i_j = X_i \cap X_j$ which is open in X_i and X_j: the union $A \cup B$ of the two open arcs, in the figure above.

We have thus a commutative diagram in the inverse category IC of topological spaces and partial homeomorphisms between open subspaces, whose horizontal arrows are invertible

$$
\begin{array}{ccc}
U_i & \xrightarrow{\ u^i\ } & X_i \\
{\scriptstyle u^i_j}\downarrow & & \downarrow{\scriptstyle w^i_j} \\
U_j & \xrightarrow[\ u^j\ ]{} & X_j
\end{array}
\qquad (4.95)
$$

The right-hand part of the diagram lives in the inverse quantaloid $\mathcal{K}[\mathbb{R}^2]$, a particular case of the inverse quantaloid $\mathcal{K}[S] = \mathsf{IC}[S]$ of 4.6.3.

We want to use this rewriting of the atlas U_I of partial homeomorphisms as an atlas X_I of partial identities, to describe the circle as embedded in $\mathbb{R}^2$, and an n-dimensional manifold as embedded in $\mathbb{R}^{2n}$.

4.6.5 Quantaloids of partial identities

Following the outline of 4.6.4, we form an inverse quantaloid $\mathcal{K}^r(n)$, whose structure is lifted by a full forgetful functor

$$|-| \colon \mathcal{K}^r(n) \to \mathcal{K}[\mathbb{R}^{2n}], \qquad (4.96)$$

with values in the inverse quantaloid $\mathcal{K}[\mathbb{R}^{2n}]$.

An object of $\mathcal{K}^r(n)$ is a pair (U, u), where U is an open subspace of $\mathbb{R}^n$ and $u \colon U \to X_u$ is a homeomorphism onto a subspace of $\mathbb{R}^{2n}$, such that the composite $U \to X_u \subset \mathbb{R}^{2n}$ is a C^r-embedding. We let $|(U, u)| = X_u$.

A morphism $\hat{a} \colon (U, u) \rightarrowtail (V, v)$ is defined by a morphism $a \colon X_u \rightarrowtail X_v$ in $\mathcal{K}[\mathbb{R}^{2n}]$, which is the partial identity on a subset $W_a \subset \mathbb{R}^{2n}$, open in X_u and in X_v.

We let $|\hat{a}| = a \colon X_u \twoheadrightarrow X_v$; but the morphism $\hat{a}$ will also be written as a, for the sake of simplicity.

Now we lift the inverse-quantaloid structure of $\mathcal{K}[\mathbb{R}^{2n}]$ to $\mathcal{K}^r(n)$.

(a) The identity $\mathrm{id}\,(U, u)$ has underlying morphism the identity of X_u.

(b) The composite of two morphisms

$$a \colon (U, u) \twoheadrightarrow (V, v), \qquad b \colon (V, v) \twoheadrightarrow (W, w),$$

has underlying morphism the composite

$$ba \colon X_u \twoheadrightarrow X_w, \qquad W_{ba} = W_a \cap W_b.$$

(c) The order relation $a \leqslant a'$ between morphisms $(U, u) \twoheadrightarrow (V, v)$ means that $W_a \subset W_{a'}$.

(d) The morphism $a^\sharp \colon (V, v) \twoheadrightarrow (U, u)$ is represented by $a^\sharp \colon X_v \twoheadrightarrow X_u$, with $W_{a^\sharp} = W_a$.

(e) A projector $e \colon (U, u) \twoheadrightarrow (U, u)$ 'is' a partial identity on an open subset of X_u.

(f) The join of a family $a_i \colon (U, u) \twoheadrightarrow (V, v)$ 'is' a partial identity on the open subset $\bigcup W_{a_i} \subset X_u$.

4.6.6 Embedded manifolds

A manifold on the inverse quantaloid $\mathcal{K}^r(n)$ is an intrinsic atlas $U = (U_i, u^i, w^i_j)$ indexed by a set I, where

(i) every (U_i, u^i) is an object of $\mathcal{K}^r(n)$,

(ii) the transition maps $w^i_j \colon (U_i, u^i) \twoheadrightarrow (U_j, u^j)$ satisfy the axioms (i)–(iii) of 3.5.1.

More explicitly:

(i′) for every $i \in I$, U_i is an open subspace of $\mathbb{R}^n$, and $u^i \colon U_i \to X_i$ is a homeomorphism onto a subspace of $\mathbb{R}^{2n}$ such that the composite $u^i \colon U_i \to X_i \subset \mathbb{R}^{2n}$ is a C^r-embedding,

(ii′) the underlying morphism $w^i_j \colon X_i \twoheadrightarrow X_j$ is the partial identity on a subset W^i_j open in X_i and in X_j, under the axioms (for $i, j, k \in I$):

$$W^i_i = X_i, \qquad W^i_j \cap W^j_k \subset W^i_k, \qquad W^i_j = W^j_i. \tag{4.97}$$

We have thus an *underlying manifold* $|U| = (X_i, (w^i_j))_I$ on $\mathcal{K}[\mathbb{R}^{2n}]$.

Moreover we have a subspace $X = \bigcup X_i$ of $\mathbb{R}^{2n}$. Then X_i is a subspace of X and we have a topological embedding

$$x^i \colon X_i \subset X, \tag{4.98}$$

which belongs to I$\mathcal{C}$ (and $\mathcal{K}[\mathbb{R}^{2n}]$) if and only if X_i is open in X.

We say that U is a C^r-*manifold embedded in* $\mathbb{R}^{2n}$ if $(X, (x^i))$ is the gluing of the underlying manifold $|U|$ in I$\mathcal{C}$. This property is characterised in Exercise (a) and the theorem below.

Without this property, a general manifold on $\mathcal{K}^r(n)$ seems to be of little interest, as shown by the following exercises.

4.6.7 Exercises and complements (Embedded C^r-manifolds)

(a) With the notation of 4.6.6, the manifold $U = (U_i, u^i, w^i_j)_I$ on $\mathcal{K}^r(n)$ is embedded in $\mathbb{R}^{2n}$ if and only if:

(i) each subset $X_i = \mathrm{Val}\, u^i$ is open in the subspace $X = \bigcup X_i$ of $\mathbb{R}^{2n}$,

(ii) $\mathrm{Def}\, w^i_j = \mathrm{Val}\, w^i_j = X_i \cap X_j$ (for all indices $i, j \in I$).

(b) Form a manifold $U = (U_i, u^i, w^i_j)_I$ on $\mathcal{K}^r(1)$, which is not embedded in $\mathbb{R}^2$. *Hints:* take I $= \{1, 2\}$.

(c) One can give a concrete example of this kind, where the gluing of the underlying manifold $|U|$ in I$\mathcal{C}$ is not locally euclidean, and is not the gluing of a manifold in I$\mathcal{C}^0$.

4.6.8 Theorem (Embedded manifolds)

Suppose we have a C^r-manifold $U = ((U_i), (u^i_j))_I$ of dimension n, and a topological embedding $u \colon X \to \mathbb{R}^{2n}$ of the gluing space of U, such that each mapping $U_i \to X \to \mathbb{R}^{2n}$ is of class C^r.

Then we can obtain U as a manifold in $\mathcal{K}^r(n)$, embedded in $\mathbb{R}^{2n}$.

Proof We start from a manifold $U = ((U_i), (u^i_j))_I$ on I$\mathcal{C}^r$: every U_i is open in $\mathbb{R}^n$, every $u^i_j \colon U_i \to U_j$ is a partial C^r-diffeomorphism between open subsets, and the axioms 3.5.1(i)–(iii) are satisfied.

By Proposition 3.5.2, its gluing in I$\mathcal{C}$ is a topological space X equipped with a family of open topological embeddings $(x^i \colon U_i \to X)_i$ fulfilling the following conditions

$$1_X = \bigvee_i x^i x^{i\sharp}, \qquad x^{j\sharp} x^i = x^i_j \qquad (i, j \in I). \tag{4.99}$$

The embedding $x^i \colon U_i \to X$ amounts to a homeomorphism $u^i \colon U_i \to X_i$, and the subspaces $X_i = \text{Val } x^i$ form an open cover of X. By hypothesis, we can assume that X is a subspace of $\mathbb{R}^{2n}$, and the composite $U_i \to X \to \mathbb{R}^{2n}$ is of class C^r.

The transition map $u_j^i \colon U_i \dashrightarrow U_j$ corresponds to a partial identity $w_j^i \colon$ $X_i \dashrightarrow X_j$ on a subset W_j^i open in X_i and X_j

$$
\begin{array}{ccc}
U_i & \xrightarrow{\ w^i\ } & X_i \\[2pt]
{\scriptstyle u_j^i}\big\downarrow & & \big\downarrow{\scriptstyle w_j^i} \\[2pt]
U_j & \xrightarrow[\ w^j\ ]{} & X_j
\end{array}
\tag{4.100}
$$

and $W_j^i = X_i \cap X_j$, by the first condition in (4.99).

We have completed the proof, representing U as an embedded manifold in the inverse quantaloid $\mathcal{K}^r(n)$, by a family (U_i, u^i) of objects of $\mathcal{K}^r(n)$, and an intrinsic atlas (U_i, u^i, w_j^i), where $w_j^i \colon X_i \dashrightarrow X_j$ is the partial identity on the subset $U_j^i = X_i \cap X_j$, open in X_i and in X_j. $\qquad\square$

4.6.9 *Arbitrary dimension

If we want a global framework for embedded manifolds of arbitrary dimension, we can consider the space $\mathbb{R}^\infty$, defined as the colimit of all inclusions $\mathbb{R}^n \subset \mathbb{R}^m$. The space $\mathbb{R}^\infty$ can be constructed as the direct sum

$$
\bigoplus_{n \in \mathbb{N}} \mathbb{R} \ = \ \bigcup_{n \in \mathbb{N}} \mathbb{R}^n,
$$

with the topology whose open sets are the subsets U such that, for every $n \in \mathbb{N}$, $U \cap \mathbb{R}^n$ is open in $\mathbb{R}^n$.

Thus every $\mathbb{R}^n$ is embedded in $\mathbb{R}^\infty$, as a closed subspace.

(a) We form an inverse quantaloid $\mathcal{K}^r(\infty)$, lifted by a full forgetful functor

$$
|-| \colon \mathcal{K}^r(\infty) \to \mathcal{K}[\mathbb{R}^\infty],
\tag{4.101}
$$

from the inverse quantaloid $\mathcal{K}[\mathbb{R}^\infty]$.

An object is a pair (U, u) where U is an open subspace of some $\mathbb{R}^n$ and $u \colon U \to X_u$ is a homeomorphism onto a subspace of $\mathbb{R}^{2n}$ (and $\mathbb{R}^\infty$) such that the composite $U \to X_u \to \mathbb{R}^{2n}$ is a C^r-embedding. We let $|(U, u)| = X_u$.

A morphism $\hat{a} \colon (U, u) \dashrightarrow (V, v)$ comes from a morphism $a \colon X_u \dashrightarrow X_v$ in $\mathcal{K}[\mathbb{R}^{2n}]$, which is the partial identity on a subset W_a of $\mathbb{R}^{2n}$, open in X_u and in X_v.

(b) Manifold and embedded manifolds on $\mathcal{K}[\mathbb{R}^\infty]$ are defined as in 4.6.6.

4.7 G-bundles and principal bundles

The topological space B is fixed throughout this section. All the bases of fibre bundles will be open subspaces of this *ground* space.

Vector bundles *on the ground* B form a subcategory Mf $\mathcal{V}_B$ of Mf $\mathcal{V}$, better suited than the whole category to develop tensor calculus.

Similarly, we have fibre bundles on a fixed ground. In particular, a G-bundle is a fibre bundle where a fixed topological group G acts on the fibres, and the transition maps are 'operated' by G.

In a principal G-bundle, the action of G on the fibre F is free and transitive, and we can identify F with the underlying topological space of G.

4.7.1 Vector bundles on a fixed ground

Reviewing vector bundles, as analysed in 4.2.4, we restrict the categories $\mathcal{V}$ and Mf $\mathcal{V}$ to a fixed ground space B, and its partial identities on open subspaces.

We form thus the category $\mathcal{V}_B$ of *trivial vector bundles* $p: U \times F \to U$ *on the ground* B, where U is an open subspace of B and F is a finite dimensional real vector space.

A morphism $(f, \overline{f}): p \twoheadrightarrow p'$ is a morphism of $\mathcal{V}$ whose map $\overline{f}: U \twoheadrightarrow U'$ is a partial identity on an open subset $W = \operatorname{Def} \overline{f} = \operatorname{Val} \overline{f}$ of $U \cap U'$ (in B)

$$
\begin{array}{ccc}
U \times F & \xrightarrow{\;f\;} & U' \times F' \\
{\scriptstyle p}\big\downarrow & & \big\downarrow{\scriptstyle p'} \\
U & \xrightarrow[\;\overline{f}\;]{} & U'
\end{array}
\qquad (4.102)
$$

Therefore $f: U \times F \twoheadrightarrow U' \times F'$ is determined by a partial map f_2 in $\mathcal{C}$ such that:

(i) $f_2: U \times F \twoheadrightarrow F'$, $\operatorname{Def} f_2 = W \times F$ (W open in $U \cap U'$),

(ii) for every $b \in W$, the mapping $f_2(b, -): F \to F'$ is $\mathbb{R}$-linear.

The partial map f is recovered letting:

$$
f(b, y) = (b, f_2(b, y)), \qquad \operatorname{Def} f = \operatorname{Def} f_2 = W \times F. \qquad (4.103)
$$

The e-subcategory $\mathcal{V}_B \subset \mathcal{V}$ is totally cohesive. Its gluing completion Mf $\mathcal{V}_B$ yields vector bundles whose basis is an open subspace of B, and their partial morphisms, with partial identities on the basis.

4.7.2 Fibre bundles on a fixed ground

Similarly, we can restrict the categories $\mathcal{B}$ and Mf $\mathcal{B}$ studied in 4.2.3 to a fixed ground space B.

The e-subcategory $\mathcal{B}_B \subset \mathcal{B}$ is described as above, with obvious modifications: the fibres are only assumed to be topological spaces, and the linearity condition 4.7.1(ii) is omitted.

Again, $\mathcal{B}_B$ is totally cohesive; its gluing completion Mf $\mathcal{B}_B$ yields fibre bundles whose basis is an open subspace of B, and their partial morphisms (with partial identities on the basis).

This framework is useful to introduce fibre bundles with a group action on the fibres. (The terminology of actions of topological groups on topological spaces was reviewed in 2.5.6.)

4.7.3 Trivial G-bundles

We have a fixed topological group G, that acts faithfully on a fixed topological space F, the *fibre*. We also fix a topological space B, the *ground*: all bases will be open subspaces of B.

We want to form an inverse category $G\mathcal{B}_{BF}$, of *trivial G-bundles* (with fibre F, on the ground B) *and partial morphisms*; the structure will be lifted by a faithful functor (as in 1.5.6)

$$G\mathcal{B}_{BF} \to \mathrm{I}\mathcal{B}_B \subset \mathcal{B}_B. \tag{4.104}$$

An object is a *trivial G-bundle* $p \colon U \times F \to U$, where U is an open subset of B; the bundle will also be written as $U \times F$. Its underlying trivial bundle is p itself, as an object of $\mathcal{B}_B$ (forgetting the action of G on the fibre).

A morphism $f_t \colon U \times F \rightarrowtail V \times F$ in $G\mathcal{B}_{BF}$ is determined by a partial map $t \colon B \rightarrowtail G$ in $\mathcal{C}$, defined on an open subset of $U \cap V$. Its underlying morphism in $\mathcal{B}_B$, still written as f_t, works as a partial identity on the ground B and acts on the fibre by t

$$f_t \colon U \times F \rightarrowtail V \times F, \qquad \mathrm{Def}\, f_t = (\mathrm{Def}\, t) \times F,$$

$$f_t(b, y) = (b, t(b)y) \qquad (b \in \mathrm{Def}\, t,\ y \in F). \tag{4.105}$$

$G\mathcal{B}_{BF}$ is a category: the composite

$$f_s f_t = f_{st} \colon U \times F \rightarrowtail V \times F \rightarrowtail W \times F,$$

is lifted from $\mathcal{B}_B$ and associated to the partial map $st \colon B \rightarrowtail G$

$$st(b) = s(b).t(b) \qquad (b \in \mathrm{Def}\, t \cap \mathrm{Def}\, s),$$

$$f_s f_t(b, y) = (b, s(b)t(b)y) \qquad (b \in \mathrm{Def}\, t \cap \mathrm{Def}\, s,\ y \in F). \tag{4.106}$$

The identity $\mathrm{id}\,(U \times F)$ is also lifted from $\mathcal{B}_B$, and associated to the partial map

$$e_U : B \rightarrowtail G, \qquad e_U(b) = 1_G \qquad (b \in U). \tag{4.107}$$

The category $G\mathcal{B}_{BF}$ is inverse, with partial inverse $f_t^\sharp : V \times F \rightarrowtail U \times F$ lifted from $I\mathcal{B}_B$, and associated to the partial map $t^\sharp : B \rightarrowtail G$ determined by inverses in G

$$
\begin{aligned}
t^\sharp(b) &= t(b)^{-1} & (b \in \mathrm{Def}\,t), \\
f_t^\sharp(b, y) &= (b, t(b)^{-1}y) & (b \in \mathrm{Def}\,t,\ y \in F).
\end{aligned}
\tag{4.108}
$$

The canonical order and linking relation of $G\mathcal{B}_{BF}$ are also lifted from $I\mathcal{B}_B$.

Let us note that, if the partial maps $t, t' : B \rightarrowtail G$ define the same partial map $f : U \times F \rightarrowtail V \times F$, then for every $b \in \mathrm{Def}\,t = \mathrm{Def}\,t'$ the operators $t(b)$ and $t'(b)$ have the same action on F, and coincide.

Therefore, the canonical order and linking relation of $G\mathcal{B}_{BF}$ correspond to the order and linking in $\mathcal{C}$

$$
\begin{aligned}
f_t \leqslant f_{t'} &\quad \Leftrightarrow \quad t \leqslant t' \text{ in } \mathcal{C}(B, G), \\
f_t \,!\, f_{t'} &\quad \Leftrightarrow \quad t \,!\, t' \text{ in } \mathcal{C}(B, G).
\end{aligned}
\tag{4.109}
$$

It follows that $G\mathcal{B}_{BF}$ is a totally cohesive inverse category: a linked family $f_{t_i} : U \times F \rightarrowtail V \times F$ has a compositive join

$$f_t = \bigvee f_{t_i} : p \rightarrowtail p', \qquad t = \bigvee t_i : B \rightarrowtail G \text{ in } \mathcal{C}. \tag{4.110}$$

An idempotent $f_t : U \times F \rightarrowtail U \times F$ is determined by a partial map $t \leqslant e_U : B \rightarrowtail G$, which means that $t(b) = 1_G$ on some open subset $\mathrm{Def}\,t \subset U$, and f_t is the partial identity on $\mathrm{Def}\,t \times F$.

4.7.4 G-bundles

The manifold completion $\mathrm{Mf}\,G\mathcal{B}_{BF}$ can be viewed as *the category of G-bundles* (with fibre F, on the ground B) *and partial morphisms*.

An object can be written as an intrinsic atlas $(U_i \times F, (t_{ij}))_I$, where

(a) U_i is open in B and $U_i \times F$ denotes the trivial G-bundle on U_i,

(b) $t_{ij} : B \rightarrowtail G$ is a partial map (the *transition action*) defined on an open subset U_{ij} of $U_i \cap U_j$,

and the following axioms are satisfied:

(i) $\quad U_{ii} = U_i,$ $\qquad\qquad t_{ii}(b) = 1_G$ $\qquad\qquad\qquad (b \in U_i),$

(ii) $\quad U_{ij} \cap U_{jk} \subset U_{ik},$ $\quad t_{jk}(b).t_{ij}(b) = t_{ik}(b)$ $\qquad (b \in U_{ij} \cap U_{jk}),$

(iii) $\quad U_{ij} = U_{ji},$ $\qquad\qquad t_{ji}(b) = t_{ij}(b)^{-1}$ $\qquad\qquad (b \in U_{ij}).$

In fact, these conditions are equivalent to saying that the associated partial maps $u_{ij}: U_i \times F \rightarrow U_j \times F$ form an intrinsic atlas on GB_{BF}.

The gluing in $\mathcal{F}$ is a fibre bundle $p: X \to U$, where $U = \cup\, U_i$ is open in B.

The morphisms of Mf GB_{BF} are very restricted, but of course one can consider morphisms of fibre bundles, in Mf $\mathcal{B}$ or Mf $\mathcal{B}_B$, which involve G-bundles.

A G-bundle is said to be *principal* if the fibre F is the topological space underlying the topological group G, acted on by G by left translations $y \mapsto gy$. If G is commutative, in additive notation, one often prefers to write this translation in the form $y \mapsto y + g$.

4.7.5 Two examples

(a) The universal covering of the circle

$$ p: \mathbb{R} \to \mathbb{S}^1, \qquad p(t) = e^{2\pi i t} = \cos 2\pi t + i.\sin 2\pi t, \qquad (4.111) $$

was considered in 4.2.6, as a fibre bundle with discrete fibres. It can be viewed as a principal $\mathbb{Z}$-bundle: the constant fibre is the discrete space $F = |\mathbb{Z}|$, acted on by the discrete group $G = \mathbb{Z}$ of the integers, by translations: $y \mapsto y + k$.

This can be realised with an atlas formed of two charts $U_i \times F$, determined by two non-total open arcs U_i of $\mathbb{S}^1$ (evenly covered by $p: \mathbb{R} \to \mathbb{S}^1$)

$$ \begin{aligned} U_1 &= \mathbb{S}^1 \setminus \{(1,0)\} = p(]0,1[), \\ U_2 &= \mathbb{S}^1 \setminus \{(-1,0)\} = p(]1/2,3/2[). \end{aligned} \qquad (4.112) $$

The transition action $t_{12}: \mathbb{S}^1 \rightarrow \mathbb{Z}$ is defined on the two connected components of $U_1 \cap U_2$

$$ U_{12} = U_1 \cap U_2 = p(]0,1/2[) \cup p(]1/2,1[), $$

$$ t_{12}(x) = 0, \quad \text{if } x \in p(]0,1/2[), \qquad (4.113) $$

$$ t_{12}(x) = 1, \quad \text{if } x \in p(]1/2,1[). $$

In this way, we are gluing half of $U_1 \times \{k\}$ on $U_2 \times \{k\}$ and the other half on $U_2 \times \{k+1\}$, as shown by the following figure

$$U_2 \times \{k+1\}$$

$$U_1 \times \{k\} \qquad\qquad\qquad U_2 \times \{k\} \qquad\qquad (4.114)$$

Of course $t_{21}(x) = -t_{12}(x)$, while t_{11} and t_{22} are constant at 0.

(b) The n-fold covering of the circle

$$p_n \colon \mathbb{S}^1 \to \mathbb{S}^1, \qquad p_n(z) = z^n \qquad\qquad (n \geqslant 2), \qquad (4.115)$$

was also considered in 4.2.6, as a fibre bundle with discrete fibres having n points.

Using the discrete group $G = \mathbb{Z}/n$, this covering can be viewed as a principal G-bundle on $\mathbb{S}^1$; the constant fibre is the discrete space $F = |\mathbb{Z}/n|$, on which G acts by translations.

We use again the previous open arcs U_1 and U_2 of $\mathbb{S}^1$, and form an atlas of two charts $U_i \times F = U_i \times |\mathbb{Z}/n|$, determined by the previous open arcs U_1 and U_2 of $\mathbb{S}^1$.

The transition action $t_{12} \colon \mathbb{S}^1 \twoheadrightarrow \mathbb{Z}/n$ is defined on the two connected components of $U_1 \cap U_2$

$$U_{12} = U_1 \cap U_2 = p(]0, 1/2[) \cup p(]1/2, 1[),$$

$$t_{12}(x) = 0, \quad \text{if } x \in p(]0, 1/2[), \qquad\qquad (4.116)$$

$$t_{12}(x) = \bar{1}, \quad \text{if } x \in p(]1/2, 1[).$$

Again, we are gluing each arc $U_1 \times \{\bar{k}\}$ on a half of $U_2 \times \{\bar{k}\}$ and a half of $U_2 \times \{\overline{k+1}\}$.

4.8 *Fundamental groupoids of smooth manifolds

This section requires some knowledge of fundamental groupoids (reviewed in 4.3.1(b)), for which we refer to [Bw], and of colimits (see Section 5.2).

We briefly mention some results of [CaG] showing that a topological space with a covering satisfying a suitable condition (in 4.8.3(i)) has a fundamental groupoid which is the gluing of a manifold of indiscrete subgroupoids. Such coverings always exist for differentiable manifolds.

4.8.1 Groupoids and partial isomorphisms

In this section, $\mathcal{G}pd$ is the inverse category *of small groupoids and partial isomorphisms*: a morphism $F \colon G \twoheadrightarrow H$ corresponds to an isomorphism $F_0 \colon \mathrm{Def}\, F \to \mathrm{Val}\, F$, from a subgroupoid of G to a subgroupoid of H.

The composite $F'F \colon G \twoheadrightarrow H \twoheadrightarrow K$ is obviously defined on the subgroupoid $\mathrm{Def}\, F'F \subset G$ consisting of the objects $x \in \mathrm{Def}\, F$ such that $F(x) \in \mathrm{Def}\, F'$, and the arrows $g \colon x \to x'$ in $\mathrm{Def}\, F$ such that $F(g) \in \mathrm{Def}\, F'$.

The partial inverse $F^\sharp \colon H \twoheadrightarrow G$ comes from the inverse isomorphism $F_0^{-1} \colon \mathrm{Val}\, F \to \mathrm{Def}\, F$. A projector $E \colon G \twoheadrightarrow G$ is a partial identity on a subgroupoid.

This inverse category is not finitely cohesive (by far), as shown below, in Exercise (b).

However, *we can consider manifolds on the inverse category $\mathcal{G}pd$ and their gluing*, as in 3.5.1.

Exercises and complements. (a) We recall that a groupoid is indiscrete, or chaotic, if it has precisely one arrow $x \to y$ for any two objects x, y.

We shall denote by $C(S)$ the indiscrete groupoid on a set (or topological space) of objects S (possibly empty).

The space X is said to be 1-connected if its fundamental groupoid is indiscrete, which means that it coincides with $C(X)$. Equivalently, X is path-connected and all its fundamental groups are trivial.

(b) Prove that two linked morphisms $F_i \colon G \twoheadrightarrow H$ in $\mathcal{G}pd$ need not have a join. (*Hints:* this already fails for groups, i.e. groupoids on one object.)

But it is easy to see that $\mathcal{G}pd$, like $\mathcal{T}$, has all upper-bounded joins, preserved by composition.

(b) Prove that the fundamental groupoid functor $\Pi_1 \colon \mathsf{Top} \to \mathsf{Gpd}$ cannot be extended to a functor $\mathcal{C} \to \mathcal{G}pd$.

4.8.2 Fundamental groupoids and open covers

Let us suppose that the topological space X has an open cover (X_i). Then $\Pi_1 X$ can be obtained as the colimit of the following diagrams of fundamental groupoids

$$
\Pi_1(X_i \cap X_j) \ \substack{\longrightarrow \\ \longrightarrow} \ \begin{array}{c} \Pi_1(X_i) \\[1em] \Pi_1(X_i) \end{array} \qquad (i, j \in I), \qquad (4.117)
$$

a sort of generalised pushout of all $\Pi_1(X_i)$ over all $\Pi_1(X_i \cap X_j)$.

This is proved in [CaG], Theorem 2.2 (with a more precise statement). The proof is essentially the same as in [Bw], Section 6.7.2, for the case $I = \{1, 2\}$, which gives an ordinary pushout: a generalisation of the well-known Seifert-van Kampen Theorem for fundamental groups.

4.8.3 Gluing fundamental groupoids

We now consider an open cover (X_i) of the space X satisfying the following condition

(i) all the subspaces $X_i \cap X_j$ are 1-connected,

which includes all X_i, of course.

This produces a manifold $C_I = (C(X_i), u_j^i)_I$ in the inverse category $\mathcal{G}pd$, where (using the notation of 4.8.1(a))

$$u_j^i \colon C(X_i) \rightarrowtail C(X_j), \qquad \mathrm{Def}\, u_j^i = C(X_i \cap X_j) = \mathrm{Val}\, u_j^i, \qquad (4.118)$$

is a partial identity on an indiscrete subgroupoid

$$\mathrm{Def}\, u_j^i = \mathrm{Val}\, u_j^i = C(X_i \cap X_j) = C(X_i) \cap C(X_j). \qquad (4.119)$$

In this situation, the result recalled in 4.8.2 means that the groupoid $\Pi_1(X)$ is the gluing of the manifold C_I, by the inclusion functors

$$u^i \colon C(X_i) \to \Pi_1(X).$$

Remarks and complements. (a) Every smooth n-dimensional manifold X has an open cover satisfying condition (i). More precisely, X has a 'good cover' (X_i), where all finite intersections $X_{i_1} \cap \ldots \cap X_{i_k}$ are diffeomorphic to $\mathbb{R}^n$ or empty. This is proved in [BoT].

(b) Every groupoid is, up to equivalence, the gluing of a manifold of indiscrete groupoids, as proved in [CaG], Proposition 6.1.

5

Complements on category theory

The previous chapters rely on enriched categories and enriched profunctors on ordered categories. This chapter and the next show how all this is a part of the general topic of enriched categories, on bases of different forms.

Further information on categories can be found in the books cited in the introduction of Section 1.4, or as specified below.

5.1 Monomorphisms and partial morphisms

We say something more on monomorphisms and epimorphisms in a category C, briefly introduced in 1.4.7. Subobjects and quotients are based on them.

Then we define pullbacks and pushouts, and use the former to build categories of partial morphisms $P_M C$, extending the obvious construction of S from Set, or C from Top.

Pushouts in Top have already been used in Section 2.6.

5.1.1 Monomorphisms and epimorphisms, continued

We already remarked that, in a category of structured sets and structure-preserving mappings, an injective mapping (of the category) is always a monomorphism, while a surjective one is an epimorphism. The converse may require a non-trivial proof, or fail. This can only be understood by working out examples and exercises, as those listed below.

A divergence appears between monos and epis: the theory of categories is selfdual, but our frameworks of structured sets are not. When we classify monos in Set, this tells us everything about the epis of Set^{op} but nothing about the epis of Set.

In fact, in all the examples below it will be easy to prove that the monomorphisms coincide with the injective morphisms. This is always true in a category concrete on Set, whose forgetful functor is representable, as proved in 5.2.3(b).

On the other hand, various problems occur with epimorphisms: classifying them in various categories of algebraic structures leads to difficult problems with no elementary solution – and no real need of it: in these categories the important epimorphisms are the regular ones (see 5.2.6).

5.1.2 Exercises and complements (Monos and epis)

A reader unfamiliar with this topic will find it useful to work out the following exercises. Solutions are given below.

(a) In Set a mono is the same as an injective mapping; an epi is the same as a surjective mapping. The category is balanced.

(b) In Top and Ab monos and epis coincide with the injective and surjective mappings of the category, respectively. Ab is balanced, and Top is not.

(c) In the category Mon, of monoids and homomorphisms, monos coincide again with the injective homomorphisms. The inclusion $\mathbb{N} \to \mathbb{Z}$ (of additive monoids) is both mono and epi in Mon, which is not balanced. Epimorphisms in Mon have no elementary characterisation; we shall see that the 'regular epimorphisms' (namely the surjective homomorphisms) are more important.

(d) In pOrd and Ord monos and epis coincide with the injective and surjective mappings of the category, respectively. These categories are not balanced.

(e) In a preordered set X, viewed as a category, all arrows are mono and epi. Saying that the category is balanced is equivalent to a precise condition on the preordering.

*(f) In the category Gp of groups all epimorphisms are surjective: a nonobvious fact, whose proof can be found in [M3], Section I.5, Exercise 5. (One cannot use the same argument as in Ab.)

Solutions. (a) If $f: X \to Y$ is a monomorphism in Set, let us suppose that $f(x) = f(x')$, with $x, x' \in X$. We consider the mappings u, v

$$\{*\} \underset{v}{\overset{u}{\rightrightarrows}} X \overset{f}{\longrightarrow} Y \qquad u(*) = x, \ \ v(*) = x'. \tag{5.1}$$

Now we have $fu = fv$, whence $u = v$ and $x = x'$, which shows that f is injective. Note that the proof works *simulating an element of X by a map* $\{*\} \to X$.

On the other hand, if $f\colon X \to Y$ is an epimorphism in Set, we define two mappings u, v with values in the set $\{0, 1\}$

$$X \xrightarrow{f} Y \overset{u}{\underset{v}{\rightrightarrows}} \{0, 1\} \tag{5.2}$$

where u is the characteristic function of the subset $f(X) \subset Y$ (with $u(y) = 1$ if and only if $y \in f(X)$) while v is the constant map $v(y) = 1$. Then $uf = vf$, whence $u = v$ and $f(X) = Y$. (A different proof can use the set $Y \times \{0, 1\}$, the disjoint union of two copies of Y; or a quotient of the latter.)

Since the invertible morphisms in Set are the bijective mappings, the category Set is balanced.

(b) For monomorphisms, the proof is similar to the previous one, in (a), making use of the singleton in Top and of the group $\mathbb{Z}$ in Ab. Note that the latter allows us to simulate an element $x \in X$ by a homomorphism $u\colon \mathbb{Z} \to X$, sending the generator 1 to x. This works because $\mathbb{Z}$ is the free abelian group on the singleton, or – in other words – because the forgetful functor Ab $\to$ Set is representable by $\mathbb{Z}$.

In Top, to prove that an epi is surjective we can proceed as in (5.2), using the indiscrete topology on $\{0, 1\}$.

In Ab we can use the quotient group $Y/f(X)$ and the homomorphisms

$$X \xrightarrow{f} Y \overset{p}{\underset{0}{\rightrightarrows}} Y/f(X) \tag{5.3}$$

with the canonical projection $p\colon Y \to Y/f(X)$ and the zero homomorphism. We are now following a different pattern: constructing arrows is fairly free in Set, somewhat less in Top, much less in categories of algebraic structures.

We conclude here that Ab is balanced, while Top is not: a bijective continuous mapping need not be invertible in Top, i.e. a homeomorphism.

(c) For monos in Mon we proceed again as in (a), making use of the additive monoid $\mathbb{N}$ (freely generated by 1).

The inclusion $\mathbb{N} \to \mathbb{Z}$, of additive monoids, is injective and mono; it is also epi, because a homomorphism $f\colon \mathbb{Z} \to M$ of monoids (in additive notation) is determined by its values on $\mathbb{N}$: if $n > 0$, then $f(n) + f(-n) = 0_M = f(-n) + f(n)$, whence $f(-n)$ is (additively) inverse to $f(n)$ in M.

(d) In pOrd and Ord every mono is injective, with the same argument as in Set, based on the (ordered) singleton.

To prove that epis are surjective in pOrd one can use a two-point set with chaotic preorder. In Ord the conclusion is the same, but the proof is more complicated: for a non-surjective monotone mapping $f\colon X \to Y$ and a point $y_0 \in Y \setminus f(X)$, one can construct an ordered set Y' where y_0 is duplicated and prove that f is not epi.

(e) The first claim is obvious. Saying that the category X is balanced means here that every arrow is an isomorphism: in other words, the preorder of X is symmetric, i.e. an equivalence relation.

5.1.3 Subobjects

Let X be an object of the category C. A subobject of X cannot be based on the notion of 'subset' (which makes no sense in a general category) but is defined as an equivalence class of monomorphisms, or better as a *selected* representative of such a class.

Given two monos m, n *with values in* X, we say that $m \prec n$ if there is a (unique) morphism u such that

$$
\begin{array}{ccc}
A & \overset{m}{\rightarrowtail} & X \\
u \downarrow & \nearrow n & \\
A' & &
\end{array}
\qquad m = nu. \qquad (5.4)
$$

We say that m, n *are equivalent*, or $m \sim n$, if $m \prec n \prec m$; in other words, there is a (unique) isomorphism u such that $m = nu$.

In every class of equivalent monos with codomain X, one is chosen and called a *subobject* of X; in the class of isomorphisms, we always choose the identity 1_X.

The subobjects of X in C form the ordered class $\mathrm{Sub}(X)$, with maximum 1_X; here, the induced *order* $m \prec n$ is also written as $m \leqslant n$. (Equalisers and kernels will always be chosen as subobjects.)

For a morphism $f\colon X \to Y$ and two subobjects $m\colon A \rightarrowtail X$, $n\colon B \rightarrowtail Y$, there can be at most one morphism $g\colon A \to B$ forming a commutative square

$$
\begin{array}{ccc}
X & \overset{f}{\longrightarrow} & Y \\
m \uparrow & & \uparrow n \\
A & \underset{g}{\longrightarrow} & B
\end{array}
\qquad fm = ng. \qquad (5.5)
$$

If this is the case, we say that g is the *restriction* of f (or *induced* by f), from m to n.

(This topic cannot be effectively managed if we define subobjects as equivalence classes of monomorphisms, as is often done.)

The category C is said to be *well powered* if every object X has a *set* of subobjects.

The usual categories of structured sets satisfy this condition: the subobjects of X can be realised as suitable subsets of the underlying set $|X|$, equipped with the restricted structure.

5.1.4 Quotients

Epimorphisms of C with a fixed domain X are dealt with in a dual way.

Their preorder, also written as $p \prec q$, means that the epimorphism p factorises through q (by a unique morphism u)

$$X \xrightarrow{p} A$$

$$p = uq, \qquad (5.6)$$

and $p \sim q$ means that $p \prec q \prec p$, or equivalently that there is a (unique) isomorphism u such that $p = uq$.

A *quotient* of X is a *selected* representative of an equivalence class of epimorphisms with domain X. They form the ordered class $\mathrm{Quo}(X)$, with maximum 1_X; again the induced order is also written as $p \leqslant q$.

For a morphism $f \colon X \to Y$ and two quotients $p \colon X \twoheadrightarrow A$, $q \colon Y \twoheadrightarrow B$, we say that the morphism $g \colon A \to B$ is *induced* by f, from p to q, if the following square commutes

$$
\begin{array}{ccc}
X & \xrightarrow{f} & Y \\
p\downarrow & & \downarrow q \\
A & \xrightarrow{g} & B
\end{array}
\qquad qf = gp. \qquad (5.7)
$$

A category is said to be *well copowered* if every object has a *set* of quotients.

This holds true in the usual categories of structured sets: the quotients of the object X can be realised as suitable quotients of the underlying set $|X|$, equipped with the induced structure.

5.1.5 Pullbacks and pushouts

(a) The *pullback* of a pair of morphisms $f \colon X_1 \to X_0 \leftarrow X_2 \colon g$ (with the same codomain) is another case of a particular limit.

It is defined as an object A equipped with two maps $u_i \colon A \to X_i$ $(i = 1, 2)$ which form a commutative square $f u_1 = g u_2$, in a universal way:

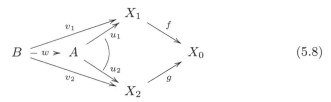

$$(5.8)$$

(i) for every triple (B, v_1, v_2) such that $f v_1 = g v_2$, there exists a unique morphism $w \colon B \to A$ such that $u_1 w = v_1$, $u_2 w = v_2$.

Pullback diagrams are often marked as above, at the corner of the solution. In Set (resp. Top, Ab) the pullback-object can be realised as a subset (resp. subspace, subgroup) of the product $X_1 \times X_2$:

$$A = \{(x_1, x_2) \in X_1 \times X_2 \mid f(x_1) = g(x_2)\}. \qquad (5.9)$$

If $f = g$, the pullback $A \rightrightarrows X_1$ of the diagram $X_1 \to X_0 \leftarrow X_1$ is called the *kernel pair* of f. In Set, it can be realised as

$$A = \{(x, x') \in X_1 \times X_1 \mid f(x) = f(x')\},$$

and amounts to the equivalence relation associated to f.

(The kernel pair should not be confused with the equaliser of the pair (f, f), which always exists and is the identity of the domain of f.)

(b) Dually, the *pushout* of a pair of morphisms (f, g) with the same domain, as below, is an object A equipped with two maps $u_i \colon X_i \to A$ $(i = 1, 2)$ which form a commutative square $u_1 f = u_2 g$, in a universal way:

$$(5.10)$$

(i*) for every triple (B, v_1, v_2) such that $v_1 f = v_2 g$, there exists a unique map $w \colon A \to B$ such that $w u_1 = v_1$, $w u_2 = v_2$.

If $f = g$, the pushout $X_1 \rightrightarrows A$ of the diagram $X_1 \leftarrow X_0 \to X_1$ is called the *cokernel pair* of f.

Pushouts can also be denoted as above, marking the corner of the solution.

5.1.6 Exercises and complements (Pullbacks and pushouts)

The following facts are important and often used.

(a) A category that has binary products and equalisers also has pullbacks. A category that has pullbacks and terminal object has finite products. Consider also the dual properties. *Hints:* for the first point, one can generalise the construction of the pullback in (5.9).

(b) (*Characterising monos*) In a category, a morphism $f\colon X \to Y$ is mono if and only if the following diagram is a pullback (the existence of general pullbacks is not assumed, here)

$$
\begin{array}{ccc}
X & \xrightarrow{\ f\ } & Y \\
{\scriptstyle 1}\big\uparrow & & \big\uparrow{\scriptstyle f} \\
X & \xrightarrow[\ 1\ ]{} & X
\end{array}
\qquad (5.11)
$$

(c) (*Preimages*) If the following square is a pullback and n is a monomorphism, so is m

$$
\begin{array}{ccc}
X & \xrightarrow{\ f\ } & Y \\
{\scriptstyle m}\big\uparrow & & \big\uparrow{\scriptstyle n} \\
\bullet & \longrightarrow & \bullet
\end{array}
\qquad (5.12)
$$

The latter is determined up to equivalence of monomorphisms (see 5.1.3): it is called the *preimage* of n along f and written as $f^*(n)$. Working with subobjects, $f^*(n)$ is determined by their choice.

(d) (*Pasting property*) If the two squares below are pullbacks, so is their 'pasting', i.e. the outer rectangle

$$
\begin{array}{ccccc}
\bullet & \longrightarrow & \bullet & \longrightarrow & \bullet \\
\big\uparrow & & \big\uparrow & & \big\uparrow \\
\bullet & \longrightarrow & \bullet & \longrightarrow & \bullet
\end{array}
\qquad (5.13)
$$

(e) In Set, if $X = X_1 \cup X_2$, then X is the pushout of X_1 and X_2 over $X_1 \cap X_2$. This fact can be extended to Top, under convenient hypotheses on the subspaces $X_i \subset X$.

(f) We have seen in Exercise 2.5.4(b) that the euclidean sphere $\mathbb{S}^n$ can be obtained as the cokernel pair, in Top, of the inclusion $\mathbb{S}^{n-1} \to \mathbb{D}^n$, with values in the compact disc.

*(g) In an abelian category, pullbacks and pushouts are characterised by conditions of exactness on an associated sequence: see [Hi], or [G10], Subsection 6.4.8.

*(h) Pushouts of fundamental groupoids have already been considered, in Section 4.8.

5.1.7 Categories of partial morphisms

As remarked in 2.1.4(f), the construction of the 'categories of partial morphisms' can be generalised. It is a well-known subject, which can be found in different forms in various papers, for instance in [RoR]. *(It is also a simplified version of the construction of a category of relations.)*

We start from a category C equipped with a wide subcategory M whose arrows are monomorphisms in C, and want to form a category P_MC where these monomorphisms have an assigned splitting: $m^\sharp m = 1$. The reader can keep in mind, as a main instance in this book, the case where C is Top and M is the wide subcategory of open topological embeddings; we want to form the category $\mathcal{C}$.

The pair (C, M) is assumed to satisfy two conditions.

(i) C is a category, and M is a wide subcategory whose arrows are monomorphisms in C, stable under the equivalence relation of monos (in 5.1.3); for each X in C, the subobjects of X belonging to M *form a set*.

(Every isomorphism of C is equivalent to an identity, and belongs to M.)

(ii) For every $f\colon X \to Y$ in C and $n\colon B \rightarrowtail Y$ in M, the preimage $m = f^*(n)\colon A \rightarrowtail X$ (see 5.1.6(c)) exists and belongs to M

$$
\begin{array}{ccc}
X & \xrightarrow{f} & Y \\
{\scriptstyle m}\big\uparrow & \searrow & \big\uparrow{\scriptstyle n} \\
A & \longrightarrow & B
\end{array}
\tag{5.14}
$$

Now, P_MC has the same objects as C. A morphism

$$\varphi = [m, f]\colon X \nrightarrow Y$$

in P_MC is an equivalence class of pairs (m, f) as in the upper row below, with m in M and f in C: we identify $[m, f] = [m', f']$ when there exists an isomorphism u in C making the following diagram commutative

$$
\begin{array}{ccccc}
X & \xleftarrow{\;m\;} & A & \xrightarrow{\;f\;} & Y \\
\big\| & & \big\downarrow{\scriptstyle u} & & \big\| \\
X & \xleftarrow[\;m'\;]{} & A' & \xrightarrow[\;f'\;]{} & Y
\end{array}
\tag{5.15}
$$

The morphism $\varphi = [m, f]$ has thus a unique *distinguished* representative, where m is a subobject of X.

The composition in $\mathsf{P}_\mathsf{M}\mathsf{C}$ is defined by a pullback in C

$$X \xleftarrow{\ m\ } A \xrightarrow{\ f\ } Y \xleftarrow{\ n\ } B \xrightarrow{\ g\ } Z \tag{5.16}$$

with n', C, f' in the pullback.

$$[n, g].[m, f] = [mn', gf'].$$

It is easy to see that the composition is well defined. The following list of exercises shows that $\mathsf{P}_\mathsf{M}\mathsf{C}$ is indeed a category, with a natural e-cohesive structure.

5.1.8 Exercises and complements

(a) For all X, Y in C, the new morphisms $X \dashrightarrow Y$ form a set. *Hints:* use the distinguished representatives mentioned above.

(b) $\mathsf{P}_\mathsf{M}\mathsf{C}$ is a category, with identities $\operatorname{id} X = [\operatorname{id} X, \operatorname{id} X]$.

(c) There is a canonical functor

$$\mathsf{C} \to \mathsf{P}_\mathsf{M}\mathsf{C},$$
$$(f \colon X \to Y) \mapsto [\operatorname{id} X, f] \colon X \dashrightarrow Y, \tag{5.17}$$

which is the identity on objects and faithful; we will identify $f = [\operatorname{id} X, f]$. (The universal property of this embedding is complicated.)

(d) Every $m \colon A \rightarrowtail X$ in M is a split mono in $\mathsf{P}_\mathsf{M}\mathsf{C}$, with a distinguished retraction $m^\sharp = [m, \operatorname{id} A] \colon X \dashrightarrow A$. Moreover $[m, f] = fm^\sharp$.

(e) The monomorphisms of $\mathsf{P}_\mathsf{M}\mathsf{C}$ are precisely those of C.

(f) $\mathsf{P}_\mathsf{M}\mathsf{C}$ is an e-category, with projectors given by the endomorphisms $mm^\sharp \colon X \dashrightarrow X$, for m in M.

5.2 Limits and colimits

The general notion of limit of a functor includes cartesian products and equalisers (dealt with in Section 1.7), pullbacks (in 5.1.5) and the classical 'projective limits'.

Dually, colimits comprise sums, coequalisers, pushouts and the classical 'injective limits' (see [M3, Bo, G10]).

5.2.1 Cones and limits

Let S be a small category and $X \colon S \to C$ a functor, that will be written in 'index notation', for $i \in S = \mathrm{Ob}\,S$ and $a \colon i \to j$ in S:

$$X \colon S \to C, \qquad i \mapsto X_i, \quad a \mapsto (X_a \colon X_i \to X_j). \tag{5.18}$$

A *cone* for X is an object A of C (the *vertex* of the cone) equipped with a family of maps $(f_i \colon A \to X_i)_{i \in S}$ in C such that the following triangles commute (for $a \colon i \to j$ in S)

$$
\begin{array}{ccc}
A & \xrightarrow{\ f_i\ } & X_i \\
 & \searrow{\scriptstyle f_j} & \big\downarrow{\scriptstyle X_a} \\
 & & X_j
\end{array}
\qquad\qquad X_a . f_i = f_j. \tag{5.19}
$$

The *limit* of $X \colon S \to C$ is a universal cone

$$(L, (u_i \colon L \to X_i)_{i \in S}).$$

This means a cone of X such that every cone $(A, (f_i \colon A \to X_i)_{i \in S})$ factorises uniquely through the former: there is a unique map $f \colon A \to L$ such that, for all $i \in S$, $u_i f = f_i$.

The solution need not exist. When it does, it is determined up to a unique coherent isomorphism, and the object L is denoted as $\mathrm{Lim}(X)$.

Dually, the *colimit* of the functor X is a universal *cocone*

$$(L', (u_i \colon X_i \to L')_{i \in S}),$$

the cocone condition being: $u_j . X_a = u_i$, for $a \colon i \to j$ in S. The object L' is denoted as $\mathrm{Colim}(X)$.

Limits and colimits can be interpreted as universal arrows: see 5.2.4.

Remarks. The categorical structure of S selects the functors $X \colon S \to C$ we are considering.

However, if X is such a functor, its cones only depend on the 'graph' underlying S (forgetting the composition of S); in particular, an identity $a = \mathrm{id}\,i$ of S gives a trivial condition (5.19) in a cone (and a trivial condition in a cocone).

5.2.2 Particular cases

The product ΠX_i of a family $(X_i)_{i \in S}$ of objects of C is the limit of the corresponding functor $X \colon S \to C$, defined on the discrete category whose objects are the elements $i \in S$ (and whose morphisms are the formal identities of these objects).

The equaliser in C of a pair of parallel morphisms $f, g\colon X_0 \to X_1$ is the limit of the obvious functor defined on the category $0 \rightrightarrows 1$ (the identities being understood).

The pullback of a pair of morphisms $f\colon X_1 \to X_0 \leftarrow X_2 : g$ (with the same codomain) is the limit of the obvious functor defined on the category $\wedge\colon 1 \to 0 \leftarrow 2$.

In fact a cone (A, u_1, u_2, u_0) essentially consists of two commutative triangles, and this is equivalent to a triple (A, u_1, u_2) forming a commutative square with f and g

$$
\begin{array}{ccc}
 & X_1 & \\
{\scriptstyle u_1} \nearrow & & \searrow {\scriptstyle f} \\
A \;\xrightarrow{\;\;\;\; u_0 \;\;\;\;}\; & & X_0 \\
{\scriptstyle u_2} \searrow & & \nearrow {\scriptstyle g} \\
 & X_2 &
\end{array}
\tag{5.20}
$$

Dually, the pushout of a pair (f, g) of morphisms with the same domain is the colimit of the obvious functor defined on the category $\vee = \wedge^{\mathrm{op}}$: $1 \leftarrow 0 \to 2$.

5.2.3 Complete categories and the preservation of limits

A category C is said to be *complete* (resp. *finitely complete*) if it has a limit for every functor $\mathsf{S} \to \mathsf{C}$ defined on a *small* category (resp. a *finite* category).

One says that a functor $F\colon \mathsf{C} \to \mathsf{D}$ *preserves the limit*

$$(L, (u_i\colon L \to X_i)_{i \in S})$$

of a functor $X\colon \mathsf{S} \to \mathsf{C}$ if the cone $(FL, (Fu_i\colon FL \to FX_i)_{i \in S})$ is the limit of the composed functor $FX\colon \mathsf{S} \to \mathsf{D}$. One says that F *preserves limits* if it preserves all the limits *which exist* in C. Analogously for the preservation of products, equalisers, etc.

One proves, by a constructive argument, that a category is complete (resp. finitely complete) if and only if it has equalisers and products (resp. finite products): see [M3], Section V.2, Theorem 1, or [G10], Theorem 2.2.4. Moreover, if C is complete, a functor $F\colon \mathsf{C} \to \mathsf{D}$ preserves all limits (resp. all finite limits) if and only if it preserves equalisers and products (resp. finite products).

Dual results hold for colimits; for instance, a category is cocomplete (resp. finitely cocomplete) if and only if it has coequalisers and sums (resp. finite sums).

The categories Set, Top, Ab are complete and cocomplete; the forgetful functor Top $\to$ Set preserves limits and colimits, while Ab $\to$ Set only preserves limits.

The category associated to a preordered set X is complete if and only if the latter has all meets; since this is equivalent to the existence of all joins (cf. 1.1.3), the category X is complete if and only if it is cocomplete. In the ordered case, this amounts to a complete lattice.

Exercises and complements. (a) A left adjoint preserves colimits, while a right adjoint preserves limits.

(b) A representable functor $C(A, -)\colon C \to$ Set (cf. 1.5.3) preserves limits and monomorphisms.

5.2.4 *Limits and colimits as universal arrows*

Consider the category C^S of functors $S \to C$ and their natural transformations (cf. 1.6.6). Letting i vary in Ob S and a in Mor S, the *diagonal functor* of C^S

$$\Delta\colon C \to C^S, \qquad \Delta(X)_i = X, \quad \Delta(X)_a = \operatorname{id} X,$$
$$(\Delta f)_i = f, \tag{5.21}$$

sends an object X of C to the constant functor at X, and a morphism $f\colon X \to Y$ to the natural transformation $\Delta f\colon \Delta X \to \Delta Y\colon S \to C$ whose components are constant at f.

(a) The limit of a diagram $X\colon S \to C$ is a universal arrow $(L, \varepsilon\colon \Delta L \to X)$ from the diagonal functor to the object X of C^S (see 1.5.7). All S-based limits in C exist if and only if the diagonal functor $\Delta\colon C \to C^S$ has a right adjoint $L\colon C^S \to C$, called the limit functor.

(a*) Dually, the colimit of X is a universal arrow $(L', \eta\colon X \to \Delta L')$ from the object X to the diagonal functor. All S-based colimits in C exist if and only if the diagonal functor $\Delta\colon C \to C^S$ has a left adjoint $L'\colon C^S \to C$, called the colimit functor.

5.2.5 *Splitting the idempotents*

An idempotent endomorphism $e\colon X \to X$ of the category C can be described as a functor $F\colon E \to C$ defined on the idempotent monoid on two elements (already used in 2.2.5(a))

$$E = \{1, \underline{e}\}, \qquad \underline{e}.\underline{e} = \underline{e}. \tag{5.22}$$

We say that the idempotent endomorphism $e\colon X \to X$ *splits* in C if it has a factorisation

$$e = mp\colon X \to A \to X, \qquad pm = 1_A. \tag{5.23}$$

Note that m is a split mono and p is a split epi. We have already seen, in 2.2.5, that every category C can be embedded in its *idempotent completion* C^E, where all idempotents split.

The reader should find it interesting (and easy) to prove the following facts.

Exercises and complements. (a) Let $e\colon X \to X$ be an endomorphism of C, associated to the functor $F\colon E \to \mathsf{C}$, $F(\underline{e}) = e$. The following conditions in the category C are equivalent:

(i) the idempotent e splits as $e = mp$, with $pm = 1_A$,

(ii) e has an epi-mono factorisation $e = mp$,

(iii) the functor F has a limit $m\colon A \to X$,

(iv) the pair of morphisms $1, e\colon X \rightrightarrows X$ has an equaliser $m\colon A \to X$,

(iii*) the functor F has a colimit $p\colon X \to A$,

(iv*) the pair $1, e\colon X \rightrightarrows X$ has a coequaliser $p\colon X \to A$.

This also proves that m is determined up to the equivalence relation of monos with values in X, and p is determined up to the equivalence relation of epis defined on X.

(b) After biproducts (in 1.7.9), this is another case where *the limit and colimit objects coincide*; here the coincidence is even necessary, as soon as our limit or colimit exists. Again (and of course) the limit-cone $m\colon A \to X$ and the colimit-cocone $p\colon X \to A$ are distinct.

(c) By (a), all idempotents split in every category with equalisers (or coequalisers). But we have seen in (2.26) that this also holds in the category $\mathsf{Rel\,Ab}$, which lacks general equalisers.

(d) If K is an inverse category and the idempotent $e\colon X \to X$ splits in K, then $e = mm^\sharp$ for a monomorphism m determined up to the equivalence relation of monos with values in X.

5.2.6 Regular subobjects and quotients

General subobjects and quotients can be of little interest, as is the case in Top (see Exercise (c), below). In the rest of this section we briefly review stronger categorical notions.

In a category, a *regular monomorphism* $m: X \to A$ is, by definition, an equaliser of some pair of maps $A \rightrightarrows A'$; it is called a *regular subobject* if it is a subobject, according to our choice (cf. 5.1.3).

Dually, a *regular epimorphism* $p: A \to X$ is a coequaliser of some pair of maps $A' \rightrightarrows A$; in particular we have the *regular quotients* of A.

Exercises and complements. These exercises show the importance of regular monos and epis; the solutions are not difficult, and left to an interested reader.

(a) In Set every monomorphism (resp. epimorphism) is regular: in fact, every subset $E \subset X$ is the equaliser of its cokernel pair, and every quotient set X/R is the coequaliser of its kernel pair.

Therefore all subobjects are regular, and can be identified with subsets. All quotients are regular, and can be identified with set-theoretical quotients (modulo an equivalence relation).

(b) In Ab and R Mod every monomorphism (resp. epimorphism) is regular (and even 'normal', as we shall see below). All subobjects are regular and can be identified with subgroups; all quotients are regular and can be identified with ordinary quotients (modulo a subgroup or submodule).

(c) In Top every subspace $E \subset X$ is the equaliser of its cokernel pair, and every quotient space X/R is the coequaliser of its kernel pair.

Here every injective continuous mapping is a monomorphism, while the regular subobjects amount to *inclusions of subspaces*: the 'general subobjects' are less important than the regular ones. Similarly, every surjective continuous mapping is an epi, but the regular quotients amount to projections on quotient spaces, in the usual sense.

The regular monomorphisms are thus the same as the topological embeddings of 1.4.9, and the regular epimorphisms are the topological projections.

(d) Similarly, in Ord, a monomorphism is any injective monotone mapping, and is a regular mono if and only if it *reflects* the ordering, so that it induces an isomorphism between the domain and an ordered subset of the codomain. Characterise regular epimorphisms.

*(e) In the usual categories of algebraic structures (more precisely, in a 'variety of algebras'), regular epimorphisms give the usual quotients of algebras, modulo a congruence; on the other hand, subalgebras correspond to general subobjects (cf. [G10], Section 4.3).

In these categories, general epis on the one hand, regular monos on the other, can be difficult to determine and — finally — are of a limited interest. Knowing a few cases should be enough: for instance, all subgroups are regular subobjects in Gp (a non-trivial fact, see [AHS], Exercise 7H), while a subsemigroup need not be a regular subobject in the category of semigroups ([AHS], Exercise 7I).

5.2.7 Kernels and cokernels

In a pointed category C, the *kernel* of a morphism $f\colon A \to B$ is defined by a universal property, as an object $\operatorname{Ker} f$ equipped with a morphism $\ker f\colon \operatorname{Ker} f \to A$, such that $f.(\ker f) = 0$ and

(i) every map h such that $fh = 0$ factorises uniquely through $\ker f$

$$\operatorname{Ker} f \xrightarrow{\ \ker f\ } A \xrightarrow{\ f\ } B \tag{5.24}$$

which means that there exists a unique morphism u such that $h = (\ker f)u$.

Equivalently, $\ker f$ is the equaliser of f and $0_{AB}\colon A \to B$; it is mono. A *normal monomorphism* is any kernel of a morphism, and is always a regular mono; a *normal subobject,* of course, is a normal monomorphism which is a subobject.

Generally, the existence of kernels does not require the existence of all equalisers. However, in a preadditive category (see 1.7.9), the equaliser of two morphisms $f, g\colon A \to B$ is the same as the kernel of the difference $f - g$, so that normal and regular monomorphisms coincide.

Dually, the *cokernel* of f is the coequaliser of f and the zero morphism $A \to B$, i.e. an object $\operatorname{Cok} f$ equipped with an (epi)morphism $\operatorname{cok} f\colon B \to \operatorname{Cok} f$, such that $(\operatorname{cok} f).f = 0$ and

(i*) every map h such that $hf = 0$ factorises uniquely through $\operatorname{cok} f$

$$A \xrightarrow{\ f\ } B \xrightarrow{\ \operatorname{cok} f\ } \operatorname{Cok} f \tag{5.25}$$

i.e. there is a unique morphism u such that $h = u(\operatorname{cok} f)$.

A *normal epimorphism* is any cokernel of a morphism, and is always a regular epi. The converse holds in a preadditive category, where $\operatorname{cok} (f - g)$ is the coequaliser of the pair f, g.

Exercises and complements. These exercises on a morphism $f\colon A \to B$ are easy and left to the reader.

(a) In $R\operatorname{Mod}$, $\operatorname{Set}_\bullet$, $\operatorname{Top}_\bullet$ and Gp the natural kernel-object is the ordinary subobject $\operatorname{Ker} f$ (the preimage of the zero subobject) and $\ker f$ is its embedding in the domain of f. However, if $u\colon K \to \operatorname{Ker} f$ is an isomorphism, also the composite $(\ker f)u\colon K \to A$ is *a* kernel of f.

*The preservation of kernels by 'exact functors' between pointed categories works *in this sense*, as any preservation of limits. Strict preservation of distinguished limits is rarely of interest.*

(b) In R Mod and Set, every subobject is normal. In Top, the normal subobjects correspond to the pointed subspaces. In Gp the normal subobjects are the usual ones (i.e. the invariant subgroups); note that *normal monomorphisms are not closed under composition* in Gp.

(c) In R Mod the natural cokernel-object is the quotient $B/f(A)$, and cok f is the canonical projection $B \to B/f(A)$. In Gp the cokernel-object is the quotient of B modulo the invariant subgroup generated by $f(A)$. In these categories all epimorphisms are normal.

(d) In Set, and Top, cokernels are easy to describe. There are non-normal epimorphisms.

(e) The category Rng of unital rings is not pointed, but the larger category Rng$'$ of possibly non-unital rings is. Normal subobjects in the latter correspond to bilateral ideals.

5.3 Tensor product of modules and Hom

As an introduction to the next section on monoidal categories, we review the definition and main properties of the tensor product and Hom-functor of R-modules – a classical topic on which the tensor calculus of vector bundles also relies. Further information, including the exactness properties of these functors, can be found in any text on Linear or Homological Algebra, like [Bou1, M1, CE].

We deal with the category R Mod of modules on a *commutative* unital ring R. We have already recalled, in 1.5.2, that the category Ab of abelian groups is canonically isomorphic to the category $\mathbb{Z}$ Mod of modules on the ring of integers, and we make no distinction between them.

The trivial R-module is written as 0.

5.3.1 Linear and multilinear maps

The hom-set R Mod(A, B) of R-homomorphisms $A \to B$ has a canonical structure of R-module, usually written as $\mathrm{Hom}_R(A, B)$, or $\mathrm{Hom}(A, B)$ when no confusion may arise.

Its operations are the pointwise sum and pointwise multiplication by scalars $\lambda \in R$:

$$(f + g)(a) = f(a) + g(a), \qquad (\lambda f)(a) = \lambda.f(a). \tag{5.26}$$

These operations are preserved by compositions: for $h\colon A' \to A$ and $k\colon B \to B'$ we have:

$$k(f+g)h = kfh + kgh, \qquad k(\lambda f)h = \lambda(kfh). \tag{5.27}$$

Now, *for a cartesian product* $A = \Pi_i A_i$ $(i = 1, ..., n)$, a mapping $\varphi\colon A \dashrightarrow B$ with values in a module B is said to be *multilinear*, or *n-linear*, on R if it is R-linear in each variable:

$$\varphi(a_1, ..., \lambda a_i' + \mu a_i'', ..., a_n)$$
$$= \lambda\varphi(a_1, ..., a_i', ..., a_n) + \mu\varphi(a_1, ..., a_i'', ..., a_n). \tag{5.28}$$

Multilinear mappings $\varphi\colon A \dashrightarrow B$ will be denoted by dot-marked arrows. They have a pointwise sum and pointwise scalar multiplication, that extend the previous ones (corresponding to the case $n = 1$).

Composing multilinear mappings is less obvious. It will be useful to remark that, given a *family* of homomorphisms $h_i\colon X_i \to A_i$ $(i = 1, ..., n)$ and a homomorphism $k\colon B \to Y$, the composite

$$k.\varphi.(\Pi_i f_i)\colon \Pi_i X_i \to Y,$$
$$(x_1, ..., x_n) \mapsto k(\varphi(f_1(x_1), ..., f_n(x_n))), \tag{5.29}$$

is multilinear. *(One could form a category, using this fact.)*

For abelian groups, a mapping $\varphi\colon \Pi_i A_i \dashrightarrow B$ is multilinear on $\mathbb{Z}$ if (and only if) it is *multiadditive*, i.e. additive in each variable.

5.3.2 Tensor product of modules

The *tensor product* of two R-modules A, B is defined by a universal property whose goal is to transform bilinear mappings into homomorphism.

Precisely, we look for a module $A \otimes_R B$ *equipped with a bilinear mapping* $\varphi_0\colon A \times B \dashrightarrow A \otimes_R B$ such that, for every bilinear mapping $\varphi\colon A \times B \dashrightarrow C$

$$
\begin{array}{ccc}
A \times B & \overset{\varphi_0}{\dashrightarrow} & A \otimes_R B \\
{\scriptstyle\varphi}\Big\downarrow & \swarrow {\scriptstyle h} & \\
C & &
\end{array}
\tag{5.30}
$$

there a unique homomorphism $h\colon A \otimes_R B \to C$ such that $\varphi = h\varphi_0$.

As always, this universal property characterises the solution, up to a unique isomorphism coherent with the structural bilinear mappings. In fact, if $\varphi_i\colon A \times B \dashrightarrow C_i$ $(i = 1, 2)$ are two solutions of our problem, there are two (determined) homomorphisms h, k such that

$$h\colon C_1 \to C_2, \qquad k\colon C_2 \to C_1, \qquad \varphi_2 = h\varphi_1, \qquad \varphi_1 = k\varphi_2.$$

Now, the homomorphism $kh: C_1 \to C_1$ satisfies

$$(kh)\varphi_1 = \varphi_1 = (\operatorname{id} C_1)\varphi_1,$$

and is the identity; similarly $hk = \operatorname{id} C_2$, which means that h and k are isomorphisms, inverse to each other.

To prove that the solution exists, we 'construct' one (by a complicated procedure, rarely used in computations)

$$A \otimes_R B = R(A \times B)/H(A, B), \qquad \varphi_0(a, b) = [(a, b)]. \qquad (5.31)$$

Here $R(A \times B)$ is the free R-module generated by the *set* $A \times B$ (see Exercise 1.5.7(b)), and its elements will be written as essentially finite R-linear combinations of pairs $(a, b) \in A \times B$.

Now, $H(A, B)$ is the submodule of $R(A \times B)$ generated by all the elements of the following types (for $a, a' \in A$, $b, b' \in B$ and $\lambda \in R$)

(i) $(a + a', b) - (a, b) - (a', b)$,

(ii) $(\lambda a, b) - \lambda(a, b)$,

(iii) $(a, b + b') - (a, b) - (a, b')$,

(iv) $(a, \lambda b) - \lambda(a, b)$.

Taking in (5.31) the quotient modulo $H(A, B)$ is precisely what is required to make the mapping φ_0 bilinear on R. We write

$$a \otimes b = \varphi_0(a, b) = [(a, b)] \in A \otimes_R B \qquad \text{(for } a \in A, b \in B). \qquad (5.32)$$

The generators (i)–(iv) of $H(A, B)$ give the bilinearity of the tensor product of elements:

$$(a + a') \otimes b = a \otimes b + a' \otimes b, \qquad (\lambda a) \otimes b = \lambda(a \otimes b),$$

$$a \otimes (b + b') = a \otimes b + a \otimes b', \qquad a \otimes (\lambda b) = \lambda(a \otimes b).$$

We often write $A \otimes B$ for $A \otimes_R B$, when no ambiguity can arise.

More generally, one can define the tensor product $\otimes A_i$ of a family $A_1, ..., A_n$ of R-modules, as a module equipped with a multilinear mapping

$$\varphi_0 \colon \prod A_i \twoheadrightarrow \otimes A_i, \qquad (5.33)$$

universal 'as above', for all n-linear mappings $\prod A_i \twoheadrightarrow C$. This will be useful, for instance to prove the associativity of the (binary) tensor product (in 5.3.5).

*For a non-commutative ring R, the tensor product $A \otimes_R B$ is defined for a right R-module A_R and a left R-module $_R B$; the result is an abelian group [Bou1, M1, CE].

More generally, one can form a double category $\mathbb{R}$ng of rings, homomorphisms and bimodules $_RA_S\colon R \rightarrowtail S$, with vertical composition the tensor product of bimodules $A \otimes_S B\colon R \rightarrowtail T$. See [GP1].*

5.3.3 Remarks, exercises and complements

(a) Every element of $A \otimes B$ can be written as a (finite) sum $\sum_i a_i \otimes b_i$; obviously, this expression is not unique (generally).

The elements of the form $a \otimes b$ form a *canonical system of generators* of $A \otimes_R B$. More generally, if a (resp. b) varies in a system of generators of A (resp. B), the elements $a \otimes b$ still form a system of generators of $A \otimes B$.

There are cases where all these generators annihilate, and $A \otimes_R B = 0$ (the trivial R-module).

(b) First, let us note that $a \otimes 0 = 0 \otimes b = 0$ (for $a \in A$ and $b \in B$). Therefore $A \otimes 0 = 0 \otimes B = 0$.

(c) Suppose now that the (commutative) ring R is an integral domain (i.e. has no proper zero-divisors). If the module A is divisible by a non-zero scalar λ (i.e. the equation $\lambda x = a$ has solutions, for every $a \in A$) and $\lambda B = 0$, then $A \otimes B = 0$.

In fact, for every $a = \lambda a' \in A$ and $b \in B$ we get $a \otimes b = (\lambda a') \otimes b = a' \otimes (\lambda b) = 0$.

(d) In the same hypothesis on R, if A is a *divisible* R-module (i.e. the equation $\lambda x = a$ has solutions, for every $a \in A$ and every scalar $\lambda \neq 0$) while B is a *torsion* R-module (every $b \in B$ is annihilated by some non-zero scalar), then $A \otimes B = 0$.

(e) Taking $R = \mathbb{Z}$, for every torsion abelian group B we have $\mathbb{Q} \otimes B = 0 = \mathbb{R} \otimes B$.

*(f) Taking $R = \mathbb{Z}$, the exactness properties of the tensor product show that $\mathbb{Z}/m \otimes \mathbb{Z}/n \cong \mathbb{Z}/d$ where d is the greatest common divisor of m, n. In particular, if m and n are coprime integers, we get $\mathbb{Z}/m \otimes \mathbb{Z}/n = 0$. (All this can be extended to modules on a principal ideal domain.)

5.3.4 Tensor product of homomorphisms

Given two homomorphisms $f\colon A \to A'$, $g\colon B \to B'$ there is a homomorphism $f \otimes g$ determined as follows on the canonical generators

$$f \otimes g\colon A \otimes_R B \to A' \otimes_R B', \qquad (f \otimes g)(a \otimes b) = f(a) \otimes g(b). \qquad (5.34)$$

In fact, the bilinear mapping

$$A \times B \to A' \otimes_R B', \qquad (a, b) \mapsto f(a) \otimes g(b), \qquad (5.35)$$

determines such a homomorphism on $A \otimes_R B$.

Plainly, this construction preserves identities and composition:

$$\mathrm{id}\, A \otimes \mathrm{id}\, B = \mathrm{id}\,(A \otimes_R B), \qquad (f'f) \otimes (g'g) = (f' \otimes g').(f \otimes g). \qquad (5.36)$$

The tensor product is thus a (covariant) functor in two variables

$$\otimes \colon R\,\mathsf{Mod} \times R\,\mathsf{Mod} \to R\,\mathsf{Mod}. \qquad (5.37)$$

This functor is *bilinear*, i.e. additive and homogeneous in each variable, with respect to the R-linear structure of morphisms defined above (in 5.3.1):

$$\begin{aligned}
(f + f') \otimes g &= f \otimes g + f' \otimes g, & (\lambda.f) \otimes g &= \lambda.(f \otimes g), \\
f \otimes (g + g') &= f \otimes g + f \otimes g', & f \otimes (\lambda.g) &= \lambda.(f \otimes g).
\end{aligned} \qquad (5.38)$$

5.3.5 Basic properties of the tensor product

The commutative ring R is fixed, and we write $A \otimes B$ for $A \otimes_R B$.

(a) The tensor product is *commutative*. More precisely, there is a canonical isomorphism determined as follows on the canonical generators:

$$A \otimes B \to B \otimes A, \qquad a \otimes b \mapsto b \otimes a. \qquad (5.39)$$

In fact, the obvious bilinear map $A \times B \to B \times A \to B \otimes A$ determines a homomorphism $A \otimes B \to B \otimes A$ that acts as above on the canonical generators; symmetrically, we have a homomorphism $B \otimes A \to A \otimes B$ that sends $b \otimes a$ to $a \otimes b$. These homomorphisms are inverse to each other (on generators, whence everywhere).

(b) The tensor product *has a unit*, the R-module R. More precisely, there is a canonical isomorphism:

$$A \otimes R \rightleftarrows A, \qquad a \otimes \lambda \mapsto \lambda a, \qquad a \mapsto a \otimes 1_R. \qquad (5.40)$$

In fact the mapping $\varphi_0 \colon A \times R \to A$, $(a, \lambda) \mapsto \lambda a$ is bilinear and satisfies the universal property that defines $A \otimes R$.

(c) The tensor product is *associative*, up to a canonical isomorphism:

$$(A \otimes B) \otimes C \to A \otimes (B \otimes C), \qquad (a \otimes b) \otimes c \mapsto a \otimes (b \otimes c). \qquad (5.41)$$

To give a simple proof, we use the ternary tensor product $A \otimes B \otimes C$ (mentioned at the end of 5.3.2), and show that $(A \otimes B) \otimes C$ is a realisation of $A \otimes B \otimes C$. Symmetrically, this also is true of $A \otimes (B \otimes C)$, so that these two realisations are isomorphic, in a coherent way.

In fact, the trilinear mapping

$$\psi_0 \colon A \times B \times C \twoheadrightarrow (A \otimes B) \otimes C, \qquad (a, b, c) \mapsto (a \otimes b) \otimes c,$$

is the composite of the mappings $\varphi_1 \times C$ and φ_2 defined below (where φ_1 and φ_2 are bilinear)

$$
\begin{array}{ccc}
A \times B \times C & \xrightarrow{\varphi_1 \times C} & (A \otimes B) \times C \\
\psi \downarrow & \swarrow \varphi' & \downarrow \varphi_2 \\
D & \xleftarrow{\quad h \quad} & (A \otimes B) \otimes C
\end{array}
\tag{5.42}
$$

It follows that ψ_0 satisfies the universal property of a ternary tensor product: a trilinear mapping $\psi \colon A \times B \times C \to D$ factorises first as $\varphi'(\varphi_1 \times C)$ where $\varphi' \colon (A \otimes B) \times C \twoheadrightarrow D$ is bilinear; then we factorise the latter as $\varphi' = h\varphi_2$, where h is a homomorphism. The uniqueness of h in the factorisation $\psi = h\psi_0$ is obvious, because the elements of type $(a \otimes b) \otimes c$ generate $(A \otimes B) \otimes C$ (by 5.3.3(a)).

(d) The tensor product is *distributive* with respect to direct sums, by a canonical isomorphism:

$$
\begin{aligned}
\left(\textstyle\bigoplus_{i \in I} A_i\right) \otimes B &\to \textstyle\bigoplus_{i \in I} (A_i \otimes B), \\
(a_i)_{i \in I} \otimes b &\mapsto (a_i \otimes b)_{i \in I}.
\end{aligned}
\tag{5.43}
$$

(e) As a consequence, there are canonical isomorphisms:

$$
\begin{aligned}
R^{(I)} \otimes B &\cong B^{(I)} = \textstyle\bigoplus_{i \in I} B, & R^m \otimes B &\cong B^m, \\
R^{(I)} \otimes R^{(J)} &\cong R^{(I \times J)}, & R^m \otimes R^n &\cong R^{mn},
\end{aligned}
\tag{5.44}
$$

where I and J are sets, m and n are natural numbers; $X^{(I)} = \bigoplus_{i \in I} X$ denotes the direct sum of a constant family of modules.

(f) *The tensor product of free modules (and all vector spaces) is thus determined:* if A and B are free on R, with bases $(a_i)_{i \in I}$ and $(b_j)_{j \in J}$, then $A \otimes B$ is free with basis

$$(a_i \otimes b_j)_{(i,j) \in I \times J}. \tag{5.45}$$

*(g) Every free R-module A is *flat*, which means that the functor $A \otimes_R -$ is exact: a consequence of (e). In particular, one proves that an abelian group A is flat if and only if it is *torsion-free*: namely, if $kx = 0$ (with $k \in \mathbb{Z}$ and $x \in A$) then $k = 0$ or $x = 0$.

*(h) For an abelian group A, the tensor product $A \otimes_{\mathbb{Z}} \mathbb{Q}$ is (readily made into) a vector space on $\mathbb{Q}$. The *rank* of the abelian group A is defined as the dimension of this vector space

$$\mathrm{rk}(A) = \dim_{\mathbb{Q}}(A \otimes_{\mathbb{Z}} \mathbb{Q}). \tag{5.46}$$

As a consequence of an important Structure Theorem of finitely generated abelian groups, an abelian group A of this kind is (non-canonically) isomorphic to a direct sum $tA \oplus \mathbb{Z}^n$, where tA is the torsion part of A and $n = \mathrm{rk}(A)$.

More precisely, the short exact sequence

$$tA \rightarrowtail A \twoheadrightarrow A/tA$$

splits, and $A/tA \cong \mathbb{Z}^n$.

5.3.6 *The functor* Hom

Another classical instrument of linear (and homological) algebra is the functor Hom (also written as Hom_R).

In fact, the set-valued hom-functor (in 1.5.3(e))

$$\mathrm{Mor} \colon R\,\mathsf{Mod}^{\mathrm{op}} \times R\,\mathsf{Mod} \to \mathsf{Set},$$

has an enriched version, essentially introduced in 5.3.1

$$\mathrm{Hom} \colon R\,\mathsf{Mod}^{\mathrm{op}} \times R\,\mathsf{Mod} \to R\,\mathsf{Mod},$$
$$(A, B) \mapsto \mathrm{Hom}(A, B), \qquad (h, k) \mapsto k. - .h. \tag{5.47}$$

The module $\mathrm{Hom}(A, B)$ is the set of R-homomorphisms $A \to B$, with the pointwise sum $f + g$ and scalar multiplication λf. Furthermore, for two R-linear mappings $h \colon A' \to A$ and $k \colon B \to B'$, the mapping

$$k. - .h \colon \mathrm{Hom}(A, B) \to \mathrm{Hom}(A', B')$$

is R-linear, as shown in (5.27).

We also note that there is a canonical functorial isomorphism

$$\mathrm{Hom}(R, -) \cong \mathrm{id} \colon R\,\mathsf{Mod} \to R\,\mathsf{Mod},$$
$$(f \colon R \to A) \mapsto f(1_R), \tag{5.48}$$

which is an 'enriched version' of the representability of the forgetful functor $R\,\mathsf{Mod} \to \mathsf{Set}$, by the module R.

5.3.7 The exponential law for sets and modules

In order to describe the relationship between tensor product and Hom, we begin with a similar, more elementary occurrence in Set.

(a) (*The exponential law for sets*) The hom-functor of Set

$$\text{Mor} \colon \mathsf{Set}^{\mathrm{op}} \times \mathsf{Set} \to \mathsf{Set},$$
$$(A, Y) \mapsto \mathsf{Set}(A, Y), \qquad (h, k) \mapsto k. - .h, \tag{5.49}$$

is related to the binary cartesian product by a natural isomorphism in the variables X, Y (depending on the 'parameter' A)

$$\varphi^A_{XY} \colon \mathsf{Set}(X \times A, Y) \to \mathsf{Set}(X, \mathsf{Set}(A, Y)),$$
$$(f \colon X \times A \to Y) \mapsto (g \colon X \to \mathsf{Set}(A, Y)), \tag{5.50}$$
$$g(x) = f(x, -) \colon A \to Y.$$

We have thus a family of adjunctions, depending on the parameter A: the functor $- \times A \colon \mathsf{Set} \to \mathsf{Set}$ is left adjoint to the functor $\mathsf{Set}(A, -) \colon \mathsf{Set} \to \mathsf{Set}$ (see Section 1.8).

The hom-sets of the category Set are often written in 'exponential notation' $Y^A = \mathsf{Set}(A, Y)$, because an indexed family $(y_a)_{a \in A}$ of Y^A is the same as a mapping $y \colon A \to Y$. With this notation, the previous family of bijections is rewritten as

$$Y^{X \times A} \cong (Y^A)^X, \tag{5.51}$$

which explains why this natural isomorphism is called an 'exponential law'.

(b) (*The exponential law for modules*) Similarly we have an *exponential law* consisting of a natural family of bijections

$$\varphi^A_{XY} \colon R\,\mathsf{Mod}(X \otimes A, Y) \to R\,\mathsf{Mod}(X, \mathrm{Hom}(A, Y)),$$
$$(f \colon X \otimes A \to Y) \mapsto (g \colon X \to \mathrm{Hom}(A, Y)), \tag{5.52}$$
$$g(x) = f(x \otimes -) \colon A \to Y,$$

saying that the endo-functor $- \otimes A$ is left adjoint to the endo-functor $\mathrm{Hom}(A, -)$ (of R-modules).

Concretely, a homomorphism $g \colon X \to \mathrm{Hom}(A, Y)$ amounts to a mapping $X \times A \twoheadrightarrow Y$ which is linear in each variable, and therefore to a homomorphism $f \colon X \otimes A \to Y$.

Note. In fact, (5.52) is a natural family of R-isomorphisms

$$\varphi^A_{XY} \colon \mathrm{Hom}(X \otimes A, Y) \to \mathrm{Hom}(X, \mathrm{Hom}(A, Y)), \tag{5.53}$$

and gives an adjunction enriched over $R\,\mathsf{Mod}$ (see 5.5.3).*

5.3.8 Exercises and complements (Tensor product of vector spaces)

The base ring is now a (commutative) field K. Modules on K are called vector spaces and have specific properties, essentially derived from the well-known theory of linear dependence in vector spaces. The category $K\mathsf{Mod}$ is often written as $K\mathsf{Vct}$.

(a) In $K\mathsf{Vct}$ every monomorphism splits, i.e. has a left inverse. In other words, every subspace of a vector space is a retract.

(b) In $K\mathsf{Vct}$ every epimorphism splits, i.e. has a right inverse. Every quotient of a vector space is a retract.

(c) For vector spaces A, B, there is a canonical homomorphism

$$u_{AB}\colon A \otimes_K B \to \mathrm{Hom}_K(A^*, B),$$
$$u_{AB}(a \otimes b)(\alpha) = \alpha(a).b \qquad (\text{for } \alpha\colon A \to K), \tag{5.54}$$

where $A^* = \mathrm{Hom}_K(A, K)$ is the *dual* K-vector space of A. This family defines a natural transformation

$$u\colon (- \otimes_K =) \to \mathrm{Hom}_K((-)^*, =)\colon K\mathsf{Vct} \times K\mathsf{Vct} \to K\mathsf{Vct}. \tag{5.55}$$

Prove that each component u_{AB} is injective. Moreover, if A and B are *finitely generated*, then u_{AB} is an isomorphism. *Hints:* use a basis of A.

It follows that the tensor product of *finitely generated* vector spaces *can be* (and often *is*) defined as: $A \otimes_K B = \mathrm{Hom}_K(A^*, B)$. This fact can also be used for vector bundles, in Section 4.2.

*(d) A reader acquainted with the basic notions of homological algebra will already know, or can easily deduce from (a) or (b), that all short exact sequences in $K\mathsf{Vct}$ split, and every additive functor $F\colon K\mathsf{Vct} \to K\mathsf{Vct}$ is exact.

5.4 Monoidal categories and closedness

A monoidal structure on a category consists of a 'well-behaved' binary product $A * B$. Enrichment on a monoidal category is considered in the next section.

Further information on the theory of monoidal categories and enrichment on them can be found in [EiK, M2, M3, Kl1, Kl2].

5.4.1 Monoidal categories

A *monoidal category* $(V, *, E)$ is a category V equipped with an object E (called *the unit object*) and a functor in two variables

$$V \times V \to V, \qquad (A, B) \mapsto A * B, \qquad (5.56)$$

often called the *tensor product* of the structure and written as $A \otimes B$.

Without entering into details, this operation is assumed to be associative *up to a natural isomorphism*

$$(A * B) * C \cong A * (B * C),$$

and the object E is assumed to be an identity, *up to natural isomorphisms*

$$E * A \cong A \cong A * E.$$

All these isomorphisms must form a 'coherent' system so that, loosely speaking, any well-formed diagram of them commutes. This allows one to forget them and write

$$(A * B) * C = A * (B * C), \qquad E * A = A = A * E.$$

A *symmetric* monoidal category is further equipped with a *symmetry isomorphism*, coherent with the other ones:

$$s_{XY} : X * Y \to Y * X. \qquad (5.57)$$

The latter *cannot be omitted*: note that $s_{XX} : X * X \to X * X$ is not the identity, in general.

The *canonical forgetful functor* of V is defined as the following representable functor (cf. 1.5.3):

$$U = V(E, -) : V \to \mathsf{Set}. \qquad (5.58)$$

This functor need not be faithful (see the examples below). It comes equipped with a natural transformation in two variables, that links the cartesian structure of Set (given by cartesian products, see below) with the monoidal structure of V

$$\varphi_{AB} : U(A) \times U(B) \to U(A * B),$$
$$\varphi(a : E \to A, \, b : E \to B) = (a * b : E = E * E \to A * B). \qquad (5.59)$$

5.4.2 Examples

(a) A category V with finite products has a symmetric monoidal structure given by the categorical product and the terminal object (the empty product); this structure is called *cartesian*.

In this case the natural transformation φ considered above, in (5.59), is a functorial isomorphism (because a representable functor preserves cartesian products, as proved in 5.2.3(b)).

In particular, we are interested in the cartesian structures of Set, Ord and Cat.

The canonical forgetful functor $\mathsf{Set}(\{*\}, -)$ is isomorphic to id Set, while $\mathsf{Ord}(\{*\}, -)$ is isomorphic to the usual forgetful functor of ordered sets, given by the underlying set: both functors are faithful.

On the other hand, the forgetful functor $\mathsf{Cat}(\mathbf{1}, -)$ is isomorphic to the functor of objects Ob : Cat $\to$ Set and is not faithful.

Let us note that, for a *pointed category* V (see 1.7.9) with finite products, the cartesian monoidal structure is often of little interest. In fact the unit object $\top$ is the zero-object 0, and the associated forgetful functor $V(0, -)$: $V \to$ Set sends every object to a singleton. (Another shortcoming of pointed cartesian structures can be found in Exercise 5.4.3(a).)

(b) In fact, various pointed categories have other monoidal structures of interest. The important symmetric monoidal structure of the category Ab of abelian groups is the usual tensor product $A \otimes B$ (reviewed in Section 5.3), with identity object $\mathbb{Z}$ and forgetful functor $\mathsf{Ab}(\mathbb{Z}, -)$ isomorphic to the usual forgetful functor of abelian groups.

Here the components $\varphi_{AB} \colon U(A) \times U(B) \to U(A \otimes B)$ of the natural transformation (5.59) are the structural bilinear mappings of the tensor product.

More generally, for a commutative unital ring R, the important (symmetric) monoidal structure of R Mod is the tensor product over R, also recalled in Section 5.3.

(c) The important symmetric monoidal structure of the category Set. of pointed sets is the *smash product* $X \wedge Y$. Writing as 0 the base-point of every pointed set, the smash product is obtained as

$$X \wedge Y = (X \times Y)/(X \times \{0\} \cup \{0\} \times Y), \tag{5.60}$$

by collapsing, in the cartesian product $X \times Y$, the subset formed by the 'cartesian axes', namely $X \times \{0\}$ and $\{0\} \times Y$ (i.e. by taking the quotient of $X \times Y$ modulo the equivalence relation that identifies all those points).

The base point of $X \wedge Y$ is obviously the class $[(0,0)]$, written again as 0, or $0_{X \wedge Y}$.

This tensor product solves a universal property similar to that of abelian groups: the set $X \wedge Y$ is characterised by having a *universal bi-pointed mapping*

$$\varphi_0 \colon X \times Y \twoheadrightarrow X \wedge Y, \qquad \varphi_0(x,y) = x \wedge y = [(x,y)], \qquad (5.61)$$

(pointed in each variable, i.e. sending all pairs $(x,0)$ and $(0,y)$ to 0).

The universal property says that every bi-pointed mapping $\varphi \colon X \times Y \twoheadrightarrow Z$ (with values in an arbitrary pointed set) can be factorised as $\varphi = f\varphi_0$, for a unique pointed mapping $f \colon X \wedge Y \to Z$.

The identity of the smash product is the set $\mathbb{S}^0 = \{-1, 1\}$, pointed at 1 (for instance).

The associativity isomorphism

$$(X \wedge Y) \wedge Z \to X \wedge (Y \wedge Z), \qquad (x \wedge y) \wedge z \mapsto x \wedge (y \wedge z), \qquad (5.62)$$

can be constructed as in 5.3.5(c), showing that the pointed sets $(X \wedge Y) \wedge Z$ and $X \wedge (Y \wedge Z)$ are realisations of a ternary tensor product

$$X \wedge Y \wedge Z = (X \times Y \times Z)/H,$$

$$H = (X \times Y \times \{0\}) \cup (X \times \{0\} \times Z) \cup (\{0\} \times Y \times Z).$$

(d) An ordered monoid M can also be interpreted as a (small, skeletal) monoidal category, with objects in M and morphisms $\lambda \to \mu$ representing the order relation $\lambda \leqslant \mu$.

The monoidal structure is given by the product $\lambda\mu$ and the unit 1 of the monoid; actually, it is a *strict* monoidal structure, where the associativity and unit isomorphisms are identities; it is symmetric if and only if the monoid is commutative.

Enrichment on an ordered monoid will be studied in Section 6.2.

5.4.3 Exponentiable objects and internal homs

In a symmetric monoidal category C, an object A is said to be *exponentiable* if the functor $- \otimes A \colon \mathsf{C} \to \mathsf{C}$ has a right adjoint, written as $(-)^A \colon \mathsf{C} \to \mathsf{C}$ or $\mathrm{Hom}(A, -)$, and called an *internal hom*. There is thus a family of bijections, natural in the variables X, Y

$$\varphi^A_{XY} \colon \mathsf{C}(X \otimes A, Y) \to \mathsf{C}(X, \mathrm{Hom}(A, Y)) \qquad (\textit{exponential law}). \quad (5.63)$$

Since adjunctions compose, all the tensor powers $A^{\otimes n}$ are also exponentiable, with

$$\mathrm{Hom}(A^{\otimes n}, -) = (\mathrm{Hom}(A, -))^n. \qquad (5.64)$$

A symmetric monoidal category is said to be *closed* if all its objects are exponentiable. There is then a functor (cf. [Kl2])

$$\mathrm{Hom}\colon \mathsf{C}^{\mathrm{op}} \times \mathsf{C} \to \mathsf{C},$$

$$(A, Y) \mapsto \mathrm{Hom}(A, Y), \qquad (h, k) \mapsto k. - .h. \tag{5.65}$$

A category with finite products is said to be *cartesian closed* if all the objects are exponentiable for the cartesian product.

Exercises and complements. (a) In a pointed category any object which is exponentiable for the cartesian product is trivial, i.e. a zero-object.

(b) A pointed cartesian closed category C is indiscrete and non-empty: it is equivalent to **1**.

*(c) A non-symmetric monoidal category can have a left and a right hom functors. This is the case of the category of cubical sets [BwH].

5.4.4 Main examples

(a) **Set** is cartesian closed, with an obvious internal hom: $\mathrm{Hom}(A, Y) = \mathsf{Set}(A, Y)$. The exponential law was proved in 5.3.7.

(b) The category **Cat** of *small* categories is cartesian closed, with the internal hom $\mathsf{Cat}(\mathsf{S}, \mathsf{C}) = \mathsf{C}^{\mathsf{S}}$ described in 1.6.6.

(c) **Ab** is symmetric monoidal closed, with respect to the usual tensor product and Hom functor. The same holds for every category $R\,\mathsf{Mod}$ of modules on a commutative unital ring.

(d) The category **Ord** of ordered sets is cartesian closed, with internal hom:

$$\mathrm{Hom}\colon \mathsf{Ord}^{\mathrm{op}} \times \mathsf{Ord} \to \mathsf{Ord}, \tag{5.66}$$

where $\mathrm{Hom}(A, Y)$, is the set of increasing mappings $A \to Y$ equipped with the pointwise order: $f \leqslant f'$ means that $f(a) \leqslant f'(a)$, for every $a \in A$.

The exponential law holds, because a mapping $X \times A \to Y$ which is monotone in each variable also respects the product-ordering on $X \times A$, applying transitivity. (Comparing this case with the previous one, the cartesian product of ordered sets appears to be the appropriate tensor product in **Ord**.)

(e) The category **Set.** of pointed sets is symmetric monoidal closed, with respect to the smash product $X \wedge A$ (cf. 5.4.2(c)), with internal hom:

$$\mathrm{Hom}\colon \mathsf{Set.}^{\mathrm{op}} \times \mathsf{Set.} \to \mathsf{Set.}, \tag{5.67}$$

where $\mathrm{Hom}(A, Y)$, is the set of pointed mappings $A \to Y$, pointed at the zero-map. The exponential law holds because a mapping $X \times A \to Y$ which is pointed in each variable amounts to a pointed map on $X \wedge A$.

(f) Other examples can be found below, and in Section 6.6.

5.4.5 *Other examples and complements*

(a) Top is not cartesian closed: the rational line $\mathbb{Q}$ (with euclidean topology) is not exponentiable, as the endo-functor $F = - \times \mathbb{Q}$ does not preserve coequalisers.

An exercise of [Bou2], Section I.5, shows that the functor F does not preserve the quotient of $\mathbb{Q}$ by the equivalence relation that identifies all the points of $\mathbb{Z}$. The proof is not easy. A characterisation of exponentiable spaces, by a property of '*quasi local compactness*', can be found in [LoR], together with many related results.

But it is well known (and not difficult to prove) that *every locally compact space A is exponentiable*: for every space Y, the space $Y^A = \mathrm{Hom}(A, Y)$ is the set of maps $\mathsf{Top}(A, Y)$ endowed with the compact-open topology.

In particular, the standard interval $\mathbb{I} = [0, 1]$ is exponentiable, with all its powers. The cylinder functor $I(X) = X \times \mathbb{I}$ is thus left adjoint to the path-functor $P(Y) = Y^{\mathbb{I}}$, and a homotopy between maps $X \to Y$ can be equivalently defined as a map $I(X) \to Y$ or a map $X \to P(Y)$.

(b) This result can be easily transferred to the category pTop of preordered spaces (see Section 4.3).

Every preordered space A that has a *locally compact topology* and an *arbitrary preorder* is exponentiable in pTop, with Y^A consisting of the set $\mathsf{pTop}(A, Y) \subset \mathsf{Top}(UA, UY)$ of *monotone continuous mappings*, equipped with the (induced) compact-open topology and the *pointwise preorder*, where $f \prec g$ means that $f(x) \prec_Y g(x)$, for all $x \in A$. (See [G8], Section 1.1.)

The directed interval $\uparrow\mathbb{I}$ is thus exponentiable in pTop, and directed homotopies therein can be equivalently defined as maps $X \times \uparrow\mathbb{I} \to Y$ or maps $X \to Y^{\uparrow\mathbb{I}}$.

(c) The category Rng of (unital) rings has a symmetric monoidal structure: the tensor product $R \otimes S$ of two rings is the tensor product $R \otimes_{\mathbb{Z}} S$ of the underlying abelian groups, with the $\mathbb{Z}$-bilinear multiplication generated by letting

$$(x \otimes y).(x' \otimes y') = xx' \otimes yy' \qquad (x, x' \in R \text{ and } y, y' \in S). \qquad (5.68)$$

The unit is the ring of integers (which is also the initial object of the category).

(d) The full subcategory $C\mathsf{Rng}$ of commutative rings inherits a monoidal structure which is co-cartesian: $R_1 \otimes R_2$ is now the categorical sum of the commutative rings R_i, with 'injections'

$$u_i \colon R_i \to R_1 \otimes R_2, \qquad u_1(x) = x \otimes 1, \quad u_2(y) = 1 \otimes y. \tag{5.69}$$

In fact, a pair of homomorphisms $f_i \colon R_i \to S$ of commutative rings has a unique $\mathbb{Z}$-linear 'extension'

$$f \colon R_1 \otimes R_2 \to S, \qquad f(x \otimes y) = f_1(x).f_2(y),$$

which is a homomorphism of rings because S is commutative.

Note that the 'injections' u_i need not be monomorphisms: $R \otimes 0$ is always the trivial ring; the same holds for $\mathbb{Z}/2 \otimes \mathbb{Z}/3$.

(e) The category Ban of Banach spaces (on the real or complex field K) and continuous linear mappings has an important symmetric monoidal closed structure, where the tensor product is the completion of the tensor product of normed vector spaces; the same is true of its wide subcategory Ban_1. The reader can find these structures in [Se].

It is interesting to note that the canonical forgetful functors yield different results:

$$\mathsf{Ban}(K, X) \cong |X|,$$
$$\mathsf{Ban}_1(K, X) \cong B_1(X) = \{x \in X \mid ||x|| \leqslant 1\}, \tag{5.70}$$

the second being (up to natural isomorphism) the standard closed ball of the Banach space X, as a set.

In both cases the internal hom-object $\mathrm{Hom}(X, Y)$ is the Banach space of *all* continuous linear mappings $X \to Y$, with its usual norm. Applying the relevant forgetful functor to the latter one gets, respectively, the set $\mathsf{Ban}(X, Y)$ or the set $B_1(\mathrm{Hom}(X, Y)) = \mathsf{Ban}_1(X, Y)$.

Enrichment on Ban or Ban_1 (cf. Section 5.5) shows as the basic idea of enriching an ordinary category C on V by *putting a V-structure on the hom-sets* $\mathsf{C}(X, Y)$ is only correct when the canonical forgetful functor of V yields the underlying set of a structured set: this is indeed the case in all the previous instances, except Ban_1.

5.5 Enrichment on a monoidal category

Enrichment on an ordered category is at the basis of our analysis of manifolds.

But the theory of enriched categories began with a different kind of enrichment, on a symmetric monoidal category $(\mathsf{V}, *, E)$. We review here this topic.

5.5.1 Preadditive categories

We start with a typical form of enrichment. As defined in 1.7.9, a preadditive category C is a category where

- every set of morphisms $\mathsf{C}(X, Y)$ is given a structure of abelian group, written as $f + f'$,

- these structures are 'consistent with composition', in the sense that all the composition mappings

$$\mathsf{C}(X, Y) \times \mathsf{C}(Y, Z) \to \mathsf{C}(X, Z), \qquad (f, g) \mapsto gf, \qquad (5.71)$$

are *additive in each variable* (and therefore bilinear on the ring $\mathbb{Z}$)

$$g(f + f') = gf + gf', \qquad (g + g')f = gf + g'f.$$

Equivalently, we can assign the composition mappings as *homomorphisms* defined on the tensor product of abelian groups

$$\mathsf{C}(X, Y) \otimes \mathsf{C}(Y, Z) \to \mathsf{C}(X, Z), \qquad f \otimes g \mapsto gf. \qquad (5.72)$$

In this way, according to a definition reviewed below, a preadditive category is the same as a category *enriched on the monoidal category* Ab (with the usual tensor product).

5.5.2 Enriched categories

A *category* C *enriched on the symmetric monoidal category* $(\mathsf{V}, *, E)$, also called a V-*category*, consists of

(a) a class of objects $\mathrm{Ob}\,\mathsf{C}$,

(b) for $x, y \in \mathrm{Ob}\,\mathsf{C}$, a V-object $\mathsf{C}(x, y)$ (the *hom-object*),

(c) for $x, y, z \in \mathrm{Ob}\,\mathsf{C}$, a V-morphism

$$c = c_{xyz} \colon \mathsf{C}(x, y) * \mathsf{C}(y, z) \to \mathsf{C}(x, z) \qquad (\text{the } composition),$$

(d) for $x \in \mathrm{Ob}\,\mathsf{C}$, a V-morphism $e = e_x \colon E \to \mathsf{C}(x, x)$ (the *unit*).

These data are to satisfy two axioms of unitarity and associativity: the following diagrams must commute in V, for all $x, y, z, w \in \mathrm{Ob}\,\mathsf{C}$ (always omitting the isomorphisms of the monoidal structure of V):

(i) (*unitarity*)

$$
\begin{array}{ccc}
\mathsf{C}(x,y) & \xrightarrow{\;e*1\;} & \mathsf{C}(x,x)*\mathsf{C}(x,y) \\[2pt]
{\scriptstyle 1*e}\downarrow & \searrow{\scriptstyle 1} & \downarrow{\scriptstyle c} \\[2pt]
\mathsf{C}(x,y)*\mathsf{C}(y,y) & \xrightarrow[\;c\;]{} & \mathsf{C}(x,y)
\end{array}
\tag{5.73}
$$

(ii) (*associativity*)

$$
\begin{array}{ccc}
\mathsf{C}(x,y)*\mathsf{C}(y,z)*\mathsf{C}(z,w) & \xrightarrow{\;c*1\;} & \mathsf{C}(x,z)*\mathsf{C}(z,w) \\[2pt]
{\scriptstyle 1*c}\downarrow & & \downarrow{\scriptstyle c} \\[2pt]
\mathsf{C}(x,y)*\mathsf{C}(y,w) & \xrightarrow[\;c\;]{} & \mathsf{C}(x,w)
\end{array}
\tag{5.74}
$$

Examples will be given in 5.5.4–5.5.6. There is an underlying category $|\mathsf{C}|$, obtained by forgetting the enriched structure via the canonical forgetful functor $|-| = \mathsf{V}(E,-)\colon \mathsf{V} \to \mathsf{Set}$ (cf. (5.58)).

Complements. For the interested reader, the formal definition of the underlying category $|\mathsf{C}|$ is as follows.

The objects are those of C. The hom-sets are the underlying sets of the hom-objects

$$|\mathsf{C}|(x,y) = |\mathsf{C}(x,y)| = \mathsf{V}(E, \mathsf{C}(x,y)). \tag{5.75}$$

Composition derives from the enriched composition, via the natural transformation φ of (5.59)

$$|\mathsf{C}(x,y)| \times |\mathsf{C}(y,z)| \xrightarrow{\;\varphi\;} |\mathsf{C}(x,y)*\mathsf{C}(y,z)| \xrightarrow{\;|c|\;} |\mathsf{C}(x,z)|. \tag{5.76}$$

Finally, the identity of the object x comes from its enriched identity $e_x \colon E \to \mathsf{C}(x,x)$

$$\operatorname{id} x = |(e_x \colon E \to \mathsf{C}(x,x))|(\operatorname{id} E), \tag{5.77}$$

where $|(e_x \colon E \to \mathsf{C}(x,x))|$ is a mapping $\mathsf{V}(E,E) \to |\mathsf{C}|(x,x)$ of sets, that we apply to $\operatorname{id} E$.

5.5.3 Enriched functors and transformations

An *enriched functor* $F \colon \mathsf{C} \to \mathsf{D}$ between V-categories, or V-*functor*, consists of a mapping $F \colon \operatorname{Ob}\mathsf{C} \to \operatorname{Ob}\mathsf{D}$ and a family of V-morphisms

$$F = F_{xy} \colon \mathsf{C}(x,y) \to \mathsf{D}(F(x), F(y)) \qquad (x,y \in \operatorname{Ob}\mathsf{C}), \tag{5.78}$$

that *preserves identities and composition*.

This means that, for $x, y, z \in \mathrm{Ob}\,\mathsf{C}$, the following diagrams commute in the category V

$$
\begin{array}{ccc}
E & \xrightarrow{\;e\;} & \mathsf{C}(x,x) \\
& \searrow{}^{e} & \downarrow{}^{F} \\
& & \mathsf{D}(Fx, Fx)
\end{array}
\tag{5.79}
$$

$$
\begin{array}{ccc}
\mathsf{C}(x,y) * \mathsf{C}(y,z) & \xrightarrow{\;c\;} & \mathsf{C}(x,z) \\
{}^{F*F}\downarrow & & \downarrow{}^{F} \\
\mathsf{D}(Fx, Fy) * \mathsf{D}(Fy, Fz) & \xrightarrow[c]{} & \mathsf{D}(Fx, Fz)
\end{array}
\tag{5.80}
$$

Again the enriched functor $F\colon \mathsf{C} \to \mathsf{D}$ has an underlying (ordinary) functor $|F|\colon |\mathsf{C}| \to |\mathsf{D}|$, with the same action on objects and the underlying action on morphisms (for $x, y \in \mathrm{Ob}\,\mathsf{C}$)

$$
|F|_{xy} \;=\; \mathsf{V}(E, F_{xy})\colon |\mathsf{C}(x,y)| \to |\mathsf{D}(F(x), F(y))|.
\tag{5.81}
$$

A V-*transformation* $\varphi\colon F \to G\colon \mathsf{C} \to \mathsf{D}$ between V-functors consists of a family of V-morphisms

$$
\varphi x\colon E \to \mathsf{D}(Fx, Gx) \qquad (x \in \mathrm{Ob}\,\mathsf{C}),
\tag{5.82}
$$

making the following diagrams commutative

$$
\begin{array}{ccc}
E * \mathsf{C}(x,y) & \xrightarrow{\;\varphi x * G\;} \mathsf{D}(Fx, Gx) * \mathsf{D}(Gx, Gy) \xrightarrow{\;c\;} \mathsf{D}(Fx, Gy) \\
\| & \qquad\qquad\qquad\qquad\qquad\qquad\qquad \| \\
\mathsf{C}(x,y) * E & \xrightarrow[F*\varphi y]{} \mathsf{D}(Fx, Fy) * \mathsf{D}(Fy, Gy) \xrightarrow[c]{} \mathsf{D}(Fx, Gy)
\end{array}
\tag{5.83}
$$

Appropriate compositions give the 2-category VCat of V-categories (see [Kl2]). One can now define V-adjunctions.

5.5.4 Examples

The following examples are rather obvious.

(a) A category enriched on Set, with respect to its cartesian structure, is nothing more than a category; an enriched functor (resp. transformation) on Set is an ordinary functor (resp. natural transformation).

(b) Enrichment on Ord, with respect to its cartesian structure, gives the structure of an ordered category, described in 2.1.1. An enriched functor amounts to an order-preserving functor.

(c) As we have seen in 5.5.1, a category enriched on Ab, with respect to the symmetric monoidal structure of the ordinary tensor product, is the same as a preadditive category; an enriched functor is an additive one (i.e. a functor that preserves the sum of parallel maps).

Ab itself has such a structure, replacing every hom-set $\mathsf{Ab}(A, B)$ by the abelian group $\mathrm{Hom}(A, B)$. The same holds for $R\,\mathsf{Mod}$ and Ban.

For a commutative ring R, one can consider R-*linear categories*, enriched on the monoidal category $(R\,\mathsf{Mod}, \otimes_R, R)$, like $R\,\mathsf{Mod}$ itself or the category of R-algebras.

(d) A category C *enriched on* $\mathsf{Set_\bullet}$, with respect to the smash product of 5.4.2(c), amounts to a category where in every hom-set $\mathsf{C}(X, Y)$ we have selected a distinguished map 0_{XY}, so that (for $f\colon X' \to X$ and $g\colon Y \to Y'$):

$$0_{XY}.f = 0_{X'Y}, \qquad g.0_{XY} = 0_{XY'}.$$

Plainly, every pointed category has such a structure. But the present notion is more general, since every full subcategory of a pointed category keeps this enrichment, even if we take out all its zero-objects.

(Some authors use the term 'pointed category' in the more general sense of a category enriched on $\mathsf{Set_\bullet}$.)

5.5.5 *Generalised metric spaces*

An unexpected example, introduced by Lawvere [Lw] in 1974, lead to far-reaching developments: a notion of Cauchy completion for enriched categories, which will be examined in various forms in Chapter 6.

The ordered set $\underline{R} = [0, \infty]^{\mathrm{op}}$ of extended positive real numbers is viewed as a strict symmetric monoidal category, and called the *Lawvere monoidal category* $\underline{R}$. An arrow $\lambda \to \mu$ is an inequality $\lambda \geqslant \mu$, the tensor product is $\lambda + \mu$, its identity is 0.

A *generalised metric space* is a small category enriched on this monoidal category.

This means a set X (of the 'objects' of the category) equipped with a (generalised) distance $d(x, y) \in \underline{R}$ (for $x, y \in X$) satisfying the following axioms (for $x, y, z \in X$):

(i) $d(x, x) = 0$ (*identity law*),

(ii) $d(x, y) + d(y, z) \geqslant d(x, z)$ (*triangle inequality*).

The unit-morphism 5.5.2(d) gives here the condition $0 \geqslant d(x, x)$, equivalent to (i); the composition morphism gives the triangle inequality. If one adds the symmetry axiom, this generalised distance becomes a *pseudometric*, in the terminology of [Bou3].

An *enriched functor* $f\colon X \to Y$ is a mapping between the underlying sets, that satisfies

$$d_X(x, x') \geqslant d_Y(f(x), f(x')) \qquad (x, x' \in X), \qquad (5.84)$$

i.e. a weak contraction.

These (non-symmetric) generalised metric spaces have been used in the book [G8], as a sort of 'directed metric spaces'. In fact, we are in the domain of *weighted* algebraic topology, an enriched version of directed algebraic topology where paths are given a weight (measuring a cost of energy, time, money, etc.), which is infinite for an illegitimate path.

5.5.6 Other examples

(a) The category Slh of semilattices and homomorphisms has a symmetric monoidal structure, where $X \otimes Y$ is equipped with a universal bi-homomorphism $X \times Y \to X \otimes Y$ (that preserves the operation in each variable).

Enrichment on $(\mathsf{Slh}, \otimes)$ amounts to a *meet-semilatticed* category (see (3.5)).

(b) Enrichment on $(\mathsf{Slh}, \times)$, endowed with the cartesian structure $X \times Y$, amounts to a meet-semilatticed category satisfying the *cartesian compositive property* (3.6).

5.6 *Two-dimensional categories and internal categories

For readers with an interest in categories, we briefly sketch four forms of 2-dimensional categories: sesquicategories, 2-categories, bicategories and double categories. A monoidal category can be viewed in this perspective, as a bicategory on a single object.

We end this chapter with a brief mention of internal categories. The alternative between local orders and global orders (on a category) is an alternative between enriched structures and internal ones.

5.6.1 Sesquicategories

The category Cat, expanded with the natural transformations as second-order morphisms $\varphi\colon F \to G$ between parallel morphisms (the functors), has higher operations, examined in 1.6.2: vertical composition and whisker composition.

Abstracting the expanded structure, a *sesquicategory* [Str] is a category C equipped with:

(a) for each pair of parallel morphisms $f, g \colon X \to Y$, a set of *cells* (or *2-cells*) $C_2(f, g)$ whose elements are written as $\varphi \colon f \to g \colon X \to Y$ (or $\varphi \colon f \to g$), so that each map f has an *identity endo-cell* id $f \colon f \to f$;

(b) a *concatenation*, or *vertical composition* of 2-cells, written as $\psi\varphi$ or $\psi.\varphi$

$$X \quad \begin{array}{c} \xrightarrow{f} \\ {\scriptstyle\downarrow \varphi} \\ \xrightarrow{} \\ {\scriptstyle\downarrow \psi} \\ \xrightarrow{h} \end{array} \quad Y \qquad \psi\varphi \colon f \to h \colon X \to Y; \qquad (5.85)$$

(c) a *whisker composition* for 2-cells and maps, or *reduced horizontal composition*, written as $k\varphi h$ (or $k{\circ}\varphi{\circ}h$, when useful)

$$X' \xrightarrow{h} X \quad \begin{array}{c} \xrightarrow{f} \\ {\scriptstyle\downarrow \varphi} \\ \xrightarrow{g} \end{array} \quad Y \xrightarrow{k} Y' \qquad (5.86)$$

$$k\varphi h \colon kfh \to kgh \colon X' \to Y'.$$

These data must satisfy the following axioms (for *associativities, identities* and *distributivity of the whisker composition*):

$$\chi(\psi\varphi) = (\chi\psi)\varphi, \qquad k'(k\varphi h)h' = (k'k)\varphi(hh'),$$

$$\varphi.\mathrm{id}\, f = \varphi = \mathrm{id}\, g.\varphi, \quad 1_Y{\circ}\varphi{\circ}1_X = \varphi, \quad k{\circ}\mathrm{id}\, f{\circ}h = \mathrm{id}\,(kfh), \qquad (5.87)$$

$$k(\psi\varphi)h = (k\psi h).(k\varphi h).$$

We also write $\varphi{\circ}h = 1_Y{\circ}\varphi{\circ}h$ and $k{\circ}\varphi = k{\circ}\varphi{\circ}1_X$. Note that each set $C(X, Y)$ is a category, under vertical composition.

5.6.2 Two-categories

A *2-category* can be defined as a sesquicategory C which satisfies the following *reduced interchange property*:

$$X \quad \begin{array}{c} \xrightarrow{f} \\ {\scriptstyle\downarrow \varphi} \\ \xrightarrow{g} \end{array} \quad Y \quad \begin{array}{c} \xrightarrow{h} \\ {\scriptstyle\downarrow \psi} \\ \xrightarrow{k} \end{array} \quad Z \qquad (\psi{\circ}g).(h{\circ}\varphi) = (k{\circ}\varphi).(\psi{\circ}f). \qquad (5.88)$$

To recover the usual definition [Be1, KlS], one defines the *horizontal composition* of 2-cells φ, ψ which are *horizontally consecutive*, as in the diagram above, using the previous identity:

$$\psi{\circ}\varphi = (\psi{\circ}g).(h{\circ}\varphi) = (k{\circ}\varphi).(\psi{\circ}f) \colon hf \to kg \colon X \to Z. \qquad (5.89)$$

The whisker composition is now a particular case: $h\varphi = \mathrm{id}\,h{\circ}\varphi$ and $\psi f = \psi{\circ}\mathrm{id}\,f$.

Then, one proves that the horizontal composition of 2-cells is associative, has identities (namely, the identity cells of identity arrows) and satisfies the *middle-four interchange property* with vertical composition (an extension of the previous reduced interchange property):

$$X \; \xrightarrow[\;\;\downarrow\psi\;\;]{\;\;\downarrow\varphi\;\;} \; Y \; \xrightarrow[\;\;\downarrow\tau\;\;]{\;\;\downarrow\sigma\;\;} \; Z \qquad\qquad (\tau.\sigma){\circ}(\psi.\varphi) = (\tau{\circ}\psi).(\sigma{\circ}\varphi). \qquad (5.90)$$

As a prime example of such a structure, Cat will also denote the *2-category* of small categories, their functors and their natural transformations.

The usual definition of a 2-category relies on the complete horizontal composition, rather than the reduced one. But in practice one usually works with the reduced horizontal composition; and there are important sesquicategories where the reduced interchange property does not hold, and one does not define a full horizontal composition.

For instance, the sesquicategory of chain complexes, chain morphisms and homotopies, over an additive category, is not a 2-category.

Adjunctions, equivalences and adjoint equivalences can be defined *inside* a 2-category: for an adjunction one should use the 'algebraic' form (c), in Definition 1.8.2.

An internal adjunction in Cat is an ordinary adjunction between small categories.

An ordered category C, as defined in 2.1.1, is the same as a 2-category where each category $C(X, Y)$ is an ordered set; it is also called a *locally ordered 2-category*. Internal adjunctions in an ordered category will be used in 6.3.4.

Remarks and complements. A 2-category is the same as a category enriched on the category Cat, with respect to its cartesian monoidal structure.

A more general notion of *bicategory* was introduced by Bénabou [Be1]; it is a laxified version of a 2-category, where the horizontal composition is associative and has units up to (coherent) vertical isomorphisms; moreover, the class of morphisms $X \to Y$ between two given objects need not be a set. A monoidal category (resp. a strict monoidal category) amounts to a bicategory (resp. a 2-category) on a single object.

5.6.3 Two-dimensional functors and universal arrows

A 2-functor $U: \mathsf{A} \to \mathsf{X}$ between 2-categories sends objects to objects, arrows to arrows and cells to cells, strictly preserving the whole structure: domains and codomains, units and compositions.

Extending universal arrows of functors (in 1.5.7), a 2-*universal arrow* from an object X of X to the 2-functor $U: \mathsf{A} \to \mathsf{X}$ is a pair

$$(A_0, h: X \to UA_0)$$

which gives an isomorphism of categories (of arrows and cells, with vertical composition):

$$\mathsf{A}_2(A_0, A) \to \mathsf{X}_2(X, UA), \qquad g \mapsto Ug.h. \tag{5.91}$$

This amounts to saying that the functor (5.91), in the variable A, is bijective on objects, full and faithful, i.e.

(i) for every A in A and every $f: X \to UA$ in X there exists a unique $g: A_0 \to A$ in A such that $f = Ug.h$,

(ii) for every pair $g, g': A_0 \to A$ in A and every cell $\varphi: Ug.h \to Ug'.h$ in X, there is a unique cell $\psi: g \to g'$ in A such that $\varphi = U\psi.h$.

(Equivalently, one can use a global universal property: for each cell $\varphi: f \to f'$: $X \to UA$ in X, there is a unique cell $\psi: g \to g'$ in A such that $\varphi = U\psi.h$. This implies that $f = Ug.h$ and $f' = Ug'.h$.)

The solution of a 2-universal problem is determined up to isomorphism.

The *2-limit* of a 2-functor is defined by a 2-universal property, extending 5.2.4. The interested reader can verify that Cat has (small) 2-limits. Further information on 2-limits and related matter can be found in [Str].

More generally, a *biuniversal arrow* from the object X to the 2-functor $U: \mathsf{A} \to \mathsf{X}$ is a pair $(A_0, h: X \to UA_0)$ such that the functor (5.91) is an *equivalence* of categories. The solution of a biuniversal problem is determined up to internal equivalence (in a 2-category).

5.6.4 Lax functors and lax colimits

In order to extend the definitions of manifold and gluing, a *lax functor* $U: I \to \mathsf{C}$ from a preordered set I to an ordered category C is a family of objects $(U_i)_{i \in I}$ and a family of morphisms $u^i_j: U_i \to U_j$ (for $i \prec j$ in I) such that:

$$u^i_i \geqslant 1_{U_i}, \qquad u^j_k.u^i_j \leqslant u^i_k \qquad (i \prec j \prec k \text{ in } I). \tag{5.92}$$

A *lax transformation* $f: U \to V: I \to \mathsf{C}$ between lax functors is a family $f_i: U_i \to V_i$ $(i \in I)$ such that:

$$
\begin{array}{ccc}
U_i & \xrightarrow{\ f_i\ } & V_i \\
{\scriptstyle u^i_j}\downarrow & \leqslant & \downarrow{\scriptstyle v^i_j} \\
U_j & \xrightarrow[\ f_j\ ]{} & V_j
\end{array}
\qquad f_j.u^i_j \leqslant v^i_j.f_i.
\tag{5.93}
$$

These transformations compose, forming a category $\mathrm{Lax}(I, \mathsf{C})$, which is pointwise ordered: $f \leqslant g$, means that $f_i \leqslant g_i$ (for $i \in I$). There is an obvious diagonal monotone functor

$$
\Delta: \mathsf{C} \to \mathrm{Lax}(I, \mathsf{C}),
\tag{5.94}
$$

where, for an object X in C, the functor $\Delta X: I \to \mathsf{C}$ is defined as:

$$
(\Delta X)_i = X, \qquad u^i_j = \mathrm{id}\, X \qquad (i \prec j \text{ in } I).
\tag{5.95}
$$

A *lax cocone* of the lax functor $U: I \to \mathsf{C}$, with vertex X, is a lax transformation $f: U \to \Delta X$.

Now, the *lax colimit* of X in C is the same as a 2-universal arrow $(X, f: U \to \Delta X)$ from the object U of $\mathrm{Lax}(I, \mathsf{C})$ to the functor Δ.

Explicitly, this means that

(i) $f_j.u^i_j \leqslant f_i$, for all $i \prec j$,

(ii) for every lax cocone $g_i: U_i \to Y$ (satisfying $g_j.u^i_j \leqslant g_i$, for $i \prec j$), there is a map $g: X \to Y$ in C such that $g_i = g.f_i$ (for $i \in I$),

(iii) if $g', g'': X \to Y$ and $g'.f_i \leqslant g''.f_i$ (for $i \in I$), then $g' \leqslant g''$.

Also here, condition (iii) gives the uniqueness of g in (ii).

Let us note that the preordered set I is a category, with an arrow $i \to j$ associated to the relation $i \prec j$; it is also a locally discrete 2-category, where the cells are reduced to identities of arrows. More generally, one defines the lax colimit (and the lax limit) of a lax functor between 2-categories.

5.6.5 Double categories

Double categories were introduced by C. Ehresmann [E3, E4]. The weak notion, called a *pseudo* (or *weak*) double category, extends bicategories and represents the general form of a 2-dimensional category; it was introduced in [GP1, GP2], together with a study of limits and adjoints in this context.

Here we briefly review the basic terminology, for the strict case.

A *double category* $\mathbb{D}$ has horizontal morphisms $f\colon X \to X'$ (with composition gf), vertical morphisms $u\colon X \nrightarrow Y$ (with composition written as $u' \cdot u$, or $u \otimes u'$), and (double) cells α, as in the diagram below

$$
\begin{array}{ccc}
X & \xrightarrow{\ f\ } & X' \\
{\scriptstyle u}\downarrow & \alpha & \downarrow{\scriptstyle v} \\
Y & \xrightarrow[\ g\ ]{} & Y'
\end{array}
\qquad\qquad \alpha\colon (u\,{\textstyle\frac{f}{g}}\,v). \qquad\qquad (5.96)
$$

A cell has a *boundary* $\alpha\colon (u\,{\textstyle\frac{f}{g}}\,v)$, written as $\alpha\colon u \to v$ or $\alpha\colon f \nrightarrow g$ when convenient.

Cells have a horizontal composition

$$
\begin{array}{ccccc}
X & \xrightarrow{\ f\ } & X' & \xrightarrow{\ f'\ } & X'' \\
{\scriptstyle u}\downarrow & \alpha & \downarrow{\scriptstyle v} & \beta & \downarrow{\scriptstyle w} \\
Y & \xrightarrow[\ g\ ]{} & Y' & \xrightarrow[\ g'\ ]{} & Y''
\end{array}
\qquad\qquad (5.97)
$$

$$
(\alpha \mid \beta)\colon (u\,{\textstyle\frac{f'f}{g'g}}\,w) \qquad\qquad (\alpha\colon u \to v,\ \beta\colon v \to w),
$$

that agrees with the horizontal composition of maps. The vertical map v is thus the *horizontal* codomain of α and the horizontal domain of β.

Symmetrically, there is a vertical composition

$$
\frac{\alpha}{\gamma}\colon (u \otimes u'\ {\textstyle\frac{f}{h}}\ v \otimes v') \qquad\qquad (\alpha\colon f \nrightarrow g,\ \gamma\colon g \nrightarrow h)
$$

when the vertical codomain g of α coincides with the vertical domain of γ.

The axioms essentially say that both laws are 'categorical', and satisfy the interchange laws. Horizontal and vertical identities, of objects and maps, are denoted as follows

$$
\begin{aligned}
&1_X\colon X \to X, &\qquad &1_u\colon u \to u, \\
&e_X\colon X \nrightarrow X, &\qquad &e_f\colon f \nrightarrow f, &\qquad &\square_X = 1_{e_A} = e_{1_A}.
\end{aligned}
\qquad (5.98)
$$

The notions of *double subcategory* and *double functor* are obvious. A 2-category amounts to a double category whose vertical arrows (for instance) are identities.

Various examples and applications of strict and weak double categories can be found in [GP1, GP2]. In particular, the theory of 2-dimensional adjunctions has its natural setting in double categories.

Complements. (a) In a *flat* double category each cell is determined by the four maps of its boundary. This is the case of the double category $\mathbb{S}$lht of 1-semilattices, homomorphisms and transfer pairs, used in 2.7.6.

(b) Horizontal transformations of vertical functors are also used in 2.7.6.

First, a *vertical functor* $F\colon \mathsf{C} \to \mathbb{D}$, defined on an ordinary category, is a functor $\mathsf{C} \to \mathrm{Ver}_0\mathbb{D}$ with values in the category of objects and vertical arrows of $\mathbb{D}$.

Now a *horizontal transformation of vertical functors* $\varphi\colon F \to G\colon \mathsf{C} \to \mathbb{D}$ assigns to every object X of C a horizontal morphism $\varphi X\colon F(X) \to G(X)$ of $\mathbb{D}$, and to every morphism $f\colon X \to Y$ in C a double cell $\varphi f\colon \varphi X \dashrightarrow \varphi Y$ as below, satisfying obvious axioms (see [E2, GP1])

$$
\begin{array}{ccc}
F(X) & \xrightarrow{\ \varphi X\ } & G(X) \\
{\scriptstyle Ff}\downarrow\bullet & \varphi f & \bullet\downarrow{\scriptstyle Gf} \\
F(Y) & \xrightarrow[\ \varphi Y\ ]{} & G(Y)
\end{array}
\tag{5.99}
$$

5.6.6 Internal categories

The last part of this section assumes a basic knowledge of simplicial objects.

(a) Let us begin by remarking that a small category C can be described as a diagram in Set having the form of a 3-truncated simplicial set

$$
C_0 \underset{e_0}{\overset{\partial_i}{\rightleftarrows}} C_1 \underset{e_i}{\overset{\partial_i}{\rightleftarrows}} C_2 \underset{e_i}{\overset{\partial_i}{\dashleftarrow}} C_3
\tag{5.100}
$$

Here $C_0 = \mathrm{Ob}\,\mathsf{C}$ is the (small) set of objects of C, $C_1 = \mathrm{Mor}\,\mathsf{C}$ is the set of morphisms, C_2 (resp. C_3) is the set of pairs (resp. triples) of consecutive morphisms.

The faces ∂_i and the degeneracies e_i are defined as follows (on an object $X \in C_0$, an arrow $f \in C_1$, a pair $(f, g) \in C_2$ and a triple $(f, g, h) \in C_3$):

$$
\begin{array}{lll}
e_0(X) = \mathrm{id}\,X, & \partial_0(f) = \mathrm{Dom}\,(f), & \partial_1(f) = \mathrm{Cod}\,(f), \\
e_0(f) = (f, 1), & e_1(f) = (1, f), & \\
\partial_0(f, g) = f, & \partial_1(f, g) = gf, & \partial_2(f, g) = g, \\
e_0(f, g) = (f, g, 1), & e_1(f, g) = (f, 1, g), & e_2(f, g) = (1, f, g), \\
\partial_0(f, g, h) = (f, g), & \partial_1(f, g, h) = (f, hg), & \partial_2(f, g, h) = (gf, h), \\
& \partial_3(f, g, h) = (g, h). &
\end{array}
$$

The simplicial identities hold, so that (5.100) is indeed a 3-truncated simplicial set:

$$\partial_i \partial_j = \partial_{j-1} \partial_i \text{ for } i < j, \qquad e_j e_i = e_i e_{j-1} \text{ for } i < j,$$

$$\partial_i e_j = e_{j-1} \partial_i \text{ for } i < j, \qquad \partial_i e_j = e_j \partial_{i-1} \text{ for } i > j+1, \qquad (5.101)$$

$$\partial_i e_j = \text{id} \quad \text{for } i = j, j+1.$$

Finally, the following two squares are pullbacks (of sets)

$$
\begin{array}{ccc}
C_2 & \xrightarrow{\partial_0} & C_1 \\
{\scriptstyle \partial_2}\downarrow & & \downarrow{\scriptstyle \partial_1} \\
C_1 & \xrightarrow{\partial_0} & C_0
\end{array}
\qquad
\begin{array}{ccc}
C_3 & \xrightarrow{\partial_0} & C_2 \\
{\scriptstyle \partial_3}\downarrow & & \downarrow{\scriptstyle \partial_2} \\
C_2 & \xrightarrow{\partial_0} & C_1
\end{array}
\qquad (5.102)
$$

The morphism $m = \partial_1 \colon C_2 \to C_1$ is called the *composition morphism* of C, or *partial multiplication*.

(b) Replacing Set with an arbitrary category X, an *internal category* C in X, or a *category object* in C, consists of a diagram (5.100) in X, where

- $C_0 = \text{Ob}\,C$ is called the *object of objects* of C,

- $C_1 = \text{Mor}\,C$ is called the *object of morphisms*,

- C_2 is called the *object of consecutive pairs of morphisms*.

The following (redundant) axioms must be satisfied:

(i) the simplicial identities (5.101) hold, so that (5.100) is a 3-truncated simplicial object in X,

(ii) the square diagrams of (5.102) are pullbacks in X.

(It is not convenient to assume that the category X has all pullbacks.)

(c) An *internal functor* $F \colon C \to D$ between internal categories in X is defined as a morphism of 3-simplicial objects. In other words we have four arrows in X

$$F_i \colon C_i \to D_i \qquad (i = 0, ..., 3), \qquad (5.103)$$

that commute with faces and degeneracies. The components F_2, F_3 are determined by the pullback condition; for X = Set we simply have

$$F_2(f, g) = (F_1 f, F_1 g), \qquad F_3(f, g, h) = (F_1 f, F_1 g, F_1 h).$$

The composition of internal functors is obvious.

(d) An *internal transformation* $\varphi \colon F \to G \colon \mathsf{C} \to \mathsf{D}$ between internal functors in X is given (or represented) by an X-morphism $\hat{\varphi}$ satisfying the following conditions

$$\hat{\varphi} \colon C_0 \to D_1, \qquad \partial_0 . \hat{\varphi} = F_0, \qquad \partial_1 . \hat{\varphi} = G_0,$$
$$m.\langle F_1, \hat{\varphi}\partial_1 \rangle = m.\langle \hat{\varphi}\partial_0, G_1 \rangle. \tag{5.104}$$

For the last condition (of *naturality*), $m = \partial_1 \colon D_2 \to D_1$ is the partial multiplication of D. The morphism $\langle F_1, \hat{\varphi}\partial_1 \rangle \colon C_1 \to D_2$ with values in the pullback D_2 is well defined because

$$\partial_1 F_1 = F_0 \partial_1 = \partial_0 \, \hat{\varphi} \, \partial_1.$$

Similarly $\langle \hat{\varphi}\partial_0, G_1 \rangle \colon C_1 \to D_2$ is legitimate because

$$\partial_1 \, \hat{\varphi} \, \partial_0 = G_0 \partial_0 = \partial_0 G_1.$$

5.6.7 Examples and complements

(a) We have already seen that an internal category in Set is a small category.

(b) As recalled in 2.8.1, a globally ordered category C (in the sense used by C. Ehresmann) is the same as an internal category in Ord, satisfying a further condition: the order induced on each hom-set $\mathsf{C}(X, Y)$ is discrete.

(c) As recalled in [GP1], an internal category in Cat is a small double category. Internal functors are the double functors; internal transformations are the horizontal ones (or the vertical ones, according to the way we are presenting the double category).

This approach to double categories masks their transpose symmetry, that interchanges horizontal and vertical arrows. But it is adequate to weak double categories, which are 'pseudo-category objects' in Cat and do not have such a symmetry.

6

Enriched categories and Cauchy completion

Enriched categories on a monoidal category have been presented in Section 5.5. Here we are mostly interested in a different form of enrichment, based on an ordered category and already used in Chapters 3 and 4, in a symmetric form.

The general form, enrichment on a bicategory, which comprises these two forms of enrichment, will only be mentioned in 6.2.1.

6.1 Ordering and upper bounds

This preliminary section reviews some terminology of ordered sets and ordered categories, from complete lattices to frames, quantales and quantaloids. Manifolds on inverse quantaloids have been dealt with in Section 4.6.

For further information, the reader is referred to [Jo1, Bo] for frames and locales, to Rosenthal's book [Ro] for quantales and quantaloids.

6.1.1 Complete upper semilattices

We already recalled (in 1.1.3) that an ordered set X is a complete upper semilattice (i.e. has all joins) if and only if it is a complete lower semilattice (i.e. has all meets), and is then called a complete lattice.

However, a *homomorphism of complete upper semilattices* $h: X \to Y$ is (only) assumed to preserve arbitrary joins; as a consequence, it also preserves the order relation and the least element $\bot = \bigvee \emptyset$, but need not preserve the greatest element $\top = \bigwedge \emptyset$, nor binary meets.

For instance, if $f: S \to T$ is a mapping of sets, its covariant extension $f_*: \mathcal{P}X \to \mathcal{P}Y$ is a homomorphism of complete upper semilattices, and need not preserve total subsets nor intersections.

Here, we are interested in the category USlh of *complete upper semi-lattices* and their homomorphisms. (The category LSlh of *complete lower semilattices* is isomorphic to USlh, by reversing order in each object.)

The free complete upper semilattice generated by the set S is the boolean algebra $\mathcal{P}(S)$, with the obvious embedding

$$\eta\colon S \to |\mathcal{P}(S)|, \qquad x \mapsto \{x\}. \tag{6.1}$$

The universal property says that every mapping $f\colon S \to |Y|$ with values in a complete lattice can be uniquely extended to a mapping $g\colon \mathcal{P}(S) \to Y$ that preserves arbitrary joins, by letting $g(A) = \bigvee_{a \in A} f(a)$, for every $A \subset S$.

The monoidal structure of USlh is examined in 6.1.6.

6.1.2 Frames and locales

A *frame* X is a complete lattice where binary meets distribute on arbitrary joins:

$$(\bigvee x_i) \wedge y = \bigvee (x_i \wedge y) \qquad \text{(for } x_i, y \in X\text{)}. \tag{6.2}$$

A *homomorphism of frames* $h\colon X \to Y$ is a mapping that preserves arbitrary joins and finite meets; as a consequence, it also preserves the order relation (obviously) and the least and greatest elements $\bot = \bigvee \emptyset$, $\top = \bigwedge \emptyset$. This defines the category Frm *of frames* (and their homomorphisms), a subcategory of USlh.

Typically, the open subsets of a topological space S form a frame $\mathcal{O}(S)$, which is a subframe of the distributive complete lattice $\mathcal{P}(S)$ of all subsets. A continuous mapping $f\colon S \to T$ gives a homomorphism of frames *in the opposite direction*, defined by the inverse image of an open subset

$$f^*\colon \mathcal{O}(T) \to \mathcal{O}(S) \qquad V \mapsto f^{-1}(V). \tag{6.3}$$

In other words, there is a contravariant functor from Top to Frm, that is often expressed as a covariant functor with values in the opposite category Loc *of locales*

$$\mathcal{O}\colon \mathsf{Top} \to \mathsf{Loc}, \qquad \mathsf{Loc} = \mathsf{Frm}^{\mathrm{op}}. \tag{6.4}$$

A *locale* is thus the same as a frame, but a *morphism of locales* $X \to Y$ is defined as a frame-homomorphism $Y \to X$. In this way, the category Loc can be viewed as a surrogate of Top: it is the framework of 'pointless topology', that studies topological properties working on the open subsets [Jo2].

A product $X = \Pi X_i$ in the category Frm is a product of ordered sets (cf. 1.7.1), with pointwise joins and finite meets, computed on the factors X_i: for a subset $\xi \subset X$ of components $\xi_i = p_i(\xi) \subset X_i$, and a finite subset $\xi' \subset X$, we have

$$(\bigvee \xi)_i = \bigvee \xi_i, \qquad (\bigwedge \xi')_i = \bigwedge \xi'_i. \tag{6.5}$$

6.1.3 Quantales

A (unital) *quantale* X [Ro] is a complete lattice equipped with an associative multiplication $x.y$ that has a unit 1 and preserves arbitrary joins in each variable

$$(\bigvee x_i).y = \bigvee x_i.y, \qquad x.(\bigvee y_i) = \bigvee x.y_i. \tag{6.6}$$

A *homomorphism of quantales* $h\colon X \to Y$ is a mapping that preserves arbitrary joins, multiplication and 1. This defines the category Qtl *of quantales.*

A frame gives a commutative idempotent quantale, letting $x.y = x \wedge y$. In the opposite category Qtl^{op} one can investigate 'noncommutative pointless topology'.

The set of endo-relations Rel Set(S) of any set S is a quantale, with respect to the canonical order of relations and their composition.

6.1.4 Quantaloids

A quantaloid is a 'many-object extension' of the notion of quantale, in the same way as categories extend monoids.

As already recalled in 3.1.5, a quantaloid C is an ordered category where every family $f_i\colon X \to Y$ $(i \in I)$ of parallel morphisms has a join, and composition distributes on them.

In other words C is an ordered category where

(i) every hom-set $\mathsf{C}(X, Y)$ is a complete (upper semi)lattice,

(ii) the composition-mappings $\mathsf{C}(X, Y) \times \mathsf{C}(Y, Z) \to \mathsf{C}(X, Z)$ preserve arbitrary joins in each variable

$$g.(\bigvee f_i) = \bigvee gf_i, \qquad (\bigvee g_j).f = \bigvee g_j f, \tag{6.7}$$

or equivalently

$$(\bigvee_j g_j).(\bigvee_i f_i) = \bigvee_{ij}(g_j.f_i). \tag{6.8}$$

6.1.5 Examples

(a) We have seen in 3.1.5 that the category Rel Set is a quantaloid, while Rel Ab is not.

(b) The ordered categories $\mathcal{S}$, $\mathcal{T}$, $\mathcal{C}$, $\mathcal{D}$, $\mathcal{C}^r$, $\mathcal{F}$, $\mathcal{V}$ studied in Chapters 3 and 4 are not quantaloids: they are uc-quantaloids, in the sense of 3.1.5. Moreover, in all of them except $\mathcal{T}$ and $\mathcal{D}$, a set of parallel morphisms is upper bounded if (and only if) it is pairwise upper bounded.

(c) In Section 4.6 we have seen various quantaloids related to the previous e-categories, like $\mathcal{C}[S]$, $\mathcal{K}[S] = I\mathcal{C}[S]$ and $\mathcal{K}^r(n)$ (in 4.6.3(a), 4.6.3(b) and 4.6.5, respectively).

(d) The category USlh of complete upper semilattices (introduced in 6.1.1) is a quantaloid, with pointwise joins of parallel morphisms

$$(\bigvee f_i)(x) = \bigvee f_i(x). \tag{6.9}$$

On the other hand, a quantaloid is the same as a category enriched on the category USlh, with respect to the symmetric monoidal structure described below.

6.1.6 A monoidal structure

The category USlh of complete upper semilattices has a symmetric monoidal closed structure, where $X \otimes Y$ is defined by the universal property of bi-homomorphisms.

The solution can be constructed as a quotient of the free upper semilattice $\mathcal{P}(X \times Y)$ (cf. 6.1.1) modulo the least congruence $\sim$ of upper semilattices that links all of the following pairs of elements of $\mathcal{P}(X \times Y)$ (where $x \in X$, $A \subset X$, $y \in Y$, $B \subset Y$)

$$A \times \{y\} \sim \{\bigvee A\} \times \{y\}, \qquad \{x\} \times B \sim \{x\} \times \{\bigvee B\}, \tag{6.10}$$

in order to force the embedding $X \times Y \to \mathcal{P}(X \times Y)$ to become a bi-homomorphism with values in the quotient $\mathcal{P}(X \times Y)/\!\!\sim\, = X \otimes Y$.

The internal Hom: $\mathsf{USlh}^{\mathrm{op}} \times \mathsf{USlh} \to \mathsf{USlh}$ is defined in the obvious way: $\mathrm{Hom}(X, Y)$ is the set of homomorphisms $\mathsf{USlh}(X, Y)$ with the pointwise order, and pointwise joins as in (6.9).

6.1.7 Proposition (Quantaloids and inverse core)

Let C *be a pre-inverse quantaloid. Then the inverse core* IC *is a sub-quantaloid of* C.

Proof We already know, from Theorem 2.4.4, that IC is an ordered sub-category of C. Take a family $u_i \colon X \to Y$ $(i \in I)$ in IC.

Letting $u = \bigvee u_i$ and $v = \bigvee u_i{}^\sharp \leqslant u^\sharp$ in C, we have

$$vu \leqslant u^\sharp u \leqslant 1_X, \qquad uv \leqslant uu^\sharp \leqslant 1_Y,$$

$$u = \bigvee{}_i (u_i.u_i{}^\sharp.u_i) \leqslant \bigvee{}_{i,j,k} (u_i.u_j{}^\sharp.u_k) = uvu \leqslant u. \tag{6.11}$$

Similarly $v = vuv$. Thus the C-morphism u has Morita inverse v in C, and belongs to IC.   □

6.2 Enrichment on a quantale: metrics and preorders

After presenting the different forms of enriched categories available in category theory, we consider the simplest case: enrichment on an ordered monoid, and in particular on a quantale.

Then we discuss enrichment on the quantales $\underline{R} = [0, \infty]^{\mathrm{op}}$ and **2**. As in Lawvere's article [Lw] and in Subsection 5.5.5 (where $\underline{R}$ and **2** are seen as strict monoidal categories), these enrichments yield generalised metric spaces and preordered sets, respectively. Cauchy completion extends the classical metric completion.

6.2.1 *Kinds of enrichment*

A category can be enriched on a *basis* of the following types (the arrows denoting inclusion):

$$\begin{array}{ccc}
\text{monoidal category} & \longrightarrow & \text{bicategory} \\
\uparrow & & \uparrow \\
\text{ordered monoid} & \longrightarrow & \text{ordered category}
\end{array} \tag{6.12}$$

- *Enrichment on a monoidal category* is the first case considered in category theory, and the more readily understood. The classical references are [EiK, Kl2]. It was presented in Section 5.5, and will be further examined in Section 6.5.

- *Enrichment on an ordered category* is one of the main topics of this book. It was investigated in a particular symmetric form, in Chapters 3 and 4, and will be studied in the general form in Sections 6.3 and 6.4.

- *Enrichment on an ordered monoid* is a basic, particular case of the previous ones, and is presented in this section.

- *Enrichment on a bicategory* unifies all these cases, and will not be examined here. It was introduced in [Bet, Wa1], and also studied in [Wa2, BetC,

BetW1, BetW2]; the origin of this topic is described in a 'Commentary' by R. Betti, in the 2020 reprint of [BetW2].

In each of these cases one has enriched functors and enriched profunctors, linked by a notion of Cauchy completeness: it was introduced by Lawvere [Lw], for enrichment on monoidal categories.

An even more general notion, enrichment on a weak double category, is now the subject of research by R. Cockett and R. Garner [CoG].

6.2.2 Ordered monoids and quantales

Let M be an ordered monoid: it is equipped with an associative multiplication $\lambda\mu$ that has a unit 1, and with an order relation $\lambda \leqslant \mu$ consistent with the multiplication

$$(\lambda \leqslant \mu \text{ and } \lambda' \leqslant \mu') \Rightarrow \lambda\lambda' \leqslant \mu\mu'. \tag{6.13}$$

M will be viewed as an ordered category on one formal object $*$; M forms the ordered set of endomorphisms $\lambda\colon * \to *$, composed by the product.

An ordered monoid M can also be viewed as a strict monoidal category, with objects $\lambda \in M$, morphisms $\lambda \leqslant_M \mu$ and tensor product $\lambda\mu$.

From now on, *we assume that M is a quantale* (i.e. a quantaloid on one object), in order to be able to compose profunctors.

The reader is invited to keep in mind the following leading example (examined in 6.2.5 and 6.2.6): the (commutative) quantale

$$\underline{R} = [0, \infty]^{\mathrm{op}} \tag{6.14}$$

of *extended positive real numbers*, ordered by the relation $\lambda \geqslant \mu$, with 'multiplication' $\lambda+\mu$. Of course we are assuming that $\infty+\lambda = \infty = \lambda+\infty$, for all $\lambda \in \underline{R}$.

This quantale was studied as a strict symmetric monoidal category in Section 5.5. Also here an enriched category on $\underline{R}$ will be a generalised metric space.

6.2.3 Enrichment on a quantale

An *M-category* (X, u), or a *category enriched on the quantale M*, is a set X (of the 'objects' of the category) equipped with a function $u\colon X \times X \to M$ satisfying the following axioms, for $x, y, z \in X$:

(i) $1_M \leqslant u(x,x)$ (*identity law*),

(ii) $u(y,z).u(x,y) \leqslant u(x,z)$ (*triangle inequality*).

We generally write the M-category as X; we can write its set of objects as $|X|$, and its structural function as u_X.

An *M-functor* $f\colon X \to Y$ is a mapping between the underlying sets, that satisfies

$$u_X(x, x') \leqslant u_Y(f(x), f(x')) \qquad (x, x' \in X). \qquad (6.15)$$

An *M-profunctor* $F\colon X \nrightarrow Y$ is a mapping satisfying

$$F\colon X \times Y \to M,$$

$$F(x', y).u(x, x') \leqslant F(x, y) \qquad (x, x' \in X,\ y \in Y), \qquad (6.16)$$

$$u(y, y').F(x, y) \leqslant F(x, y') \qquad (x \in X,\ y, y' \in Y).$$

Assuming that the object-sets $|X|$ and $|Y|$ are disjoint, to simplify things, an enriched profunctor $F\colon X \nrightarrow Y$ amounts to extending the enriched categories X, Y to the disjoint union $|X| + |Y|$, with a structural function u_F that extends u_X and u_Y (and is trivial 'backwards')

$$u_F(x, y) = F(x, y) \in M, \qquad u_F(y, x) = \min M \qquad (x \in X,\ y \in Y).$$

The composite of $F\colon X \nrightarrow Y$ with a consecutive profunctor $G\colon Y \nrightarrow Z$ is computed by a least upper bound in the quantale M

$$(GF)(x, z) = \bigvee_{y \in Y} (G(y, z).F(x, y)). \qquad (6.17)$$

We have thus the category $\mathrm{Prf}(M)$ of M-categories and M-profunctors. It is an *ordered* category: if $F, G\colon X \nrightarrow Y$, the relation $F \leqslant G$ is defined pointwise, by the order of M

$$F(x, y) \leqslant G(x, y) \qquad (\text{for } x \in X,\ y \in Y), \qquad (6.18)$$

and can be viewed as an *M-transformation* $F \to G$.

6.2.4 Adjoint profunctors and Cauchy completeness

Two M-profunctors $F\colon X \nrightarrow Y$ and $G\colon Y \nrightarrow X$ are *adjoint*, with *F left adjoint to G* (written as $F \dashv G$) if

$$\mathrm{id}\, X \leqslant GF, \qquad FG \leqslant \mathrm{id}\, Y, \qquad (6.19)$$

which means that they are adjoint in the ordered category $\mathrm{Prf}(M)$ (cf. 5.6.2). Equivalently:

$$u_X(x, x') \leqslant \bigvee_{y \in Y} (G(y, x').F(x, y)) \qquad (x, x' \in X),$$
$$F(x, y').G(y, x) \leqslant u_Y(y, y') \qquad (x \in X,\ y, y' \in Y). \qquad (6.20)$$

Every functor $f\colon X \to Y$ has two associated adjoint profunctors $f_* \dashv f^*$

$$f_*\colon X \nrightarrow Y, \qquad f_*(x,y) = u_Y(f(x),y),$$
$$f^*\colon Y \nrightarrow X, \qquad f^*(y,x) = u_Y(y,f(x)), \qquad (6.21)$$

as proved in Exercise (a).

Now, the M-category Y is said to be *Cauchy complete* if, for every M-category X and every pair of adjoint profunctors $F \dashv G$, as in (6.19), there exists an M-functor $f\colon X \to Y$ that produces the adjunction

$$F(x,y) = u_Y(f(x),y), \qquad G(y,x) = u_Y(y,f(x)). \qquad (6.22)$$

Exercises and complements. (a) Prove the adjunction $f_* \dashv f^*$, in (6.21).

(b) Classify the M-structures of the singleton, for a general quantale M and for Rel Set(S).

(c) Unless otherwise specified, the singleton $\{*\}$ will be equipped with the *canonical M-structure*, with $u(*,*) = 1_M$.

Then, a functor $\{*\} \to X$ is an object of X.

*(d) Theorem 6.4.5 will prove, in the more general context of categories enriched on quantaloids, that an M-category Y is Cauchy-complete if (and only if) it satisfies the previous condition for all pairs of adjoint profunctors

$$F\colon \{*\} \rightleftarrows Y \colon G,$$

for the canonical M-structure of the singleton.

6.2.5 Generalised metric spaces

We come now to the main example, the *Lawvere quantale* $\underline{R} = [0,\infty]^{\mathrm{op}}$ of extended positive real numbers, ordered by the relation $\lambda \geqslant \mu$, with (commutative) multiplication $\lambda + \mu$. A join in $\underline{R}$ is a meet in $[0,\infty]$, and will be written as $\bigwedge \lambda_i$, with respect to the natural order.

As in 5.5.5, a small category enriched on this quantale gives a generalised metric space: a set X (of the 'objects' of the category) equipped with a (generalised) distance $d(x,y) \in \underline{R}$ (for $x,y \in X$) satisfying the following axioms (for $x,y,z \in X$):

(i) $d(x,x) = 0$ (*identity law*),

(ii) $d(x,y) + d(y,z) \geqslant d(x,z)$ (*triangle inequality*).

(We replace $0 \geqslant d(x,x)$ with the equivalent form (i).) Here the singleton has a unique structure, the canonical one with $d(*,*) = 0$.

An *enriched functor* $f\colon X \to Y$ is a mapping between the underlying sets, that satisfies

$$d_X(x, x') \geqslant d_Y(f(x), f(x')) \qquad (x, x' \in X), \tag{6.23}$$

i.e. a weak contraction.

An *enriched profunctor* $F\colon X \to Y$ is a mapping such that (for $x, x' \in X$ and $y, y' \in Y$)

$$F\colon X \times Y \to \underline{R},$$
$$d(x', x) + F(x, y) + d(y, y') \geqslant F(x', y'). \tag{6.24}$$

Assuming that X and Y are disjoint, this amounts to extending the metrics of X and Y to their disjoint union, adding new distances

$$d_F(x, y) = F(x, y) \in \underline{R}, \qquad d_F(y, x) = \infty \qquad (x \in X, \, y \in Y),$$

in a consistent way, i.e. satisfying (ii).

The composite of $F\colon X \to Y$ with a consecutive enriched profunctor $G\colon Y \to Z$ is computed by a greatest lower bound in the complete lattice $[0, \infty]$ (a supremum in $[0, \infty]^{\mathrm{op}}$)

$$(GF)(x, z) = \bigwedge_{y \in Y} (F(x, y) + G(y, z)). \tag{6.25}$$

We have thus the category $\mathrm{Prf}(\underline{R})$ of generalised metric spaces and profunctors. It is an *ordered* category: if $F, G\colon X \to Y$, the relation $F \geqslant G$ is defined pointwise by the order of $\underline{R}$

$$F(x, y) \geqslant G(x, y) \qquad \text{(for } x \in X, \, y \in Y). \tag{6.26}$$

This should be viewed as an $\underline{R}$-*transformation* $F \to G$ (according to the order relation $\lambda \geqslant \mu$ of $\underline{R}$).

6.2.6 Adjoint profunctors and symmetric distance

Two profunctors of generalised metric spaces $F\colon X \to Y$ and $G\colon Y \to X$ are *adjoint*, with F *left adjoint to* G $(F \dashv G)$, if:

$$\mathrm{id}\, X \geqslant GF, \qquad FG \geqslant \mathrm{id}\, Y. \tag{6.27}$$

Equivalently:

$$d_X(x, x') \geqslant \bigwedge_{y \in Y} F(x, y) + G(y, x') \qquad (x, x' \in X),$$
$$G(y, x) + F(x, y') \geqslant d_Y(y, y') \qquad (x, x' \in X, \, y, y' \in Y). \tag{6.28}$$

Every functor $f\colon X \to Y$ has two associated adjoint profunctors $f_* \dashv f^*$ (by Exercise 6.2.4(a))

$$f_*\colon X \nrightarrow Y, \qquad f_*(x,y) = d_Y(f(x), y),$$
$$f^*\colon Y \nrightarrow X, \qquad f^*(y, x) = d_Y(y, f(x)). \tag{6.29}$$

A generalised metric space X has a *symmetric distance*

$$\underline{d}(x, y) = d(x, y) + d(y, x), \tag{6.30}$$

which can be called a 'pseudometric' (as in Bourbaki [Bou3]), as it takes values in $[0, \infty]$ and the separation axiom is not assumed: $\underline{d}(x, y) = 0$ need not imply $x = y$.

A sequence $(x_n)_{n \in \mathbb{N}}$ in X is said to be a *Cauchy sequence* in X, or *to converge to a point* $x \in X$, if this holds with respect to the symmetric distance. (Uniqueness of limits can fail.)

As in the ordinary case, every convergent sequence is Cauchy. A generalised metric space is said to be *Cauchy complete* if the converse holds: every Cauchy sequence converges.

6.2.7 Lawvere Completeness Theorem [Lw]

The generalised metric space Y is Cauchy complete if and only if it is Cauchy complete as an enriched category:

(i) for every generalised metric space X and every pair of adjoint profunctors

$$F\colon X \nrightarrow Y, \qquad G\colon Y \nrightarrow X, \qquad F \dashv G, \tag{6.31}$$

there exists an enriched functor $f\colon X \to Y$ (i.e. a weak contraction) that produces the adjunction

$$F(x, y) = d_Y(f(x), y), \qquad G(y, x) = d_Y(y, f(x)). \tag{6.32}$$

Proof By 6.2.4(d) (a particular case of Theorem 6.4.5), condition (i) can be equivalently expressed letting X be the singleton metric space $\{*\}$.

We suppose that this condition is satisfied and let $(y_n)_{n \in \mathbb{N}}$ be a Cauchy sequence in Y. We construct a pair of profunctors

$$F\colon \{*\} \nrightarrow Y, \qquad F(y) = F(*, y) = \lim_n d(y_n, y),$$
$$G\colon Y \nrightarrow \{*\}, \qquad G(y) = G(y, *) = \lim_n d(y, y_n). \tag{6.33}$$

They are adjoint, with $F \dashv G$:

$$(GF)(*,*) = \bigwedge_y (F(y) + G(y)) = \bigwedge_y (\lim_n d(y_n, y) + \lim_n d(y, y_n))$$
$$= \bigwedge_y \lim_n (d(y_n, y) + d(y, y_n))$$
$$\leqslant \lim_{m,n} (d(y_n, y_m) + d(y_m, y_n)) = 0,$$
$$FG(y, y') = G(y, *) + F(*, y') = \lim_n (d(y, y_n) + d(y_n, y')) \geqslant d_Y(y, y').$$

By hypothesis, there is a functor $f \colon \{*\} \to Y$, i.e. a point $\overline{y} \in Y$, such that for all $y \in Y$

$$F(y) = f_*(y) = d_Y(\overline{y}, y), \qquad G(y) = f^*(y) = d_Y(y, \overline{y}). \tag{6.34}$$

Therefore (y_n) converges to $\overline{y}$ in the symmetric distance $\underline{d}$

$$\lim_n d(y_n, \overline{y}) = F(\overline{y}) = f_*(\overline{y}) = 0,$$
$$\lim_n d(\overline{y}, y_n) = G(\overline{y}) = f^*(\overline{y}) = 0.$$

Conversely, we suppose that Y is Cauchy complete as a generalised metric space, and let $F \dashv G$ be a pair of adjoint profunctors (with $X = \{*\}$). Because of the relation

$$0 = d(*, *) \geqslant \bigwedge_{y \in Y} F(y) + G(y),$$

we can choose, for every $n > 0$, a point $y_n \in Y$ such that

$$F(y_n) + G(y_n) < 1/n.$$

The sequence (y_n) is Cauchy in Y, because

$$d(y_n, y_m) + d(y_m, y_n) \leqslant G(y_n, *) + F(*, y_m) + G(y_m, *) + F(*, y_n)$$
$$< 1/n + 1/m.$$

Let $\overline{y}$ be any limit of the sequence (y_n) in Y. Then $F(\overline{y}) = 0 = G(\overline{y})$, because, for $n > 0$

$$F(\overline{y}) \leqslant F(y_n) + d(y_n, \overline{y}) < 1/n + d(y_n, \overline{y}),$$
$$G(\overline{y}) \leqslant d(\overline{y}, y_n) + G(y_n) < d(\overline{y}, y_n) + 1/n.$$

The pair of adjoint profunctors $f_* \dashv f^*$ associated to the functor $f = \overline{y} \colon \{*\} \to Y$ coincides thus with $F \dashv G$

$$f_*(y) = d_Y(\overline{y}, y) \leqslant G(\overline{y}) + F(y) = F(y) \leqslant F(\overline{y}) + d_Y(\overline{y}, y) = d_Y(\overline{y}, y),$$
$$f^*(y) = d_Y(y, \overline{y}) \leqslant G(y) + F(\overline{y}) = G(y) \leqslant d_Y(y, \overline{y}) + G(\overline{y}) = d_Y(y, \overline{y}).$$

$$\square$$

6.2.8 Enrichment on the quantale 2

As a more elementary example, we examine the lattice $\mathbf{2} = \{0 < 1\}$, viewed as a commutative quantale with multiplication $\lambda \wedge \mu = \min(\lambda, \mu)$ and unit 1.

A small category enriched on $\mathbf{2}$ is the same as a preordered set. In fact, such an enrichment means a set X equipped with a function $u(x, y) \in \{0, 1\}$ (defined on the set $X \times X$) which we interpret as the truth-value $[x \prec y]$ of a relation $x \prec y$ in X.

The enrichment axioms amount to the reflexivity and transitivity of this relation (taking into account that $\lambda \geqslant 1$ means $\lambda = 1$)

(i) $[x \prec x] = 1$,   (for $x \in X$),

(ii) $[x \prec y] \wedge [y \prec z] \leqslant [x \prec z]$ (for $x, y, z \in X$).

An enriched functor $f \colon X \to Y$ amounts to a monotone mapping

$$[x \prec x']_X \leqslant [f(x) \prec f(x')]_Y \qquad \text{(for } x, x' \in X). \tag{6.35}$$

An enriched profunctor $F \colon X \nrightarrow Y$ is a monotone mapping $X^{\mathrm{op}} \times Y \to \mathbf{2}$. In other words:

$$(x' \prec x \text{ in } X, \ y \prec y' \text{ in } Y) \ \Rightarrow \ F(x, y) \leqslant F(x', y') \text{ in } \mathbf{2}. \tag{6.36}$$

Assuming that X and Y are disjoint, this amounts to an extension of the preorders of X and Y to the disjoint union $F = X \cup Y$, where no element of Y precedes any element of X. We are adding new instances

$$[x \prec y]_F = F(x, y) \qquad [y \prec x]_F = 0 \qquad \text{(for } x \in X, y \in Y). \tag{6.37}$$

The order-preserving property (6.36) ensures that the extended relation is transitive (for $x, x' \in X$ and $y, y' \in Y$):

$$[x' \prec x]_X \wedge [x \prec y]_F \wedge [y \prec y']_Y \leqslant [x' \prec y']_F.$$

The composite of $F \colon X \nrightarrow Y$ with an enriched profunctor $G \colon Y \nrightarrow Z$ is

$$[x \prec z]_{GF} = \bigvee_{y \in Y} ([x \prec y]_F \wedge [y \prec z]_G), \tag{6.38}$$

which means that $x \prec z$ if and only if there is some $y \in Y$ such that $x \prec y$ in F and $y \prec z$ in G.

The identity profunctor of the preordered set X is $[x \prec x']_X$.

We have thus the category $\mathrm{Prf}(\mathbf{2})$ of preordered sets and profunctors between them. It is an *ordered* category: if $F, G \colon X \nrightarrow Y$, the relation $F \leqslant G$ is defined pointwise by the order of $\mathbf{2}$

$$F(x, y) \leqslant G(x, y) \qquad \text{(for } x \in X, \ y \in Y). \tag{6.39}$$

6.2.9 Adjoint profunctors of preordered sets

Two profunctors $F\colon X \nrightarrow Y$ and $G\colon Y \nrightarrow X$ of preordered sets are adjoint, with F left adjoint to G (written as $F \dashv G$) if:

$$\mathrm{id}\, X \leqslant GF, \qquad FG \leqslant \mathrm{id}\, Y. \tag{6.40}$$

Equivalently (leaving the universal quantifiers understood):

$$\begin{aligned} [x \prec x']_X &\leqslant \bigvee_{y \in Y} F(x,y) \wedge G(y,x'), \\ G(y,x) \wedge F(x,y') &\leqslant [y \prec y']_Y. \end{aligned} \tag{6.41}$$

The first inequality above is equivalent to saying that: if $x \prec x'$ in X, there exists some element $y(x,x') \in Y$ such that

$$F(x, y(x,x')) = G(y(x,x'), x') = 1.$$

Every functor $f\colon X \to Y$ has two associated adjoint profunctors $f_* \dashv f^*$

$$\begin{aligned} f_*\colon X \nrightarrow Y, &\qquad f_*(x,y) = [f(x) \prec y]_Y, \\ f^*\colon Y \nrightarrow X, &\qquad f^*(y,x) = [y \prec f(x)]_Y. \end{aligned} \tag{6.42}$$

In fact, if $x \prec x'$ in X, we have $f(x) \prec f(x')$ in Y, and we can take $y(x,x') = f(x)$. Furthermore, if $y \prec f(x) \prec y'$, we obviously have $y \prec y'$.

An exercise. (a) *Every preordered set is Cauchy-complete*, as a category enriched on **2**.

In other words, given two adjoint profunctors $F \dashv G$ as in (6.40), we can define a monotone mapping $f\colon X \to Y$ such that the associated adjoint profunctors $f_* \dashv f^*$ coincide with F and G.

6.3 Enrichment on a quantaloid

We extend the previous section, replacing the quantale M by a quantaloid C (cf. 6.1.4).

More generally, we begin by defining categories, functors and profunctors enriched on an ordered category C. Then, assuming that C is a quantaloid, all profunctors can be composed.

All this is related to the notion of manifold on a totally cohesive e-category, explored in Chapters 3 and 4.

6.3.1 Enrichment on an ordered category

A (small) *category U enriched* on the ordered category C, or C-*category*, is a family $(U_i)_{i \in I}$ of objects of C, indexed by a set I and equipped with morphisms $u_j^i \colon U_i \to U_j$ (of C) such that (for $i, j, k \in I$)

(i) $1_{U_i} \leqslant u_i^i$ *(identity law),*

(ii) $u_k^j.u_j^i \leqslant u_k^i$ *(composition law, or triangle inequality).*

(Equivalently, we are considering a lax functor $U \colon C(I) \to C$, where $C(I)$ is the indiscrete category on the set I.)

Extending the notation of a symmetric manifold, in Chapter 3, the C-category U can be written as $((U_i), (u_j^i))_I$, or as U_I. When convenient, U_I can be viewed as a *chiral manifold* on C; in this terminology, U_i is a *chart* of U, the morphism u_j^i is a *transition map*, and the family (u_j^i) is an *intrinsic atlas* of U.

In particular, we have:

$$u_i^i.u_i^i = u_i^i, \qquad u_j^i.u_i^i = u_j^i = u_j^j.u_j^i. \tag{6.43}$$

A C-*functor* $f \colon U \to V$, where $V = ((V_h), (v_k^h))_H$ is an enriched category indexed by the set H, is a mapping $f \colon I \to H$ between the set of indices of U and V, such that we have, in C

$$U_i = V_{fi}, \qquad u_j^i \leqslant v_{fj}^{fi} \qquad \text{(for } i, j \in I\text{)}. \tag{6.44}$$

It is also called an *enriched functor* on C. (This notion seems to be of little interest in the context of local structures; but we have already seen that it is important in the examples of Section 6.2.)

Composition of enriched functors is obvious; the identity of U_I is id I. We say that f is a *fully faithful* C-functor if $u_j^i = v_{fj}^{fi}$, for all $i, j \in I$.

Given C-functors $f, g \colon U \to V$ (so that $U_i = V_{fi} = V_{gi}$), a *transformation* $f \to g$ is actually a condition

$$\text{id } U_i = \text{id } V_{fi} = \text{id } V_{gi} \leqslant v_{gi}^{fi} \qquad \text{(for all } i \in I\text{)}, \tag{6.45}$$

that we write in the form $f \prec g$ (*f precedes g*). This relation is a preorder, and we say that f and g are *isomorphic* C-functors ($f \cong g$) when $f \prec g \prec f$, that is

$$1 \leqslant v_{gi}^{fi}, \qquad 1 \leqslant v_{fi}^{gi} \qquad \text{(for all } i \in I\text{)}. \tag{6.46}$$

We have thus the preordered category Enr(C), of C-categories and C-functors, and the associated ordered category Enr(C)$/\cong$. We write as Enr$^\circ$(C) the category Enr(C) with opposite preorder: $f \prec {}^{op}g$ if $g \prec f$.

6.3.2 Comments and examples

(a) (*The Enrichment and the Manifold Interpretations*) We shall freely interchange the terminology of enriched categories (where enriched functors are the main arrows) and that of manifolds (where we are interested in profunctors).

In a sense, we are outlining two different theories with the same technical basis, from which they diverge to different applications.

The intersection of these two theories is examined in 6.4.8

(b) Given two enriched functors $f, g \colon U_I \to V_H$, the condition $f \prec g$ is equivalent to $u_j^i \leqslant v_{gj}^{fi}$ (for $i, j \in I$). In fact, if $f \prec g$:

$$u_j^i = u_j^i \operatorname{id} U_i \leqslant v_{gj}^{gi} . v_{gi}^{fi} \leqslant v_{gj}^{fi}.$$

Conversely, the relation $u_j^i \leqslant v_{gj}^{fi}$ gives $1_{U_i} \leqslant u_i^i \leqslant v_{gi}^{fi}$.

(c) (*Trivial manifolds*) For an object W of C we write as $\hat{W} = (W, \operatorname{id} W)$ the C-category indexed by the singleton $\{*\}$, with intrinsic atlas

$$W_* = W, \qquad u_*^* = \operatorname{id} W. \tag{6.47}$$

It will be called the *trivial* C-*category* on the object W, or the *trivial manifold* on the chart W.

An enriched functor $f \colon \hat{W} \to U_I$ is the same as picking out an index $i = f(*) \in I$, with $W = U_i$ (the condition $1_W \leqslant u_i^i$ being automatically satisfied).

(d) If C is the quantale **2** of Subsection 6.2.8, an enriched functor $f \colon X \to Y$ is a monotone mapping of preordered spaces. The condition $f \prec g$ means that $f(x) \prec g(x)$, for all $x \in X$.

(e) If C is the Lawvere quantale $\underline{R} = [0, \infty]^{\mathrm{op}}$ of 6.2.5, an enriched functor $f \colon X \to Y$ is a weak contraction of generalised metric spaces. The condition $f \prec g$ means that $d_Y(fx, gx) = 0$ for all $x \in X$; the condition $f \cong g$ means that the symmetric distance $\underline{d}_Y(fx, gx)$ is always zero.

6.3.3 Profunctors

With the notation above, a *profunctor* $a = (a_h^i) \colon U \nrightarrow V$ (on C) is a family of morphisms $a_h^i \colon U_i \to V_h$ of C (for $i \in I$, $h \in H$) such that, for all $i, j \in I$ and $h, k \in H$:

(i) $a_h^j . u_j^i \leqslant a_h^i$, $\quad v_k^h . a_h^i \leqslant a_k^i$ $\quad$ (*profunctor laws*, or *triangle inequalities*).

Here also, we have

$$a_h^i . u_i^i = a_h^i = v_h^h . a_h^i. \tag{6.48}$$

The composite of $a \colon U \nrightarrow V$ with a second profunctor

$$b = (b_m^h) \colon V \nrightarrow W, \qquad W = ((W_m), (w_n^m))_M,$$

'should' be defined by joins in the ordered set $\mathsf{C}(U_i, W_m)$

$$(ba)_m^i = \bigvee_{h \in H} b_m^h . a_h^i \colon U_i \to W_m, \qquad (6.49)$$

which do exist if each ordered hom-set of C is a complete lattice.

We assume, for the rest of this section, that C *is a quantaloid* (cf. 6.1.4); the composition of profunctors is then well defined and associative, with identities $\mathrm{id}\, U$ defined as follows (and working as identities because of (6.48))

$$(\mathrm{id}\, U)_j^i = u_j^i \colon U_i \to U_j. \qquad (6.50)$$

We have defined the ordered category $\mathrm{Prf}(\mathsf{C})$ of C-categories and C-profunctors. For two profunctors $a, b \colon U \nrightarrow V$, the order relation $a \leqslant b$ derives from the ordering of C, as

$$a_h^i \leqslant b_h^i \quad \text{for all indices } i, h. \qquad (6.51)$$

Assuming that the sets I and H are disjoint, to simplify things, a profunctor $a \colon U \nrightarrow V$ amounts to extending the manifolds U and V to the disjoint union $I + H$, adding new transition morphisms a_h^i, and trivial transition morphisms from H-indices to I-indices

$$a_h^i \in \mathsf{C}(U_i, V_h), \qquad a_i^h = \min \mathsf{C}(V_h, U_i) \qquad (i \in I, \, h \in H).$$

The profunctor laws (i) say that the new changes a_h^i are consistent with the old ones, while the trivial transitions a_i^h automatically are.

In the Manifold Interpretation, $\mathrm{Prf}(\mathsf{C})$ can be written as $\mathrm{chMf}\,(C)$, and called the ordered category *of chiral manifolds and profunctors on* C.

6.3.4 Adjoint profunctors

Two C-profunctors $a = (a_h^i) \colon U \nrightarrow V$ and $b = (b_i^h) \colon V \nrightarrow U$ are *adjoint*, with $a \dashv b$ (*a left adjoint to* b) if

$$
\begin{aligned}
1_U &\leqslant ba \colon U \nrightarrow U && \text{(the \textit{unit} of the adjunction)}, \\
ab &\leqslant 1_V \colon V \nrightarrow V && \text{(the \textit{counit} of the adjunction)},
\end{aligned}
\qquad (6.52)
$$

which means that they are adjoint in the ordered category $\mathrm{Prf}(\mathsf{C})$ (cf. 5.6.2). Equivalently:

$$
\begin{aligned}
u_j^i &\leqslant \bigvee_{h \in H} b_j^h . a_h^i \colon U_i \to U_j && (i, j \in I), \\
a_k^i . b_i^h &\leqslant v_k^h \colon V_h \to V_k && (i, j \in I, \, h, k \in H).
\end{aligned}
\qquad (6.53)
$$

The following properties are straightforward.

(a) If $a \dashv b$, then $a = aba$ and $b = bab$.

(b) If $a \dashv b$ and $a' \dashv b'$ (for $a, a' \colon U \nrightarrow V$), then $a \leqslant a'$ if and only if $b' \leqslant b$.

(c) If $a \dashv b$, then a strictly determines b, and conversely.

(d) Adjunctions of profunctors compose: if $a \dashv b$ (with $a \colon U \nrightarrow V$) and $c \dashv d$ (with $c \colon V \nrightarrow W$), then $ca \dashv bd$.

We write as $\mathsf{AdjPrf}(\mathsf{C})$ the ordered category of C-categories and adjoint C-profunctors, where a morphism $a \colon U \nrightarrow V$ is a profunctor which has a (unique) right adjoint. It is a wide subcategory of $\mathrm{Prf}(\mathsf{C})$

$$\mathsf{AdjPrf}(\mathsf{C}) \subset \mathrm{Prf}(\mathsf{C}), \qquad (6.54)$$

and $a \leqslant a'$ is equivalent to $b \geqslant b'$ (if $a \dashv b$ and $a' \dashv b'$).

6.3.5 Theorem

(a) Every C-functor $f \colon U_I \to V_H$ has two associated profunctors, with the following components (for $i \in I$, $h \in H$):

$$
\begin{aligned}
f_* &\colon U \nrightarrow V, & (f_*)_h^i &= v_h^{fi} \colon U_i = V_{fi} \to V_h, \\
f^* &\colon V \nrightarrow U, & (f^*)_i^h &= v_{fi}^h \colon V_h \to V_{fi} = U_i,
\end{aligned}
\qquad (6.55)
$$

and they are adjoint: $f_* \dashv f^*$.

(b) The functor f is fully faithful if and only if the unit of the adjunction is invertible, i.e. $f^* f_* = 1_U$.

(c) The procedures $f \mapsto f_*$ and $f \mapsto f^*$ are functorial (co- and contravariantly): in other words, given a consecutive C-functor $g \colon V_H \to W_M$ we have

$$(gf)_* = g_* f_*, \qquad (gf)^* = f^* g^*. \qquad (6.56)$$

(d) For C-functors $f, g \colon U \to V$, we have:

$$
\begin{aligned}
f \prec g &\quad \Leftrightarrow \quad g_* \leqslant f_* &\quad \Leftrightarrow \quad f^* \leqslant g^*, \\
f \cong g &\quad \Leftrightarrow \quad f_* = g_* &\quad \Leftrightarrow \quad f^* = g^*.
\end{aligned}
\qquad (6.57)
$$

Proof In the following formulas, it is understood that $i, j \in I$, $h, k \in H$ and $m, n \in M$; the universal quantifier 'for all', on these indices, is also understood as convenient.

(a) The data of f_* do form a profunctor $U \nrightarrow V$, because

$$v_h^{fj}.u_j^i \leqslant v_h^{fj}.v_{fj}^{fi} \leqslant v_h^{fi}, \qquad v_k^h.v_h^{fi} \leqslant v_k^{fi}.$$

Similarly f^* is a profunctor. They are adjoint because

$$u_j^i \leqslant v_{fj}^{fi} = \bigvee_{h \in H} v_{fj}^h.v_h^{fi}, \qquad v_k^{fi}.v_{fi}^h \leqslant v_k^h. \qquad (6.58)$$

(The equality in the left formula above comes from taking $h = fi$ and applying (6.48).)

(b) Both properties amount to saying that the left inequality in (6.58) is an equality (for all $i, j \in I$).

(c) In the present hypotheses, and letting $W = ((W_m), (w_n^m))_M$, we have

$$(g_*f_*)_m^i = \bigvee_{h \in H} w_m^{gh}.v_h^{fi} \leqslant \bigvee_{h \in H} w_m^{gh}.w_{gh}^{gfi} \leqslant w_m^{gfi},$$

$$(f^*g^*)_i^m = \bigvee_{h \in H} v_{fi}^h.w_{gh}^m \leqslant \bigvee_{h \in H} w_{gfi}^{gh}.w_{gh}^m \leqslant w_{gfi}^m,$$

which means that $g_*f_* \leqslant (gf)_*$ and $f^*g^* \leqslant (gf)^*$.

But $g_*f_* \dashv f^*g^*$, by composing adjunctions (cf. 6.3.4(d)); it follows that $g_*f_* = (gf)_*$ by 6.3.4(b).

(d) We only have to prove that $f \prec g$ is equivalent to $f^* \leqslant g^*$, i.e. $v_{fi}^h \leqslant v_{gi}^h$ (for all indices h and i).
Indeed, if $1 \leqslant v_{gi}^{fi}$ then

$$v_{fi}^h \leqslant v_{gi}^{fi}.v_{fi}^h \leqslant v_{gi}^h.$$

Conversely, if $v_{fi}^h \leqslant v_{gi}^h$ (for all h) then $1 \leqslant v_{fi}^{fi} \leqslant v_{gi}^{fi}$. $\qquad \square$

6.3.6 A synopsis

We summarise the main points developed so far.

(a) Enr(C) is the preordered category of C-categories and C-functors (see 6.3.1). We recall that, for $f, g \colon U_I \to V_H$, $f \prec g$ means that $v_{gi}^{fi} \geqslant \text{id}\, U_i$ (for $i \in I$); this makes sense, because here $U_i = V_{fi} = V_{gi}$. The associated equivalence relation is written as $f \cong g$.

(b) Prf(C) is the ordered category of C-categories and C-profunctors (see 6.3.3). We recall that, for profunctors $a, b \colon U_I \nrightarrow V_H$, the relation $a \leqslant b$ means that $a_h^i \leqslant b_h^i$ (for all indices i, h).

(c) AdjPrf(C) $\subset$ Prf(C) is the ordered category of C-categories and adjoint C-profunctors (see 6.3.4), embedded in Prf(C) as a wide subcategory.

(d) There is another monotone functor, which is the identity on objects (by Theorem 6.3.5)

$$\mathrm{Enr}^\circ(\mathsf{C}) \to \mathrm{AdjPrf}(\mathsf{C}), \qquad f \mapsto f_*, \tag{6.59}$$

$$f \prec^{\mathrm{op}} g \quad \Leftrightarrow \quad f_* \leqslant g_* \quad \Leftrightarrow \quad g* \leqslant f^*,$$
$$f \cong g \quad \Leftrightarrow \quad f_* = g_* \quad \Leftrightarrow \quad f* = g*.$$

Thus the ordered category $\mathrm{Enr}^\circ(\mathsf{C})/\cong$ is embedded as a wide ordered subcategory of $\mathrm{AdjPrf}(\mathsf{C})$ and $\mathrm{Prf}(\mathsf{C})$.

(e) The quantaloid C will be said to be Cauchy-complete (in Section 6.4) if the functor (6.59) is full, which means that the associated embedding

$$\mathrm{Enr}^\circ(\mathsf{C})/\cong \; \to \; \mathrm{AdjPrf}(\mathsf{C})$$

is the identity.

6.3.7 *A double category

The preordered category $\mathrm{Enr}(\mathsf{C})$ and the ordered category $\mathrm{Prf}(\mathsf{C})$ can be amalgamated in a strict double category $\mathbb{Prf}(\mathsf{C})$, similar to the weak double category $\mathbb{Cat}$ of small categories, functors and profunctors [GP1].

The objects are all categories enriched on C. The horizontal arrows form the category $\mathrm{Enr}(\mathsf{C})$, the vertical ones the category $\mathrm{Prf}(\mathsf{C})$.

A double cell is *flat*, i.e. a condition on its boundary

$$
\begin{array}{ccc}
U_I & \overset{f}{\longrightarrow} & X_J \\
a \downarrow & \leqslant & \downarrow b \\
V_H & \underset{g}{\longrightarrow} & Y_K
\end{array}
\qquad a_h^i \leqslant b_{gh}^{fi}, \text{ for all indices } i, h. \tag{6.60}
$$

The horizontal composition is obvious. For the vertical one, given a vertically consecutive cell

$$
\begin{array}{ccc}
V_H & \overset{g}{\longrightarrow} & Y_K \\
c \downarrow & \leqslant & \downarrow d \\
W_M & \underset{r}{\longrightarrow} & Z_N
\end{array}
\qquad c_m^h \leqslant d_{rm}^{gh}, \text{ for all indices } h, m, \tag{6.61}
$$

we have:

$$
\begin{aligned}
(ca)_m^i = \bigvee_{h \in H} b_m^h a_h^i &\leqslant \bigvee_h d_{rm}^{gh} b_{gh}^{fi} \\
&\leqslant \bigvee_{k \in K} d_{rm}^k b_k^{fi} = (db)_{rm}^{fi}.
\end{aligned}
\tag{6.62}
$$

In this double category:

- a *special* cell from $f = \operatorname{id} U$ to $g = \operatorname{id} V$, as below at the left, amounts to the relation $a \leqslant b$,

$$
\begin{array}{ccc}
U_I \xrightarrow{\ 1\ } U_I & \qquad & U_I \xrightarrow{\ f\ } X_J \\
{\scriptstyle a}\downarrow \quad \leqslant \quad \downarrow{\scriptstyle b} & & {\scriptstyle 1}\downarrow \quad \downarrow \quad \downarrow{\scriptstyle 1} \\
V_H \xrightarrow[\ 1\ ]{} V_H & & U_I \xrightarrow[\ g\ ]{} X_J
\end{array}
\qquad (6.63)
$$

- a *globular* cell from $a = \operatorname{id} U$ to $b = \operatorname{id} X$, as above at the right, amounts to the relation $u^i_{i'} \leqslant x^{fi}_{gi'}$ (for all $i, i' \in I$), which is equivalent to $f \prec g$, by 6.3.2(b).

6.4 Cauchy completion on a quantaloid

The notion of Cauchy completeness for a category enriched on a quantale, examined in Section 6.2, is now extended to enrichment on a quantaloid.

V is always a category enriched on the quantaloid C.

6.4.1 Definition

The C-category V is said to be *Cauchy complete* if:

(i) every pair of adjoint C-profunctors $a \colon U \rightleftarrows V \colon b$ coincides with the pair $f_* \dashv f^*$ of a suitable C-functor $f \colon U \to V$.

After characterising this property, we prove that, when the quantaloid C is small:

- the Cauchy completion of a C-category V always exists,
- it is constructed by adding to V all its profunctor-redundant charts (cf. 6.4.3),
- and is *profunctor-isomorphic* to V.

Cauchy completion is important in the theory of enriched categories, and has concrete instances like the completion of metric spaces (as we have seen in Section 6.2), or the sheafification of presheaves [Wa1].

On the other hand, when the natural morphisms are the profunctors this procedure seems to be of little interest. This was the case of the local structures considered in Section 4.6 as symmetric enriched categories on an e-cohesive quantaloid: replacing an intrinsic atlas by the associated maximal atlas (with all the possible redundant charts) just gives an isomorphic object.

Now, the cohesive categories $\mathcal{C}^r$, $\mathcal{B}$, $\mathcal{V}$, $\mathcal{D}_0$, $\mathcal{F}$, $\mathcal{V}$ etc., investigated in Sections 4.1–4.5, are not quantaloids, and some of them (like $\mathcal{F}$ and $\mathcal{V}$) are not even small. There seems to be no point in extending to enrichment on an e-cohesive category a notion of Cauchy completion that would produce — if anything — a new structure that can only be distinguished at the light of morphisms we are not interested in.

6.4.2 Charts and profunctors

Let $V = V_H$ be an enriched category, or chiral manifold, on the quantaloid C.

An index $k \in H$ amounts to a mapping $k \colon \{*\} \to H$ and gives a C-functor $f(k) \colon \hat{V}_k \to V$, defined on the trivial manifold $\hat{V}_k$ (as we have already seen, in 6.3.2(c)).

Therefore every chart V_k, or more precisely every index $k \in H$, gives a profunctor-adjunction $\hat{V}_k \rightleftarrows V$, associated to the functor $f(k)$

$$f(k)_* = (v_h^k)_{h \in H} \colon \hat{V}_k \nrightarrow V, \qquad f(k)^* = (v_k^h)_{h \in H} \colon V \nrightarrow \hat{V}_k. \qquad (6.64)$$

We say that two indices $k, k' \in H$ *essentially give the same chart* in V, and write $k \sim k'$, if their adjoint profunctors coincide, which means that:

(i) $V_k = V_{k'}$, $v_h^k = v_h^{k'}$, $v_k^h = v_{k'}^h$ (for all $h \in H$).

Letting $I = H \setminus \{k\}$ and $J = H \setminus \{k'\}$, one can then prove that the obvious functors $V_I \rightleftarrows V_J$ (that interchange k' and k) are inverse to each other.

Exercises and complements. (a) The relation $k \sim k'$ can equivalently be written as:

(ii) $V_k = V_{k'}$, $v_k^k = v_{k'}^k = v_k^{k'} = v_{k'}^{k'}$.

*(b) The C-category V is said to be *skeletal* [Wa1] if the relation $k \sim k'$ is the identity. Every C-category V has a skeleton, isomorphic to V in Prf(C). (A more general situation of this kind will be examined in the next subsection.)

6.4.3 Full subcategories and redundant charts

Let $V = ((V_h), (v_k^h))_H$ be a C-category. Restricting all the indices to a subset $I \subset H$ we obtain a full C-subcategory V_I of V.

Then the inclusion $I \subset H$ gives a fully faithful functor $f \colon V_I \to V$ and an adjunction $f_* = a \dashv b = f^*$ with $ba = \operatorname{id} V_I$ (which is the restriction of the pair of identity profunctors $V \rightleftarrows V$)

$$a: V_I \rightleftarrows V : b, \qquad a_h^i = v_h^i, \qquad b_i^h = v_i^h,$$
$$v_j^i = \bigvee_{h \in H} v_j^h . v_h^i, \qquad \bigvee_{i \in I} v_k^i . v_i^h \leqslant v_k^h \quad (i, j \in I, \, h, k \in H). \tag{6.65}$$

The indices of $H \setminus I$ (or the corresponding charts) will be said to be *profunctor-redundant* in V if the adjunction above is an isomorphism of $\operatorname{Prf}(\mathsf{C})$, i.e. $ab = \operatorname{id} V$. This is equivalent to saying that

$$v_k^h = \bigvee_{i \in I} v_k^i . v_i^h : V_h \to V_k, \quad \text{for all } h, k \in H \setminus I. \tag{6.66}$$

In fact, if $h, k \in H$ and $h \in I$ or $k \in I$ we already have that $\bigvee_{i \in I} v_k^i . v_i^h = v_k^h$, as above.

We simply speak of *redundant charts*. We also say that V_I is obtained from V_H *by discarding redundant charts*, and that V_H is obtained from V_I *by adding* (consistent) *redundant charts*. As a particular case, this certainly holds if the subset $I \subset H$ meets all the classes of the equivalence relation 6.4.2(i); then we can say that V_I is obtained from V *by discarding repeated charts*.

We are particularly interested in the case where there is *one* discarded index, say $\infty \in H$, and $I = H \setminus \{\infty\}$.

Given a C-category V_I on I, one can extend it to a C-category V_H indexed by the disjoint union $H = I \cup \{\infty\}$, *by adding a chart*. The supplementary data consist of a C-object V_∞ together with morphisms (for $i \in I$)

$$v_i^\infty : V_\infty \to V_i, \qquad v_\infty^i : V_i \to V_\infty, \qquad v_\infty^\infty : V_\infty \to V_\infty, \tag{6.67}$$

such that the extended triangle inequalities hold on $H = I \cup \{\infty\}$ *and* $v_\infty^\infty \geqslant 1$.

The added chart V_∞ is redundant if and only if

$$v_\infty^\infty = \bigvee_{i \in I} v_\infty^i . v_i^\infty : V_\infty \to V_\infty. \tag{6.68}$$

All this is stated in the following theorem.

Note. This terminology is influenced by the Manifold Interpretation, where we are only interested in profunctors.

In the Enriched Category Interpretation, profunctor-redundant indices should not be viewed as redundant, generally. For instance, for a generalised metric space X, the metric completion adds points which are profunctor-redundant with respect to X (see Theorem 6.4.7), but not redundant at all from the usual 'point of view' of the main morphisms, the weak contractions.

6.4.4 *Theorem* (Adding a redundant chart)

Let W be an object of C and $\hat{W}$ the one-index enriched category (defined in 6.3.2(c)). Let $V = ((V_h), (v_k^h))_H$ be an arbitrary C-category indexed by the set H.

(a) An adjunction $a\colon \hat{W} \rightleftarrows V \colon b$ ($a \dashv b$) of profunctors amounts to assigning two families of C-morphisms such that (for $h, k \in H$)

$$a_h \colon W \to V_h, \qquad\qquad b^h \colon V_h \to W,$$

$$v_k^h . a_h \leqslant a_k, \qquad\qquad b^k . v_k^h \leqslant b^h, \qquad\qquad (6.69)$$

$$1_W \leqslant \bigvee_{h \in H} b^h a_h, \qquad a_k b^h \leqslant v_k^h \colon V_h \to V_k.$$

(b) This is equivalent to extending V to a C-category V' indexed by the disjoint union $H' = H \cup \{\infty\}$, with a new (consistent) chart $V_\infty = W$ that is profunctor-redundant in V' (as defined above), i.e.

$$v_\infty^\infty = \bigvee_{h \in H} b^h a_h. \qquad\qquad (6.70)$$

(c) The adjunction $a\colon \hat{W} \rightleftarrows V \colon b$ ($a \dashv b$) of profunctors comes from a functor $\hat{W} \to V$ if and only if there exists an index $k \in H$ such that $W = V_k$ and

$$a_h = v_h^k \colon V_k \to V_h, \qquad b^h = v_k^h \colon V_h \to V_k \qquad \text{(for every } h \in H), \quad (6.71)$$

which means that the redundant chart to be added is already in V. In this case the functor $\hat{W} \to V$ is given by the mapping $k\colon \{*\} \to H$.

Proof (a) is trivial. The points (b) and (c) have been proved in 6.4.3, with

$$v_h^\infty = a_h, \qquad v_\infty^h = b^h, \qquad v_\infty^\infty = \bigvee_{h \in H} b^h a_h.$$

$\square$

6.4.5 *Theorem* (Cauchy completeness, I)

A C-category V is Cauchy-complete if (and only if) it satisfies the completeness condition 6.4.1(i) for all pairs of adjoint profunctors $a\colon \hat{W} \rightleftarrows V \colon b$, where W is an arbitrary object of C and $\hat{W}$ is the associated one-index enriched category (cf. 6.3.2).

Proof We follow the argument given in [Bet], page 12, for enrichment on a bicategory; it extends the previous proof of Lawvere for the monoidal case, in [Lw].

We suppose that the C-category $V = ((V_h), (v_k^h))_H$ satisfies the restricted condition on trivial manifolds, and take an adjunction $a\colon U \rightleftarrows V \colon b$, with an arbitrary C-category $U = ((U_i), (u_j^i))_I$.

Fixing an index $i \in I$, the trivial enriched category $\hat{W}$ of the object $W = U_i$ gives a functor $i\colon \hat{W} \to U$ (determined by the mapping $i\colon \{*\} \to I$) and the corresponding adjunction

$$i_*\colon \hat{W} \rightleftarrows U \colon i^*,$$
$$(i_*)_j = u_j^i\colon U_i \to U_j, \quad (i^*)^j = u_i^j\colon U_j \to U_i = W \qquad (j \in I). \tag{6.72}$$

Its composite with the adjunction $a \dashv b$ is computed as follows

$$ai_*\colon \hat{W} \rightleftarrows V \colon i^*b, \qquad\qquad v_k^h.a_h^j.u_j^i \leqslant a_k^i,$$
$$(ai_*)_k = \bigvee\nolimits_{j \in I} a_k^j u_j^i = a_k^i\colon U_i \to V_k, \tag{6.73}$$
$$(i^*b)^k = \bigvee\nolimits_{j \in I} u_i^j b_j^k = b_i^k\colon V_k \to U_i \qquad (k \in H).$$

By hypothesis, the composed adjunction is the outcome of a functor $h\colon \hat{W} \to V$ (determined by an index $h \in H$, as a mapping $h\colon \{*\} \to H$). Therefore $V_h = W (= U_i)$ and, for $k \in H$

$$a_k^i = (h_*)_k = v_k^h\colon V_h \to V_k, \qquad b_i^k = (h^*)^k = v_h^k\colon V_k \to V_h = W. \tag{6.74}$$

Let $f\colon I \to H$ be the mapping that sends every $i \in I$ to the element $f(i) = h \in H$ determined above. This mapping is a C-functor, because $U_i = V_{fi}$ and

$$u_j^i \leqslant u_j^i.v_{fi}^{fi} = u_j^i.b_i^{fi} \leqslant b_j^{fi} = v_{fj}^{fi}. \tag{6.75}$$

Finally the adjoint pair $f_* \dashv f^*$ coincides with the original pair $a \dashv b$

$$(f_*)_k^i = v_k^{fi} = a_k^i, (f^*)_i^k = v_{fi}^k = b_i^k \qquad (i \in I,\, k \in H), \tag{6.76}$$

because of (6.74). □

6.4.6 Corollary (Cauchy completeness, II)

A C-category V is Cauchy-complete if and only if it already contains every profunctor-redundant chart that one can add.

Proof It is a consequence of Theorems 6.4.4 and 6.4.5. □

6.4.7 Theorem and Definition (Cauchy completion [BetW1])

Let us suppose that the ground quantaloid C is small.

Every C-*category V has a* Cauchy completion, *namely a Cauchy complete* C-*category U equipped with a* C-*functor*

$$f \colon V \to U, \tag{6.77}$$

that is universal in the following sense:

(i) every functor $g \colon V \to W$ with values in a Cauchy complete C-*category factorises as $g \cong ef$, by a functor $e \colon U \to W$ determined up to isomorphism (in the sense of 6.3.1).*

Moreover V and U are profunctor-isomorphic, via the adjunction (f_*, f^*) associated to f.

Note. The completion U is constructed as the 'maximal atlas' of the original C-category V, by putting together all the *profunctor-redundant charts* of V. This is why we want C to be small.

Proof Let $V = ((V_h), (v_k^h))_H$. We want to define a new C-category

$$U = ((U_i), (u_j^i))_I, \tag{6.78}$$

where I is the *set* of all the redundant charts of V.

As we have seen in 6.4.4, each redundant chart $i \in I$ is an adjunction of profunctors starting from a trivial manifold $\hat{U}_i$

$$a^i \colon \hat{U}_i \rightleftarrows V \colon b_i \qquad (a^i \dashv b_i).$$

This consists of one object U_i of C and two families of C-morphisms (indexed by H)

$$a_h^i \colon U_i \to V_h, \qquad b_i^h \colon V_h \to U_i \qquad (h \in H), \tag{6.79}$$

such that, for $h, k \in H$:

$$
\begin{aligned}
v_k^h . a_h^i \leqslant a_k^i, && b_i^k . v_k^h \leqslant b_i^h, \\
1_{U_i} \leqslant \bigvee_{h \in H} b_i^h . a_h^i \colon U_i \to U_i, && a_h^i . b_i^h \leqslant v_k^h \colon V_h \to V_k.
\end{aligned}
\tag{6.80}
$$

Now we define the transition morphisms of U and check their coherence conditions (for $i, j, r \in I$)

$$u_j^i = \bigvee_{h \in H} b_j^h . a_h^i \colon U_i \to U_j,$$

$$
\begin{aligned}
u_j^r . u_r^i = \bigvee_{h, k \in H} (b_j^k . a_k^r)(b_r^h . a_h^i) &\leqslant \bigvee_{h, k \in H} b_j^k . v_k^h . a_h^i \\
&\leqslant \bigvee_{h \in H} b_i^h . a_h^i = u_j^i,
\end{aligned}
\tag{6.81}
$$

$$u_i^i = \bigvee_{h \in H} b_i^h . a_h^i \geqslant 1_{U_i}.$$

There is a fully faithful functor $f \colon V \to U$ that sends the index $h \in H$ to the chart $f(h)$ already present in V, with $U_{fh} = V_t$, $a_s^{fh} = v_s^h$ and $b_{fh}^s = v_h^s$ (for $s \in H$). In fact

$$u_{fk}^{fh} = \bigvee_{s \in H} b_{fk}^s . a_s^{fh} = \bigvee_{s \in H} v_k^s . v_s^h = v_k^h.$$

It will be useful to note that

$$\begin{aligned} u_{fh}^i &= \bigvee_{k \in H} b_{fh}^k . a_k^i = \bigvee_{k \in H} v_h^k . a_k^i = a_h^i, \\ u_j^{fh} &= \bigvee_{k \in H} b_j^k . a_k^{fh} = \bigvee_{k \in H} b_j^k . v_k^h = b_j^h. \end{aligned} \tag{6.82}$$

We have already seen (in Theorem 6.3.5) that this fully faithful functor $f \colon V \to U$ produces an adjunction $f_* = c \dashv d = f^*$ with $dc = 1$

$$c \colon V \rightleftarrows U \colon d, \qquad c_i^h = u_i^{fh}, \qquad d_h^i = u_{fh}^i \qquad (h \in H,\ i \in I). \tag{6.83}$$

Actually we have here a profunctor-isomorphism, i.e. $cd = 1_U$, as follows from (6.82)

$$\bigvee_{h \in H} c_j^h . d_h^i = \bigvee_h u_j^{fh} . u_{fh}^i = \bigvee_h b_j^h . a_h^i = u_j^i.$$

Take now a functor $g \colon V \to W$ with values in a Cauchy complete C-category $((W_m), (w_n^m))_M$, so that $V_h = W_{g(h)}$ and $v_k^h \leqslant w_{gk}^{gh}$.

To define a functor $e \colon U \to W$, we note that every redundant chart $i \in I$ of V determines a redundant chart $g[i] \colon \hat{U}_i \rightleftarrows W$, composing the adjunctions

$$(a^i \dashv b_i) \colon \hat{U}_i \rightleftarrows V, \qquad (g_* \dashv g^*) \colon V \rightleftarrows W.$$

Since W is Cauchy complete, this redundant chart is already present in W, and we can choose an index $e(i) \in M$ such that $(e(i)_*, e(i)^*) = g[i]$. This means that $W_{ei} = U_i$ and, for $m \in M$

$$w_h^{ei} = \bigvee_{h \in H} w_m^{gh} . a_h^i \colon U_i \to W_m, \qquad w_{ei}^m = \bigvee_{h \in H} b_i^h . w_{gh}^m \colon W_m \to U_i.$$

Now, e is a functor $U \to W$ because:

$$\begin{aligned} w_{ej}^{ei} &\geqslant \bigvee_{h \in H} w_{ej}^{gh} . w_{gh}^{ei} \geqslant \bigvee_h b_j^h . w_{gh}^{gh} . w_{gh}^{gh} . a_h^i \\ &\geqslant \bigvee_h b_j^h . v_h^h . a_h^i = \bigvee_h b_j^h . a_h^i = u_j^i. \end{aligned}$$

To verify that $ef \cong g$, i.e. $(ef)_* = g_*$, we prove that $(ef)_* \leqslant g_*$ and $(ef)^* \leqslant g^*$. Indeed, every index $h \in H$ is sent to the chart $f(h) = i \in I$ with $U_i = V_h$, $a_k^i = v_k^h$ and $b_i^k = v_h^k$ so that

$$((ef)_*)_m^h = w_m^{ei} = \bigvee_{k \in H} w_m^{gk} . a_k^i = \bigvee_k w_m^{gk} . v_k^h \leqslant \bigvee_k w_m^{gk} . w_{gk}^{gh} = w_m^{gh},$$

$$((ef)^*)_h^m = w_{ei}^m = \bigvee_{k \in H} b_i^k . w_{gk}^m = \bigvee_k v_h^k . w_{gk}^m \leqslant \bigvee_k w_{gh}^{gk} . w_{gk}^m = w_{gh}^m.$$

The uniqueness of the profunctor e_* (associated to the functor e) is obvious, because f_* is invertible. $\qquad\square$

6.4.8 Enriched categories and symmetry

Let C be an inverse quantaloid. The enriched categories (or chiral manifolds) on C include the symmetric manifolds on C studied in Chapters 3 and 4 (on a more general basis), and we can consider their interplay.

(a) First, C is a quantaloid, and we have at our disposition the theory of (small) enriched categories on C, or C-categories, outlined in 6.3.6. In particular, we have two embeddings of ordered categories

$$\text{Enr}^\circ(\mathsf{C})/\cong \ \subset \ \text{AdjPrf}(\mathsf{C}) \ \subset \ \text{Prf}(\mathsf{C}), \tag{6.84}$$

and C is Cauchy-complete if and only if the first inclusion is the identity.

Let us note that, since C is inverse, every C-category U_I is unital: $u_i^i \geqslant 1$ means $u_i^i = 1$ $(i \in I)$.

(b) Second, C is a totally cohesive inverse category (with indiscrete linking), and we have at our disposition the theory of manifolds, or symmetric enriched categories on C, developed in Chapter 3.

Prf(C) can be written as chMf(C), and called the ordered category *of chiral manifolds and profunctors on* C. It contains two subcategories

$$\text{IMf}(\mathsf{C}) \ \subset \ \text{Mf}(\mathsf{C}) \ \subset \ \text{chMf}(\mathsf{C}) = \text{Prf}(\mathsf{C}). \tag{6.85}$$

IMf(C) is the category of (symmetric) manifolds and symmetric profunctors, determined in chMf(C) as follows. A manifold U_I is a chiral manifold that satisfies the symmetry law of 3.5.1

$$u_i^j = (u_j^i)^\sharp \qquad (i,j \in I). \tag{6.86}$$

A symmetric profunctor $a \colon U_I \nrightarrow V_H$ is a profunctor which satisfies the symmetry laws of 3.5.3

$$a_k^i a_h^{i\,\sharp} \leqslant v_k^h, \qquad a_h^{j\,\sharp} a_h^i \leqslant u_j^i \qquad (i,j \in I,\ h,k \in H), \tag{6.87}$$

also expressed as the left and right linking laws:

$$a_k^i \underline{e}(a_h^i) = v_k^h a_h^i \qquad \underline{e}^*(a_h^j) a_h^i = a_h^j u_j^i, \qquad (i,j \in I,\ h,k \in H). \tag{6.88}$$

Mf(C) is the category of (symmetric) manifolds and linked profunctors. It has the same objects as IMf(C) and more general morphisms, the left linked profunctors of 3.6.3, satisfying the equivalent properties (for $i \in I$ and $h,k \in H$)

$$a_k^i a_h^{i\,\sharp} \leqslant v_k^h, \tag{6.89}$$

$$a_k^i \underline{e}(a_h^i) = v_k^h a_h^i. \tag{6.90}$$

We already remarked, in 6.4.1, that Cauchy completion is of a marginal interest for manifolds.

6.5 *Cauchy completion on a symmetric monoidal category

This section gives some hints at the basic form of Cauchy completeness, for enrichment on symmetric monoidal categories, introduced by Lawvere in 1974 [Lw]; the reader can find there a very interesting presentation of this topic. Further information can be found in many papers dealing with the more general case of enrichment on a bicategory [Bet, Wa1, Wa2, BetC, BetW1, BetW2].

Categories enriched on a symmetric monoidal category and enriched functors were dealt with in Section 5.5. Enriched profunctors in this context are presented now. Their theory goes beyond the natural bounds of this book: associativity of composition only works up to isomorphism, forming a bicategory.

V is always a symmetric monoidal category.

6.5.1 Profunctors of categories

We begin by considering ordinary categories, which correspond to enrichment on Set, with respect to its cartesian monoidal structure. Their profunctors were introduced by Bénabou [Be2] and Lawvere [Lw], under the names of *distributors* and *bimodules*, respectively.

A *profunctor* between two categories C and D consists of a functor

$$F\colon \mathsf{C}^{\mathrm{op}} \times \mathsf{D} \to \mathsf{Set},$$

which we view as an arrow $F\colon \mathsf{C} \nrightarrow \mathsf{D}$. (We use the opposite direction with respect to the original one, to be consistent with morphism of manifolds.)

Concretely we are assigning

(a) a set $F(x, y)$, for every x in C and y in D,

(b) a mapping $F(f, g)\colon F(x, y) \to F(x', y')$, for every $f\colon x' \to x$ in C and $g\colon y \to y'$ in D,

so that the following axioms hold (for x in C, y in D, f', f consecutive in C and g, g' consecutive in D)

(i) $F(1_x, 1_y) = \mathrm{id}\, F(x, y)$,

(ii) $F(f', g').F(f, g) = F(ff', g'g)$.

This amounts to assigning new arrows $u\colon x \nrightarrow y$ from the objects of C to those of D, the elements $u \in F(x, y)$, forming a larger category $\mathsf{F} = \mathsf{C} +_F \mathsf{D}$ that contains C and D as disjoint subcategories and has no arrows from the objects of D to those of C.

The new composition in the category F is:

$$guf = F(f, g)(u) \qquad \text{(for } f \colon x' \to x \text{ in C, } g \colon y \to y' \text{ in D),} \qquad (6.91)$$

$$
\begin{array}{ccc}
x' \xrightarrow{\ f\ } x & & \text{C} \\
\downarrow{\scriptstyle u} & & \downarrow{\scriptstyle F} \\
y \xrightarrow{\ g\ } y' & & \text{D}
\end{array}
$$

Associativity and unitarity follow from (i) and (ii). We remark that $F(y, x) = \emptyset$ is the initial object of Set (for every x in C and y in D).

This representation is also a reason to view F as directed from C to D.

Profunctors compose, by colimits in Set. Given a second profunctor $G \colon D \to E$, we define $(GF)(x, z)$ as a quotient of a disjoint union of sets

$$U = \Sigma_y\, F(x, y) \times G(y, z) \qquad \text{(for } y \text{ in D).} \qquad (6.92)$$

The quotient set $(GF)(x, z) = U/R$ is the coequaliser in Set of the following mappings σ, τ

$$\Sigma_{y', y''}\, F(x, y') \times D(y', y'') \times G(y'', z) \;\rightrightarrows\; \Sigma_y\, F(x, y) \times G(y, z),$$

$$\sigma(u, g, v) = (gu \colon x \to y'',\ v \colon y'' \to z), \qquad (6.93)$$

$$\tau(u, g, v) = (u \colon x \to y',\ vg \colon y' \to z),$$

so that R is the equivalence relation generated by identifying each $\sigma(u, g, v)$ to the corresponding $\tau(u, g, v)$.

*The colimit we are using is called a *coend* in category theory, and written as $\int^y F(x, y) \times G(y, z)$.*

The composition is associative, *up* to natural bijections produced by the universal property of colimits. The identity profunctor of the category C is defined as the hom-functor itself (cf. 1.5.3(e))

$$\text{Mor}_C \colon C^{op} \times C \to \text{Set,}$$

and acts as a unit, up to natural bijections. We have thus a bicategory (see 5.6.2) of (small) categories and profunctors.

A functor $F \colon C \to D$ has two associated profunctors:

$$F_* = \text{Mor}_D(F^{op} \times \text{id}\,D) \colon C^{op} \times D \to \text{Set,}$$

$$F_*(x, y) = D(F(x), y),$$

$$F^* = \text{Mor}_D(\text{id}\,D^{op} \times F) \colon D^{op} \times C \to \text{Set,} \qquad (6.94)$$

$$F^*(y, x) = D(y, F(x)).$$

A natural transformation $\varphi \colon F \to G \colon \mathsf{C} \to \mathsf{D}$ gives two transformations of profunctors

$$\varphi_* \colon G_* \to F_* \colon \mathsf{C} \nrightarrow \mathsf{D}, \qquad \varphi^* \colon F^* \to G^* \colon \mathsf{D} \nrightarrow \mathsf{C}. \tag{6.95}$$

Note the contravariant behaviour of the mapping $\varphi \mapsto \varphi_*$, as in 6.3.5(d).

6.5.2 Enriched profunctors

In the general case, an *enriched profunctor* $F \colon \mathsf{C} \nrightarrow \mathsf{D}$ between V-categories, or V-*profunctor*, consists of:

(a) a family of V-objects $F(x, y)$, for x in C and y in D,

(b) a family of V-morphisms $c \colon \mathsf{C}(x', x) * F(x, y) * \mathsf{D}(y, y') \to F(x', y')$, called *compositions*, for x, x' in C and y, y' in D,

so that the axioms of unitarity and associativity hold true: the following diagrams commute (composition in C and D is also denoted by c)

$$
\begin{array}{ccc}
E * F(x, y) * E & \xrightarrow{\;e*1*e\;} & \mathsf{C}(x, x) * F(x, y) * \mathsf{D}(y, y) \\
 & \searrow{\scriptstyle 1} & \downarrow{\scriptstyle c} \\
 & & F(x, y)
\end{array}
\tag{6.96}
$$

$$
\mathsf{C}(x'', x') * \mathsf{C}(x', x) * F(x, y) * \mathsf{D}(y, y') * \mathsf{D}(y', y'')
$$

$$
\begin{array}{ccc}
{\scriptstyle c*1*c} \swarrow & & \searrow {\scriptstyle 1*c*1} \\
\mathsf{C}(x'', x) * F(x, y) * \mathsf{D}(y, y'') & & \mathsf{C}(x'', x') * F(x', y') * \mathsf{D}(y', y'') \\
{\scriptstyle c} \searrow & & \swarrow {\scriptstyle c} \\
 & F(x'', y'') &
\end{array}
\tag{6.97}
$$

When V is Set, or Ord, or Ab (or any symmetric monoidal category made concrete by its canonical forgetful functor) we can write the V-morphisms c on elements, as defining a composition mapping $c(f, u, g) = guf$. The axioms (6.96) and (6.97) are then translated as follows (assuming that all compositions are legitimate):

$$1_y\, u\, 1_x = u, \tag{6.98}$$

$$g'(guf)f' = (g'g)u(ff'). \tag{6.99}$$

6.5.3 Composing enriched profunctors

Suppose that our ground category V has all colimits, preserved by the functors $- * V$ (for V in V).

Given two V-profunctors $F \colon C \nrightarrow D$ and $G \colon D \nrightarrow E$, one defines $(GF)(x, z)$ as the coequaliser in V of the two morphisms σ, τ defined by the following commutative squares (via the universal properties of the sum in their domain)

$$
\begin{array}{ccc}
F(x,y') * D(y',y'') * G(y'',z) & \xrightarrow{\ c*1\ } & F(x,y'') * G(y'',z) \\
{\scriptstyle i_{y'y''}} \downarrow & & \downarrow {\scriptstyle i_{y''}} \\
\textstyle\sum_{y'y''} F(x,y') * D(y',y'') * G(y'',z) & \underset{\tau}{\overset{\sigma}{\rightrightarrows}} & \textstyle\sum_y F(x,y) * G(y,z) \\
{\scriptstyle i_{y'y''}} \uparrow & & \uparrow {\scriptstyle i_{y'}} \\
F(x,y') * D(y',y'') * G(y'',z) & \xrightarrow[\ 1*c\ ]{} & F(x,y') * G(y',z)
\end{array}
$$

The V-morphisms

$$
c \colon C(x',x) * (GF)(x,z) * E(z,z') \to (GF)(x',z'), \qquad (6.100)
$$

are defined using the universal property of the previous colimit (preserved by tensoring).

Again, we get a bicategory $\mathrm{Prf}(V)$.

In particular, if V is a quantale, the parallel arrows σ, τ are necessarily equal, and $(GF)(x, z)$ is computed as a join of products in V (a 'matrix multiplication'), as in Section 6.2

$$
(GF)(x,z) = \bigvee\nolimits_y F(x,y) * G(y,z). \qquad (6.101)
$$

6.5.4 Cauchy completeness

One goes on, adapting to the present context the treatment of enriched profunctors in Section 6.3.

After defining V-transformations of V-profunctors, adjunctions of V-profunctors and the adjoint V-profunctors $F_* \dashv F^*$ associated to a V-functor $F \colon C \to D$, the V-category D is said to be *Cauchy complete* if

(i) every pair of adjoint V-profunctors $C \rightleftarrows D$ is isomorphic to the pair $F_* \dashv F^*$ of a suitable V-functor $F \colon C \to D$.

In the particular case $V = \mathsf{Set}$, an ordinary category D is Cauchy complete if and only if all its idempotent endomorphisms split: see 5.2.5, and [BoD] for a proof.

6.6 *Cohesive categories as enriched categories

This section deals with another relationship between cohesive categories and enrichment.

Cohesive sets abstract the properties of the hom-sets $C(X, Y)$ of a cohesive category C. On the other hand, a cohesive category can be defined as a category enriched on the category Coh of cohesive sets, with respect to a suitable symmetric monoidal structure of the latter.

6.6.1 Definition

A *cohesive set* will be a set X equipped with two binary relations $x \leqslant y$ and $x\,!\,y$, satisfying the following axioms:

(chs.1) $\leqslant$ is an order relation on X,

(chs.2) ! is a proximity relation (i.e. reflexive and symmetric), called the *linking* or *compatibility* relation,

(chs.3) if $x \leqslant x'$, $y \leqslant y'$ and $x'\,!\,y'$ then $x\,!\,y$,

(chs.4) if $x\,!\,y$, then the meet $x \wedge y$ exists in X, and is called a *linked meet*.

A *cohesive mapping* $f \colon X \to Y$ is a mapping between cohesive sets which preserves the order, the linking and binary linked meets. Cohesive mappings are closed under composition, and we write as Coh the *category of cohesive sets and cohesive mappings*.

In a cohesive set X a subset ξ is said to be *linked* if $x\,!\,y$, for all $x, y \in \xi$. Every finite non-empty linked subset ξ has a meet $\bigwedge \xi$ in X, which will also be called a *linked meet*.

6.6.2 Definition

A *finitely cohesive* (resp. *σ-cohesive, totally cohesive*) *set* will be a cohesive set X such that every finite (resp. countable, arbitrary) linked subset ξ has a join $\bigvee \xi$ in X (called a *linked join*). Moreover *linked binary meets must distribute over such joins*: in other words, if ξ is a finite (resp. countable, arbitrary) linked subset of X and $y\,!\,\xi$, then

(i) $y\,!\,\bigvee \xi$, $y \wedge (\bigvee \xi) = \bigvee_{x \in \xi} (y \wedge x)$.

A *finitely cohesive* (resp. *σ-cohesive, totally cohesive*) *mapping* $f \colon X \to Y$ is a mapping between finitely cohesive (resp. σ-cohesive, totally cohesive) sets which preserves the order relation, the linking relation, binary linked meets and the joins of finite (resp. countable, arbitrary) linked subsets. We write as fCoh (resp. σCoh, tCoh) the category of such objects and maps.

We still use, as in 3.2.3, the cardinal bound ρ as a general substitute for the terms: *finite* (or finitely), *countable* (or countably, or σ-), *arbitrary* (or totally). We speak thus of the category $\rho\mathsf{Coh}$ of *ρ-cohesive sets and ρ-cohesive mappings*, and of a *ρ-subset* of a set.

In all these cases the linking relation is determined by the order: $x\,!\,y$, if and only if the pair x, y is upper bounded.

6.6.3 Examples

Every hom-set $\mathsf{C}(X, Y)$ of a cohesive (resp. totally cohesive) category is a cohesive (resp. totally cohesive) set.

In particular, the set $\mathcal{S}(X, Y)$ of partial mappings from the set X to the set Y is totally cohesive, with the order $f \leqslant g$, and linking relation $f\,!\,g$, defined in 3.1.1.

The same holds for the set $\mathcal{C}(X, Y)$ of partial continuous mappings, between topological spaces X and Y, defined on an open subspace of X. Or for the set $\mathcal{C}^r(U, V)$ of partial C^r-mappings between euclidean open spaces U, V, defined on an open subspace of U.

6.6.4 Exercises

(a) Let X be a ρ-cohesive set. A non-empty ρ-subset $\xi \subset X$ is linked if and only if it has some upper bound (e.g. $\vee\,\xi$), if and only if it is pairwise upper bounded.

If ξ and η are ρ-subsets of morphisms and $\xi\,!\,\eta$, then $\vee\,\xi\,!\,\vee\,\eta$ and

$$(\vee\,\xi) \wedge (\vee\,\eta) = \vee\, x \wedge y \qquad (\text{for } x \in \xi,\, y \in \eta). \tag{6.102}$$

(b) A cohesive set X is finitely cohesive if and only if it satisfies the following two conditions:

(chs.5a) the ordered set X has a least element 0 (automatically preserved by binary meets),

(chs.5b) every linked pair $x\,!\,y$, has a join $x \vee y$, and linked binary meets distribute over such joins.

(c) A finitely cohesive set is σ-cohesive if and only if:

(chs.5c) every increasing sequence (x_n) in X, obviously linked, has a join $\vee\, x_n$; linked binary meets distribute over increasing countable joins.

6.6.5 The monoidal structure

The category Coh has arbitrary cartesian products. In particular, the cartesian product $X \times Y$ of two cohesive sets has the componentwise structure:

$$(x, y) \leqslant (x', y') \quad \Leftrightarrow \quad x \leqslant x' \text{ in } X \text{ and } y \leqslant y' \text{ in } Y, \tag{6.103}$$

$$(x, y) \,!\, (x', y') \quad \Leftrightarrow \quad x \,!\, x' \text{ in } X \text{ and } y \,!\, y' \text{ in } Y. \tag{6.104}$$

Now we can imitate the construction of the tensor product of modules (in Section 5.3) or pointed sets (in 5.4.2).

We consider a mapping $f \colon X \times Y \twoheadrightarrow Z$ (with values in a cohesive set) that is *linked in each variable*. Then f does preserve the order relation, because of the transitive property

$$(x, y) \leqslant (x', y') \quad \Rightarrow \quad (x, y) \leqslant (x', y) \leqslant (x', y')$$
$$\Rightarrow \quad f(x, y) \leqslant f(x', y) \leqslant f(x', y'),$$

but need not take (6.104) to the linking relation of Z.

We are thus lead to consider a finer proximity relation on the *set $X \times Y$*:

$$(x, y) \,!!\, (x', y') \quad \Leftrightarrow$$
$$(x \,!\, x' \text{ in } X \text{ and } y = y') \text{ or } (x = x' \text{ in } X \text{ and } y \,!\, y' \text{ in } Y). \tag{6.105}$$

This relation is certainly preserved by our mapping f, but need not satisfy axiom (chs.3).

Finally, we define the *tensor product $X \otimes Y$* of two cohesive sets as the set $X \times Y$, with the product order (6.103) and the linking relation generated by (6.105):

$$(x, y) \,!_{\otimes}\, (x', y') \quad \Leftrightarrow$$
$$(x \,!\, x' \text{ and } \exists y'' \colon y, y' \leqslant y'') \text{ or } (y \,!\, y' \text{ and } \exists x'' \colon x, x' \leqslant x''). \tag{6.106}$$

More precisely, $!_{\otimes}$ is the least relation that contains (6.105) and makes the ordered set $X \times Y$ into a cohesive set. It follows easily that an order-preserving mapping $f \colon X \times Y \to Z$ is linked in each variable if and only if it is a cohesive map $X \otimes Y \to Z$.

It is now easy to verify that Coh is a symmetric monoidal closed category (cf. 5.4.3), with this tensor product and the singleton $\{*\}$ as a unit. The internal hom-functor $\mathrm{Hom}(X, Y)$ has the pointwise structure

$$f \leqslant f' \quad \Leftrightarrow \quad \forall x \in X,\ f(x) \leqslant f(x') \text{ in } Y,$$
$$f \,!\, f' \quad \Leftrightarrow \quad \forall x \in X,\ f(x) \,!\, f(x') \text{ in } Y. \tag{6.107}$$

The subcategory ρCoh of ρ-cohesive sets has a restricted structure of this kind.

Finally, a cohesive (resp. ρ-cohesive) category amounts to a category enriched on Coh (resp. ρCoh).

6.7 *Cohesive categories and Ehresmann's global orders

We end by comparing cohesive categories with part of the local structures of Charles Ehresmann. Essentially, we are dealing with *enriched structures* on the one hand, and *internal structures* on the other. Technically, a globally ordered category C will be turned into an e-cohesive category P_MC of partial maps, which gives back C as its subcategory of total morphisms; all this under suitable conditions.

As in 2.8.1, we mostly refer to Andrée C. Ehresmann's comments in Volume [E6] ('Structures locales') of C. Ehresmann's works, where his terminology is presented making use of internal categories.

6.7.1 Introducing the comparison

(a) We have seen that a locally ordered category is an *enriched category* on the category Ord of ordered sets.

On the other hand, forgetting questions of size, a globally ordered category (i.e. an ordered category in Ehresmann's sense) is an *internal category* in Ord satisfying a further condition: all hom-sets inherit a discrete order (see 5.6.7).

(b) We have also seen, in Section 6.6, that a totally cohesive category is the same as an *enriched category* on the monoidal category tCoh of totally cohesive sets. We recall below, in 6.7.3, that Ehresmann's *inductive* (resp. *local*) categories are *internal categories* in two categories of 'inductive' (resp. 'local') sets, reviewed in 6.7.2, and closely related to tCoh.

(c) The relationship with e-cohesive categories is more complex: we introduce in 6.7.4 a modified and simplified version of an inductive category, which we prove to be equivalent to an e-cohesive category with a 'good' splitting of projectors.

(d) Within these bounds, the present gluing completion theorem (in Section 3.6), restricted to totally cohesive e-categories, seems to correspond to Ehresmann's 'théorème d'élargissement complet d'un foncteur local' [E2]. The connections at the level of cohesive or prj-cohesive categories should be much more involved.

6.7.2 Inductive sets

We begin by recalling Ehresmann's terminology related to 'conditioned' complete lattices.

An *inductive set* is an ordered set where every upper bounded subset has a join, or equivalently every non-empty subset has a meet. An *inductive mapping* between these objects has to preserve upper bounded joins and meets of upper bounded pairs. We write their category as Ind (Ehresmann's notation is $\mathcal{I}$).

One can define a monoidal structure on the category Ind (proceeding as in 6.1.6, for USlh). A category enriched on Ind is a uc-quantaloid (in the sense of 3.1.5) with an additional condition: composition has to preserve, in each variable, all meets of upper bounded pairs of morphisms.

More particularly, an *Ehresmann local set* is an inductive set satisfying a condition similar to 6.6.2(i): if the set $\xi \cup \{y\}$ is upper bounded in X then the pair $(y, \vee \xi)$ is (obviously) upper bounded and

$$y \wedge (\vee \xi) = \vee_{x \in \xi} y \wedge x. \tag{6.108}$$

An Ehresmann local set is thus a sort of conditioned frame, while a frame is the same as an Ehresmann local set with a maximum. (In fact, Ehresmann's term inspired the present meaning of 'locale' in pointless topology, recalled in 6.1.2.)

The full subcategory of Ind determined by Ehresmann local sets will be written as ELoc. In an inductive set, we use the term *upper bounded join* (resp. *meet*) to mean the join (resp. meet) of an upper bounded subset.

Notes. (a) Ehresmann deals with *inductive classes* and *local classes*, rather than sets.

(b) The category tCoh of totally coherent sets is a subcategory of ELoc.

(c) A hom-set $\mathcal{T}(X, Y)$ is a local set, finitely cohesive and not totally cohesive, in general: see Exercise 3.2.6(g).

6.7.3 Ehresmann's inductive categories

(a) A small *inductive category* ([E6] p. 374) is a category C equipped with the structure of an internal category in the category Ind of inductive sets, satisfying two additional conditions:

(i) the order induced on each hom-set $C(X, Y)$ is discrete,

(ii) if $k \subset gf$ there exist $f' \subset f$ and $g' \subset g$ such that $k = g'f'$.

More explicitly, a category internal in Ind (see 5.6.6) is the same as a globally ordered category where:

(a) Ob C and Mor C are inductive sets,

(b) the structural mappings of C are inductive mappings

$$\text{Dom} : \text{Mor C} \to \text{Ob C}, \qquad \text{Cod} : \text{Mor C} \to \text{Ob C},$$

$$\text{id} : \text{Ob C} \to \text{Mor C}, \qquad c : \text{Mor}_2\, \text{C} \to \text{Mor C}.$$

Thus Ob C is a retract of Mor C, and its structure is determined by the latter.

(The set Mor_2 C of consecutive pairs of morphisms is the following pullback

$$\begin{array}{ccc} \text{Mor}_2\,\text{C} & \longrightarrow & \text{Mor C} \\ \downarrow & & \downarrow {\scriptstyle \text{Cod}} \\ \text{Mor C} & \underset{\text{Dom}}{\longrightarrow} & \text{Ob C} \end{array} \qquad (6.109)$$

in the category Ind.)

An *inductive functor* $F : \text{C} \to \text{D}$ is an internal functor in Ind: in other words, it is an ordinary functor which preserves upper bounded joins and upper bounded binary meets of morphisms (and of objects, as a consequence).

(b) An *inductive category* is defined in the same way, but we allow Ob C and Mor C to be classes, and only require each hom-class $\text{C}(X, Y)$ to be a set.

We have already seen in 2.8.1 the main examples we are interested in:

- the category Top with its global order 'of open inclusions',
- its substructure C^rTop.

(c) A *small local category* ([E2], [E6] p. 400) is a category C equipped with the structure of an internal category in the category ELoc, satisfying the conditions (i), (ii).

Again, in an Ehresmann local category, Ob C and Mor C are classes.

6.7.4 From global orders to e-categories

Let C be a category with global order $\subset$ (as defined in 2.8.1). We want to form an e-category $\text{P}_\text{M}\text{C}$ *of partial morphisms*, as in Section 5.1, with respect to a subcategory M determined by C.

(The reader can keep in mind the case where C is Top and M is the wide subcategory of open topological embeddings.)

We assume three conditions on C.

(i) If $f\colon X \to Y$ is in C and $X' \subset X$ (i.e. $1_{X'} \subset 1_X$), there exists a (unique) $f' \subset f$ with $\mathrm{Dom}\, f' = X'$ and $\mathrm{Cod}\, f' = Y$, called the *restriction of f on X'.*

This is similar to condition 2.8.3(v), for a groupoid.

(ii) If $X' \subset X$, the restriction $h\colon X' \to X$ of $\mathrm{id}\, X$ satisfies $\mathrm{id}\, X' \subset h \subset \mathrm{id}\, X$. This restriction will be called an *inclusion,* or *inclusion morphism,* and written as $X' \rightarrowtail X$. The inclusions with values in a given object X *form a set* (in bijective correspondence with the set of objects $X' \subset X$).

Thus, inclusions are monomorphisms and form a wide subcategory of C (see 6.7.5(a)).

(iii) The wide subcategory M of all monomorphisms equivalent to inclusions is closed under preimages.

More precisely, condition (iii) means that, given $f\colon X \to Y$ and an inclusion $k\colon Y' \rightarrowtail Y$, the pullback of k along f exists, and we can fix it as an inclusion $h = f^*(k)\colon X' \rightarrowtail X$

$$
\begin{array}{ccc}
X & \xrightarrow{\ f\ } & Y \\[2pt]
h\big\uparrow & & \big\uparrow k \\[2pt]
X' & \xrightarrow[\ f'\ ]{} & Y'
\end{array}
\tag{6.110}
$$

The pair (C, M) satisfies thus the conditions 5.1.7(i), (ii), and we can form the e-cohesive category $\mathsf{P_M C}$.

The main properties of the global order, in the present situation, are described in the following statement.

6.7.5 Proposition

Let C be a category with global order, satisfying the conditions 6.7.4(i), (ii).

(a) The inclusions of C are monomorphisms, and form a wide subcategory of C.

(b) The following square with vertical inclusions is commutative if and only if $f' \subset f$

$$
\begin{array}{ccc}
X & \xrightarrow{\ f\ } & Y \\[2pt]
h\big\uparrow & & \big\uparrow k \\[2pt]
X' & \xrightarrow[\ f'\ ]{} & Y'
\end{array}
\tag{6.111}
$$

(c) If $h\colon X' \rightarrowtail X$ *and* $k\colon X'' \rightarrowtail X$ *are inclusions,* $X' \subset X''$ *if and only if* $h \subset k$, *if and only if* $h \prec k$ *in the preorder of monomorphisms. The latter is an order relation, for inclusions.*

(d) In particular, if the inclusion $h\colon X' \rightarrowtail X$ *is an isomorphism, then* $X' = X$ *and* $h = 1_X$.

Proof (a) Let $h\colon X' \to X$ be an inclusion and suppose that $hu_1 = hu_2 = f$. Therefore $u_i = 1_{X'}u_i \subset hu_i = f$; since u_1 and u_2 are parallel morphisms, they coincide.

As to composition, if $k\colon X'' \rightarrowtail X'$ is also an inclusion, we have $hk\colon X'' \to X$ with $hk \subset h.\mathrm{id}\, X' \subset \mathrm{id}\, X$. Therefore kh is an inclusion.

(b) We have $1_{X'} \subset h \subset 1_X$ and $1_{Y'} \subset k \subset 1_Y$.

Therefore, if the square (6.111) commutes

$$f' = 1_{Y'}f' \subset kf' = fh \subset f1_X = f.$$

Conversely, if $f' \subset f$, then $f' = f'1_{X'} \subset fh$, and $kf' \subset 1_Y(fh) = fh$; finally $kf' = fh$.

(c) First, if $X' \subset X''$, then $h \subset 1_{X'} \subset 1_{X''} \subset k$.

Second, if $h \subset k$ then $X' \subset X''$. The restriction k' of k on X' gives the commutative triangle below

$$(6.112)$$

where $k' \subset k$ and $h = k'$, showing that $h \prec k$.

Finally, if $h \prec k$, we form the right square above, where h' is an inclusion, by (a), and $X' \subset X''$. □

6.7.6 *From e-categories to global orders*

Conversely, we start now from an e-cohesive category A, whose morphisms are written as dot-marked arrows $a\colon X \twoheadrightarrow Y$, and want to form a globally ordered category $\mathsf{C} = \mathrm{Prp}(\mathsf{A})$ of *proper morphisms*. The reader can keep in mind the case $\mathsf{A} = \mathcal{C}$.

Loosely speaking, we suppose that all projectors e of A split (see 5.2.5), and there exists a 'good' choice of the splitting mono (which in $\mathcal{C}$ would be given by the subspace $\mathrm{Def}\, e$).

More precisely, we assume three conditions on A.

(i) Each projector $e\colon X \twoheadrightarrow X$ of A has an assigned splitting

$$e = h_e h_e{}^\sharp, \qquad h_e{}^\sharp h_e = 1.$$

The split monomorphism h_e is called the *inclusion* of the projector e and denoted as $h_e\colon H_e \rightarrowtail X$.

As a consequence, the inclusion h_e belongs to the inverse core IA, and $h_e{}^\sharp$ is its partial inverse, called the *retraction* of e. Moreover, the order $e \leqslant e'$ in $\mathsf{Prj}\,(X)$ corresponds to the preorder $h_e \prec h_{e'}$ of inclusions, which is thus an ordering: two equivalent inclusions coincide.

(ii) The inclusion of the projector $\mathrm{id}\,X$ is the identity, and any inclusion $X \rightarrowtail X$ is the identity.

As a consequence, if an inclusion $h_e\colon H_e \rightarrowtail X$ is epi, then $e = 1_X$ and $h_e = 1_X$. We also note that the second point in condition (ii) is not redundant: in $\mathcal{C}$ the partial identity $e\colon \mathbb{R} \twoheadrightarrow \mathbb{R}$ defined on the open interval $]0,1[$ can also be split as $e = hh^\sharp$, where $h\colon \mathbb{R} \to \mathbb{R}$ is a topological embedding onto $]0,1[$.

(iii) Inclusions are closed under composition, forming a wide subcategory M_0.

Now we let $\mathsf{C} = \mathsf{Prp}\,(\mathsf{A})$ be the wide subcategory of A formed by the *proper morphisms* $f\colon X \to Y$, or *total morphisms*, defined by the property $\underline{e}(f) = 1$. It is indeed a subcategory: if a and b are proper, the relation $a.\underline{e}(ba) = \underline{e}(b).a = a$ shows that $\underline{e}(ba) \geqslant \underline{e}(a) = 1$.

We define a global order $f' \subset f$ on C. On the objects, we let $X' \subset X$ if there is a projector e of X whose inclusion h_e is defined on $H_e = X'$. Then, given two proper morphisms $f'\colon X' \to Y'$ and $f\colon X \to Y$, we define $f' \subset f$ if $X' \subset X$, $Y' \subset Y$ and the inclusions $h\colon X' \rightarrowtail X$ and $k\colon Y' \rightarrowtail Y$ give a commutative square

$$
\begin{array}{ccc}
X & \xrightarrow{\ f\ } & Y \\[2pt]
{\scriptstyle h}\big\uparrow & & \big\uparrow{\scriptstyle k} \\[2pt]
X' & \xrightarrow[\ f'\ ]{} & Y'
\end{array}
\qquad\qquad (6.113)
$$

We verify below that this is indeed a global order relation on C.

This only works because we are using *distinguished* splittings of projectors, satisfying (ii) and (iii): in particular, the axiom 2.8.1(iii)

$$(f' \subset f,\ \mathrm{Dom}\,f = \mathrm{Dom}\,f',\ \mathrm{Cod}\,f = \mathrm{Cod}\,f') \;\Rightarrow\; f' = f,$$

follows from condition (ii): if $\mathrm{Dom}\,f = \mathrm{Dom}\,f'$ and $\mathrm{Cod}\,f = \mathrm{Cod}\,f'$, the inclusions h, k of diagram (6.113) are identities.

6.7.7 Theorem

If A *is an e-cohesive category satisfying conditions 6.7.6(i)–(iii), the previous construction give a globally ordered category* C *(see 2.8.1) that satisfies conditions 6.7.4(i)–(iii).*

Proof First we examine the relation $X' \subset X$, defined by the existence of a projector e of X with inclusion $h_e \colon X' \rightarrowtail X$. We begin by noting that, for a given object X, the 'smaller' objects $X' \subset X$ form a set, corresponding to the set of projectors of X in A (a subset of $\mathsf{A}(X, X)$).

The relation $X \subset X$ follows from 6.7.6(ii). As to transitivity, if $X'' \subset X' \subset X$ we have inclusions $k \colon X'' \rightarrowtail X'$ and $h \colon X' \rightarrowtail X$ whose composite $hk \colon X'' \rightarrowtail X$ is assumed to be an inclusion. The relation is antisymmetric: if $X' \subset X \subset X'$, the composed inclusion $X' \rightarrowtail X'$ is the identity, therefore the inclusion $k \colon X \rightarrowtail X'$ is epi, and must be the identity, by 6.7.6(ii).

It is now easy to see that the relation $f' \subset f$ on the morphisms of C, defined in (6.113), is an order relation and satisfies Ehresmann's axioms of global orders (in 2.8.1(i)–(iii)).

Now we verify conditions 6.7.4(i)–(iii).

First, if $f \colon X \to Y$ is in C and $X' \subset X$, the restriction of f on X' is the composite $fh \colon X' \to Y$ with the inclusion $h \colon X' \rightarrowtail X$.

Second, an inclusion $h \colon X' \rightarrowtail X$ satisfies $\mathrm{id}\, X' \subset h$, by the commutative square

$$\begin{array}{ccc} X' & \xrightarrow{\ h\ } & X \\ {\scriptstyle 1}\big\uparrow & & \big\uparrow{\scriptstyle h} \\ X' & \xrightarrow[\ 1\ ]{} & X' \end{array} \tag{6.114}$$

Suppose we have $f \colon X \to Y$ in C and an inclusion $k \colon Y' \rightarrowtail Y$, corresponding to a projector $e' = kk^\sharp$ of A. We want to form a pullback

$$\begin{array}{ccc} X & \xrightarrow{\ f\ } & Y \\ {\scriptstyle h}\big\uparrow & \searrow & \big\uparrow{\scriptstyle k} \\ X' & \xrightarrow[\ f'\ ]{} & Y' \end{array} \tag{6.115}$$

In the e-cohesive category A, $e'f \leqslant f$, and there exists $e \in \mathsf{Prj}\,(X)$ such that $e'f = fe$. We let $h = h_e$ and $f' = k^\sharp fh$ (in A), forming a commutative square in A: $kf' = e'fh = feh = fh$.

The morphism f' is proper:

$$\underline{e}(f') \geqslant \underline{e}(kf') = \underline{e}(fh) = 1_{X'}.$$

Finally we prove that this square is a pullback in C. Take two morphism u, v so that $fu = kv$

$$ \text{(6.116)} $$

Let $a = h^{\sharp}u$ in A. Then $f'a = v$, because:

$$ k(f'a) = f(hh^{\sharp})u = feu = e'fu = (kk^{\sharp})fu = (kk^{\sharp})kv = kv. $$

This also shows that a is a proper morphism:

$$ \underline{e}(a) \geqslant \underline{e}(f'a) = \underline{e}(kv) = 1_Z. $$

The commutativity of the upper triangle follows from axiom (ECH.2):

$$ ha = hh^{\sharp}u = u\underline{e}(hh^{\sharp}u) = u, $$

because $\underline{e}(hh^{\sharp}u) = \underline{e}(h^{\sharp}u) = \underline{e}(a) = 1.$ $\square$

6.7.8 Theorem

The two procedures above, in 6.7.4 and 6.7.6, are inverse to each other, up to isomorphism.

Proof (a) Let C be a category with global order, satisfying conditions 6.7.4(i)–(iii). Let $A = P_MC$, where M is the wide subcategory of C containing each monomorphism equivalent to an inclusion $h: X' \rightarrowtail X$ (a restriction of 1_X).

A is an e-category, with projectors $e = mm^{\sharp}: X \twoheadrightarrow X$ for m in M, as defined in 5.1.8(f). Moreover, each projector e has a distinguished splitting $h_e h_e{}^{\sharp}$, where h_e is an inclusion of C. The condition 6.7.6(ii) is also satisfied: if $h: X' \rightarrowtail X$ is an inclusion and $X' = X$ then $h = 1_X$. Finally, inclusions are closed under composition, as proved in 6.7.4.

Now Prp (A) coincides with C, and the new order $\subseteq$ is the same.

On the objects, $X' \subseteq X$ if there is in A a projector e of X with inclusion $h_e: X' \rightarrowtail X$. Then $h_e \subset 1_X$ and $X' \subset X$; conversely, if $X' \subset X$ in C, the inclusion $h: X' \rightarrowtail X$ shows that $X' \subseteq X$.

On the morphisms, the order relations coincide, because of the characterisation 6.7.5(b).

(b) Let A be an e-category with an assigned splitting of projectors, satisfying conditions 6.7.6(i)–(iii).

By Theorem 6.7.7, the wide subcategory $C = \mathrm{Prp}\,(A)$ has a global order, defined in 6.7.6, that satisfies conditions 6.7.4(i)–(iii).

Let $B = P_M C$, where M is the wide subcategory of C containing every monomorphism equivalent to the 'inclusion' of a projector (produced by the assigned splitting).

Then A is canonically isomorphic to B, sending the morphism $a \colon X \rightarrow Y$ of A to the morphism $[h, f] \colon X \rightarrow Y$ of B, where $h \colon A \rightarrowtail X$ is the inclusion of C associated to the projector $\underline{e}(a)$, and $f = ah \colon A \rightarrow Y$.   □

7
Solutions of the exercises

Easy exercises and exercises marked with * may be left to the reader.

7.1 Exercises of Chapter 1

7.1.1 Solutions of 1.1.4

(a) Assuming that meets distribute over joins, we have:

$$
\begin{aligned}
(x \vee y) \wedge (x \vee z) &= ((x \vee y) \wedge x) \vee ((x \vee y) \wedge z) \\
&= x \vee (x \wedge z) \vee (y \wedge z) = x \vee (y \wedge z).
\end{aligned}
$$

(b) If x'' is also a complement of x in X, we have

$$
x' = x' \wedge 1 = x' \wedge (x \vee x'') = (x' \wedge x) \vee (x' \wedge x'') = 0 \vee (x' \wedge x'') = x' \wedge x'',
$$

whence $x' \leqslant x''$, and symmetrically $x'' \leqslant x'$.

(c) The properties of the first two lines in (1.9) are easily verified. The last line is a consequence.

(d) An easy consequence of the previous point. The condition $x = \bigvee A = \min(U(A))$ means that $x \in U(A) \cap LU(A)$. The condition $x = \bigwedge(U(A)) = \max(LU(A))$ means that $x \in LU(A) \cap ULU(A)$. But we have seen that $ULU(A) = U(A)$.

(e) Let X be totally ordered, and therefore a lattice. Since x and y play the same role in axiom 1.1.3(D), we can suppose that $x \leqslant y$. Then $(x \wedge z) \leqslant (y \wedge z)$ and

$$
(x \vee y) \wedge z = y \wedge z = (x \wedge z) \vee (y \wedge z).
$$

(f), (g) A consequence of (d). The subset A is upper bounded in X if and only if $U(A)$ is non-empty. A is upper bounded and non-empty if and only if $U(A)$ is non-empty and lower bounded.

(j) In the abelian group $\mathbb{Z}^2$ the diagonal Δ has various complements, like $\mathbb{Z} \oplus 0$ and $0 \oplus \mathbb{Z}$. The same argument works for any non-trivial abelian group $A = B^2$.

7.1.2 Solutions of 1.1.9

(a) We know that, in the lattice $\mathrm{Sub}(A)$, $H \wedge K = H \cap K$ and $H \vee K = H + K$ (by Exercise 1.1.4(i)).

Now, for H, K, L in $\mathrm{Sub}(A)$, with $H \subset L$, we only have to verify that:

$$(H + K) \cap L \subset H + (K \cap L),$$

as the other inclusion is obvious.

We take an element $x = h + k \in L$, with $h \in H$, $k \in K$; then $k = (h + k) - h \in L$ (because $H \subset L$), and $x = h + k \in H + (K \cap L)$.

7.1.3 Solutions of 1.2.3

(d) We can assume that $X \neq \emptyset$.

Let $f \colon X \to X$ be a mapping. For every $x \in X$ we choose an element $\overline{x}$ such that $f(x) = f(\overline{x})$; we also fix one of these distinguished elements, $x_0 = \overline{x}_0$. Now a partial inverse $g \colon X \to X$ of f can be defined as follows, on $y \in X$:

- if there is some $x \in X$ such that $y = f(x)$, we let $g(y) = \overline{x}$,
- otherwise we let $g(y) = x_0$.

(e) If X is empty or a singleton, $S(X)$ is the trivial group, and an inverse monoid. Otherwise, $S(X)$ is not inverse: any pair $f, g \colon X \to X$ of distinct costant mappings is a regular pair; but (f, f) is also.

7.1.4 Solutions of 1.3.3

(a) The inclusion $a \subset aa^\sharp a$ is obvious (and already holds for relations of sets). If $(x, y) \in aa^\sharp a$ there are $x', y' \in A$ such that the pairs (x, y'), (x', y') and (x', y) belong to a, and then

$$(x, y) = (x, y') - (x', y') + (x', y) \in a.$$

(This also works for the endo-relations of a group: the commutativity of the sum in A has not been used.)

(b) For a set X having two distinct elements x, y, it is easy to find an endo-relation a which is properly contained in $aa^\sharp a$. For instance, $a = \{(x, x), (x, y), (y, y)\}$, since $(y, x) \in aa^\sharp a$.

7.1.5 Solutions of 1.4.8

(a) For $x \in X$ we let $u(p(x)) = q(x)$. The mapping u is well defined, since $p(x) = p(x')$ gives $nq(x) = f(x) = f(x') = nq(x')$ and $q(x) = q(x')$.

Diagram (1.39) commutes. Moreover $up = q$ and $nu = m$ show that u is epi and mono, whence a bijection.

(b) One simply uses the factorisation of the underlying mapping in Set, putting on the intermediate set A the unique structure (of abelian group, or R-module, or pointed set) consistent with p and m (the structure for which they belong to the category we are examining).

(c) If $f \colon X \to Y$ is mono and epi in C, the factorisations $f = f.1_X = 1_Y.f$ are epi-mono; essential uniqueness forces f to be an isomorphism.

(d) We already know that Top and pOrd are not balanced.

7.1.6 Solutions of 1.5.7

(a) If $(A, \eta \colon X \to UA)$ and $(B, \eta' \colon X \to UB)$ are universal arrows from X to U, the A-maps $g \colon A \to B$ and $g' \colon B \to A$ such that $Ug.\eta = \eta'$ and $Ug'.\eta' = \eta$ are inverse to each other. In fact

$$U(g'g).\eta = Ug'.\eta' = \eta = U(\mathrm{id}\,A).\eta,$$

whence $g'g = \mathrm{id}\,A$; similarly, $gg' = \mathrm{id}\,B$.

(b) The *free R-module RX* on a set X is a direct sum of copies of R (viewed as an R-module):

$$RX = \bigoplus_{x \in X} R$$
$$= \{\lambda \colon X \to |R| \mid \lambda(x) = 0, \text{ except for a finite subset of } X\}. \tag{7.1}$$

An element λ is thus a *quasi null* mapping $\lambda \colon X \to |R|$; it can be written as a quasi null family $(\lambda_x)_{x \in X}$ of scalars. These families are added componentwise, and multiplied componentwise by any scalar.

The set X has a canonical mapping, called the insertion of X as the *canonical basis* of RX

$$\eta \colon X \to U(RX),$$
$$(\eta(x))_x = 1_R, \qquad (\eta(x))_y = 0_R \ \text{(for } y \neq x). \tag{7.2}$$

The element $\eta(x)$ is often written as e_x, and can be identified with x (*unless R is the trivial ring!*). Each quasi null family (λ_x) of scalars is an essentially finite linear combination $\Sigma_{x \in X} \lambda_x e_x$.

The universal property of the pair (RX, η) says that each mapping

$f\colon X \to U(A)$ with values in an R-module can be uniquely extended to an R-homomorphism $g\colon RX \to A$, so that $g\eta = f$. In fact the homomorphism g is the R-linear extension of f, computed as

$$g(\textstyle\sum_{x\in X}\lambda_x e_x) = \sum_{x\in X}\lambda_x f(x), \tag{7.3}$$

by an essentially finite linear combination in A.

Note. If R is the trivial ring, all R-modules are trivial (i.e. singletons) and the mapping η is constant: the trivial R-module is free *on each set*. Otherwise, the universal mapping $\eta\colon X \to URX$ is injective.

(c) In a group G, the *subgroup of commutators* $[G,G]$ is generated by all *commutators*

$$[x,y] = xyx^{-1}y^{-1} \qquad (x,y \in G). \tag{7.4}$$

Plainly, G is commutative if and only if this subgroup is trivial. It is easy to see that this subgroup is always normal in G: the set of commutators is invariant under inner automorphisms of G.

The universal arrow from the group G to the embedding $\mathsf{Ab} \to \mathsf{Gp}$ is the canonical projection on the *abelianised* quotient group G^{ab}

$$\eta\colon G \to G^{\mathrm{ab}} = G/[G,G]. \tag{7.5}$$

In fact, every homomorphism $f\colon G \to A$ with values in an abelian group takes all commutators of G to 0_A, and induces a (unique) homomorphism $g\colon G^{\mathrm{ab}} \to A$ such that $g\eta = f$.

7.1.7 Solutions of 1.6.1

(a) The component $\hat{\lambda}_A\colon A \to A$ on the left module A is defined as $\hat{\lambda}_A(x) = \lambda x$, the multiplication by λ.

(b) Let φ be a natural endo-transformation of $\mathrm{id}\,(R\,\mathsf{Mod})$. The homomorphism $\varphi_R\colon R \to R$ of left R-modules is the left multiplication by the scalar $\lambda = \varphi_R(1)$, which commutes with any other $\mu \in R$, since $\lambda\mu = \varphi_R(\mu) = \mu\varphi_R(1) = \mu\lambda$.

For every left R-module A, the naturality of φ on the homomorphism $f\colon R \to A$ that sends 1 to an element $a \in A$ gives the relation: $\varphi_A(a) = \varphi_A(f(1)) = f(\varphi_R(1)) = f(\lambda) = \lambda a$.

*(c) Proving that the operations of C are also determined by the corresponding natural transformations requires topics investigated later on. The multiplication in C just amounts to the vertical composition of these transformations, by Exercise 1.6.2(b).

Less trivially, the sum in C amounts to the sum of our transformations: this rests on the (easy) fact that $R\,\mathsf{Mod}$ is an additive category (see 1.7.9), and the (more complex) fact that this structure on $R\,\mathsf{Mod}$ is uniquely determined (only mentioned in 1.7.9).

7.1.8 Solutions of 1.6.2

(a) If all the components φX are invertible in D, their inverses form a natural transformation which is inverse to φ. The converse is obvious.

(b) Obviously $\hat{\lambda}\hat{\mu} = (\lambda\mu)\hat{\ }$. The last claim follows from 1.6.1(b).

7.1.9 Solutions of 1.6.6

(b) The non-identity arrow of $\mathbf{2}$ is written as $\iota\colon 0 \to 1$.

A natural transformation $\varphi\colon F \to G\colon \mathsf{C} \to \mathsf{D}$ gives a functor

$$\Phi\colon \mathsf{C} \times \mathbf{2} \to \mathsf{D}, \qquad \Phi(-,0) = F, \quad \Phi(-,1) = G,$$
$$\Phi(f\colon X \to X', \iota\colon 0 \to 1) = \varphi(f)\colon F(X) \to G(X'), \tag{7.6}$$

where $\varphi(f)$ is defined in (1.48): the diagonal of the 'naturality square' of φ on the morphism f.

Similarly, we have a functor

$$\Phi'\colon \mathsf{C} \to \mathsf{D}^{\mathbf{2}}, \qquad \Phi'(X) = \varphi_X\colon F(X) \to G(X),$$
$$\Phi'(f\colon X \to X') = (Ff, Gf)\colon \varphi_X \to \varphi_{X'}. \tag{7.7}$$

Both procedures are invertible.

7.1.10 Solutions of 1.7.4

(c) The Hausdorff space Y has a closed diagonal $\Delta \subset Y \times Y$. The equaliser of f, g is the preimage of Δ by the continuous mapping $(f, g)\colon X \to Y \times Y$.

(f) Let $m\colon E \to X$ be the equaliser of $f, g\colon X \to Y$. Given two morphisms $u, v\colon Z \to E$ such that $mu = mv\colon Z \to X$, this morphism h gives $fh = fmu = gmu = gh$; thus h factors uniquely through m, and $u = v$.

(g) If $m\colon A \to X$ and $p\colon X \to A$ are such that $pm = \mathrm{id}\,A$, it is easy to see that m is the equaliser of mp and $\mathrm{id}\,X$.

7.1.11 Solutions of 1.7.8

(d) For two homomorphisms $f, g \colon X \to Y$ in Ab we take the quotient Y/H, modulo the subgroup

$$H = \{f(x) - g(x) \mid x \in X\}.$$

This is the same as the quotient of Y modulo the congruence of abelian groups spanned by the equivalence relation R used in (a). Similarly, in Gp, we take the quotient Y/H modulo the normal subgroup of Y generated by the elements $f(x).g(x)^{-1}$ (for $x \in X$).

(e) In Set$_\bullet$ or Top$_\bullet$ we take the coequaliser in Set or Top, with the base-point determined by the projection $Y \to Y/R$.

(f) In Top the solution is the standard circle $\mathbb{S}^1$, with euclidean topology; the details are written out in 2.5.2(a).

In Set we take the same set. In pOrd we take the same set with the chaotic preorder. In Ord, the associated ordered set gives the singleton.

7.1.12 Solutions of 1.7.9

(a) The cartesian projection $p_i \colon X \to X_i$ of a product admits, as a section, the morphism $f_i \colon X_i \to X$ of components

$$p_i f_i = \operatorname{id} X_i, \qquad p_j f_i = 0 \colon X_i \to X_j \quad \text{(for } i \neq j\text{)}.$$

(b) The morphism $f \colon X_1 + X_2 \to X_1 \times X_2$ is determined by the canonical injections and projections letting

$$p_i f u_i = \operatorname{id} X_i, \qquad p_j f u_i = 0 \colon X_i \to X_j \quad \text{(for } i \neq j\text{)}.$$

(One can begin by constructing two morphisms $f_i \colon X_i \to X_1 \times X_2$, and then take $f = [f_1, f_2]$.)

*(c) It is easy to verify that a product $X \times Y$ in Ban$_1$ is a direct sum $X \oplus Y$ of vector spaces with the l_∞-norm, while the sum $X + Y$ is the same vector space with the l_1-norm

$$||(x, y)||_\infty = \max (||x||, ||y||), \qquad ||(x, y)||_1 = ||x|| + ||y||,$$

$$||(x, y)||_\infty \leqslant ||(x, y)||_1 \leqslant 2||(x, y)||_\infty.$$

The same holds in Ban. The morphism $f \colon X + Y \to X \times Y$ which is the identity of the underlying vector spaces is in Ban$_1$; it is an isomorphism in Ban, because $||f^{-1}|| \leqslant 2$, but not in Ban$_1$ (except in trivial cases).

We have thus proved that Ban$_1$ and Ban have binary products and sums, which are biproducts in Ban. In fact, Ban_1 does not have biproducts, as verified below, in (e).

*(d) For Banach spaces X, Y, the set $\mathsf{Ban}(X, Y)$ of continuous linear mappings is a linear subspace of the vector space $K \operatorname{Mod}(X, Y)$, on the scalar field K ($\mathbb{R}$ or $\mathbb{C}$). It is actually a Banach space, with the usual norm

$$||f|| = \sup\{||f(x)|| \mid x \in X, ||x|| \leqslant 1\}.$$

*(e) Of course $\mathsf{Ban}_1(X, Y)$ is not a linear subspace of $\mathsf{Ban}(X, Y)$, but we want to show that Ban_1 *cannot* have an additive structure: using the scalar field K of as a Banach space, we prove that the product $(K \times K, p_1, p_2)$ in Ban_1 cannot be a biproduct.

Suppose for a contradiction that $(K \times K, u_1, u_2)$ is a sum in Ban_1, with $p_i u_i = \operatorname{id} K$ and $p_j u_i = 0 \colon K \to K$ (for $i \neq j$ in $\{1, 2\}$). Then the homomorphisms $u_1, u_2 \colon K \to K \times K$ are the first and second injection, respectively. There should be a morphism $h \colon K \times K \to K$ of cocomponents $\operatorname{id} K$ and $\operatorname{id} K$. By linearity

$$h(x, y) = h(x, 0) + h(0, y) = hu_1(x) + hu_2(y) = x + y,$$

and h is not in Ban_1. (For instance, $|h(1, 1)| = 2 > ||(1, 1)||_\infty = 1$.)

7.1.13 Solutions of 1.8.5

(a) By 1.5.7(c), the reflector $\mathsf{Gp} \to \mathsf{Ab}$ is the abelianisation functor $(-)^{\text{ab}}$.

(b) An abelian group A has a *torsion subgroup* $\mathsf{t}A$, formed by the elements of finite period; A is a torsion group if $\mathsf{t}A = A$. The coreflector takes the abelian group A to its torsion subgroup $\mathsf{t}A$, and the counit of the adjunction is the embedding $\varepsilon A \colon \mathsf{t}A \to A$.

(c) An abelian group A is *torsion-free* if $\mathsf{t}A = 0$. The unit of the adjunction is the canonical projection $\eta A \colon A \to A/\mathsf{t}A$.

7.1.14 Solutions of 1.8.8

(a) We let x vary in X, and y in Y.

(i) $\Rightarrow$ (ii). Assuming (i), $f(x) \leqslant f(x)$ gives $x \leqslant g(f(x))$; moreover, if $x \leqslant g(y)$ then $f(x) \leqslant y$.

(ii) $\Rightarrow$ (iii). Assuming (ii), f is necessarily increasing. From

$$f(x) \in \{y \in Y \mid x \leqslant g(y)\}$$

we get $x \leqslant g(f(x))$; moreover $f(g(y)) = \min\{t \in Y \mid g(y) \leqslant g(t)\} \leqslant y$.

(iii) $\Rightarrow$ (i). We assume (iii). From $f(x) \leqslant y$ we get $x \leqslant g(f(x)) \leqslant g(y)$. From $x \leqslant g(y)$ we get $f(x) \leqslant f(g(y)) \leqslant y$.

(b) The right adjoint to the embedding $i\colon \mathbb{Z} \to \mathbb{R}$ is the integral-part function, or *floor* function

$$[-]\colon \mathbb{R} \to \mathbb{Z}, \qquad [x] = \max\{k \in \mathbb{Z} \mid k \leqslant x\}. \tag{7.8}$$

The left adjoint is the *ceiling* function

$$\min\{k \in \mathbb{Z} \mid k \geqslant x\} = -[-x], \tag{7.9}$$

related (here) to the right adjoint by the anti-isomorphism $x \mapsto (-x)$ of the real and integral lines.

(c) An irrational number has no 'best' rational approximation, lower or upper. For instance, the subset $\{q \in \mathbb{Q} \mid q \leqslant \sqrt{2}\}$ has no maximum, while $\{q \in \mathbb{Q} \mid q \geqslant \sqrt{2}\}$ has no minimum.

(d) In fact, for $X \subset A$ and $Y \subset B$:

$$X \subset f^* f_*(X), \qquad f_* f^*(Y) = Y \cap f(A) \subset Y. \tag{7.10}$$

7.2 Exercises of Chapter 2

7.2.1 Solutions of 2.1.6

(a) In $\mathcal{S}$, a sum $\sum_{i \in I} X_i$ is realised as in Set, with the same (everywhere defined) injections $u_i\colon X_i \to X$ (see 1.7.6(a)).

Given a family $f_i\colon X_i \nrightarrow Y$ of partial mappings, we let

$$f\colon X \nrightarrow Y, \qquad f(x,i) = f_i(x), \tag{7.11}$$

where $\operatorname{Def} f = \bigcup_i u_i(\operatorname{Def} f_i)$, so that $f u_i = f_i$, for all $i \in I$. As $X = \bigcup_i u_i X_i$, this is the only solution.

The coequaliser of two partial mappings $f, g\colon X \nrightarrow Y$

$$\begin{array}{ccc}
X & \overset{f}{\underset{g}{\rightrightarrows}} & Y \xrightarrow{\;p\;} Y/R \\
 & & \downarrow{\scriptstyle h} \quad \swarrow{\scriptstyle k} \\
 & & Z
\end{array} \tag{7.12}$$

can be constructed as in Set, even though f, g are partial mappings. We take the projection

$$p\colon Y \to Y/R, \tag{7.13}$$

modulo the equivalence relation spanned by the pairs $(f(x), g(x)) \in Y^2$, for $x \in \operatorname{Def} f \cap \operatorname{Def} g$.

Explicitly, $(y', y'') \in R$ if and only if there exists a finite sequence $y_0, ..., y_n$ in Y where $y_0 = y'$, $y_n = y''$ and each pair of consecutive points is of the form $(f(x), g(x))$, or $(g(x), f(x))$, or (y, y).

Given a partial mapping $h \colon Y \dashrightarrow Z$ such that $hf = hg$, Def h is saturated for R (see 1.4.9), and h is constant on each equivalence class of R contained in Def h. We define $k \colon Y/R \dashrightarrow Z$ on $p(\text{Def } h)$, letting $k([y]) = h(y)$. Thus $kp = h$, and this is clearly the only solution. (Note that p is also epi in $\mathcal{S}$.)

(b) In $\mathcal{C}$, a sum $\Sigma_{i \in I} X_i$ is realised as in Top. The proof is as in (a), applying Lemma 2.1.7(a): f is defined on the open subset Def f, and continuous on each subspace of its open cover $(u_i(\text{Def } f_i))_{i \in I}$.

Coequalisers too are constructed as in (a), putting on Y/R the quotient topology. The partial map h is continuous on the open subset $V = \text{Def } h \subset Y$, saturated for p; therefore k is defined on an open subset $p(V)$ of Y/R; if W is open in Z, $h^{-1}(W) = p^{-1}k^{-1}(W)$ is open in Y and $k^{-1}(W)$ is open in Y/R.

(c) In $\mathcal{D}$, a *finite* sum $\Sigma_{i \in I} X_i$ is realised as in Top. The proof works as in (b), applying Lemma 2.1.7(b): Def f is a closed subspace of X, and $(u_i(\text{Def } f_i))_{i \in I}$ is a finite closed cover of Def f.

Coequalisers are constructed as in (b). The partial map h is continuous on the closed subset $D = \text{Def } h \subset Y$, saturated for p; therefore k is defined on a closed subset $p(D)$ of Y/R; if C is closed in Z, $h^{-1}(C) = p^{-1}k^{-1}(C)$ is closed in Y and $k^{-1}(C)$ is closed in Y/R.

*(d) Take a product $\prod_i (X_i, \overline{x}_i) = (\prod_i X_i, \overline{x})$ in Top$_\bullet$, and suppose that each base point $\overline{x}_i$ is closed in X_i and not contained in any proper open subset.

The base point $\overline{x} = (\overline{x}_i)$ is closed in $\prod_i X_i$, because $\{\overline{x}\} = \bigcap p_i^{-1}\{\overline{x}_i\}$ is an intersection of closed subsets. Moreover, if $U = \prod_i U_i$ is a basic open set of $\prod_i X_i$ (see 1.7.2(a)) that contains $\overline{x}$, each factor U_i contains $\overline{x}_i$ and is total.

As to equalisers, if (X, x_0) belongs to Top$'_\bullet$, any pointed subspace (E, x_0) also does. First, the point x_0 is closed in X, and therefore in E; second, if U is open in X, and the open subset $U \cap E$ of E contains x_0, also U does: then $U = X$ and $U \cap E = E$.

*(e) The proof is partially similar to the previous one, but based on a finite set of indices I.

The base point $\overline{x} = (\overline{x}_i)$ is open in $\prod_{i \in I} X_i$, because $\{\overline{x}\} = \bigcap p_i^{-1}\{\overline{x}_i\}$ is a finite intersection of open subsets. Moreover, if $U = \prod_i U_i$ is a non-empty basic open set of $\prod_i X_i$, then each U_i contains $\overline{x}_i$ and U contains $\overline{x}$.

As to equalisers, if (X, x_0) belongs to $\mathsf{Top}''_\bullet$, any pointed subspace (E, x_0) also does. First, the point x_0 is open in X, and therefore in E; second, if U is open in X and $U \cap E$ is not empty, then $U \neq \emptyset$ contains x_0, and also $U \cap E$ does.

7.2.2 Solutions of 2.2.2

(a) From Exercise 1.4.7(c).

(b) Suppose we have two epi-mono factorisations $a = mp = nq$ in C

$$
\begin{array}{ccccc}
X & \xrightarrow{\ p\ } & A & \xrightarrow{\ m\ } & Y \\
\| & & \downarrow u & & \| \\
X & \xrightarrow{\ q\ } & B & \xrightarrow{\ n\ } & Y
\end{array}
\qquad (7.14)
$$

Defining $u = n^\sharp m \colon A \to B$ we get a commutative diagram:

$$up = n^\sharp mp = n^\sharp nq = q, \quad (nu)p = nn^\sharp mp = nn^\sharp nq = nq = mp.$$

In fact we have proved more: in every category where all monomorphisms split, epi-mono factorisations are essentially unique. Dually, the same holds in every category where all epimorphisms split.

7.2.3 Solutions of 2.2.3

(a) Given a third relation $c \colon Z \nrightarrow W$, the ternary composites $c(ba)$ and $(cb)a$ are both computed as the set of all pairs $(x, w) \in X \times W$ for which:

- there are $y \in Y$ and $z \in Z$ such that: $(x, y) \in a$, $(y, z) \in b$, $(z, w) \in c$.

The rest is obvious.

(b) For a family of relations $a_i \colon X_i \nrightarrow Y$, we can view each a_i as a subset of $X \times Y$, and define the relation $a \colon X \nrightarrow Y$ as the union of these subsets. This is the only relation $a \colon X \nrightarrow Y$ such that $au_i = a_i$, for all indices i.

7.2.4 Solutions of 2.2.5

(a) We let E be the idempotent monoid on two elements

$$E = \{1, \underline{e}\}, \qquad \underline{e}.\underline{e} = \underline{e}, \qquad (7.15)$$

so that a functor $F \colon E \to \mathsf{C}$ amounts to an idempotent endomorphism $e \colon X \to X$ in C. Natural transformations of these functors are the morphisms considered above.

(The full embedding $\mathsf{C} \to \mathsf{C}^E$ is the 'diagonal functor'; the general case is in 5.2.4.)

*(b) For every idempotent endomorphism $f\colon Y \to Y$ of D, we choose a splitting

$$f = m_f p_f\colon Y \twoheadrightarrow \operatorname{Im} f \rightarrowtail Y, \qquad p_f m_f = 1_{\operatorname{Im} f},$$

under the constraint that the splitting of an identity is trivial: $1_Y = 1_Y 1_Y$.
    Then we define $G\colon \mathsf{C}^E \to \mathsf{D}$ letting

$$G(e) = \operatorname{Im} F(e),$$

$$G(a\colon e \to e') = p_{F(e')}\, F(a)\, m_{F(e)}\colon \operatorname{Im} F(e) \to \operatorname{Im} F(e'),$$

where $e\colon X \to X$ and $e'\colon X' \to X'$ are idempotents of C.

7.2.5 Solutions of 2.2.6

(a) In C^P, the object e (an idempotent of C) is isomorphic to the object $e' = \underline{e}(e) = e^\sharp e$ (a projector of C), by the inverse morphisms

$$e'\colon e \to e', \qquad e\colon e' \to e \qquad (ee' = e\colon e \to e,\ e'e = e'\colon e' \to e').$$

(b) The full embedding $\mathsf{C}^P \to \mathsf{C}^E$ is essentially surjective on the objects, whence an equivalence (by 1.6.3).

(c) P is the monoid E of (7.15), equipped with its unique involution: $\underline{e}^\sharp = \underline{e}$.

7.2.6 Solutions of 2.3.5

(a) The equivalence $F\colon \mathsf{Set}_{\scriptscriptstyle\bullet} \to \mathcal{S}$ (in (2.5)) restricts to an equivalence between $\mathcal{I}$ and the wide subcategory of $\mathsf{Set}_{\scriptscriptstyle\bullet}$ of pointed mappings $f\colon (X, x_0) \to (Y, y_0)$ which are injective outside of $f^{-1}\{y_0\}$.

7.2.7 Solutions of 2.4.3

(a) If $e \leqslant 1_X$ and $f \leqslant 1_X$, it follows that ef is idempotent; but then $ef = ef.ef \leqslant f.e$, and similarly $fe \leqslant ef$.

(b) Given a partial mapping $e\colon X \rightarrowtail X$ with $e \leqslant 1_X$, we have $e(x) = x$ for all $x \in \operatorname{Def} e$. Thus e is the identity on $\operatorname{Def} e$, and idempotent.

(c) Take a constant endo-mapping of a set with at least two elements.

7.2.8 Solutions of 2.4.6

(a) For instance the increasing real function $f(x) = x - 1$ is smaller than
id $\mathbb{R}$, and not idempotent.

(b) An endo-relation $e \leqslant 1_X$ is a partial identity on a subset Def $e = $ Val e
of X, and is a symmetric idempotent. The inverse core $I(\mathrm{Rel\,Set})$ is the
wide subcategory of injective, single-valued relations, or partial bijections,
and coincides with $\mathcal{I} = I\mathcal{S}$.

   An endo-relation $a \geqslant 1_X$ amounts to a subset of $X \times X$ that contains the
diagonal Δ. It need not be idempotent: if $X = \{x, y, z\}$ has three elements,
the endo-relation $a = \Delta \cup \{(x, y), (y, z)\}$ is not idempotent, because $(x, z) \in aa$.

(c) Again, an endo-relation $e \leqslant 1_A$ is a partial identity on a subgroup
Def $e = $ Val e, and is a symmetric idempotent. Here this fact can also be
easily deduced from the regular involution: $e = ee^\sharp e \leqslant e^\sharp e \leqslant e$, whence
$e = e^\sharp e = \underline{e}(e)$. The inverse core $I(\mathrm{Rel\,Ab})$ is the wide subcategory of
injective, single-valued relations of abelian groups, i.e. partial isomorphisms
between subgroups.

   Here the condition $a = be$ (with $e \leqslant 1$) is not equivalent to $a \leqslant b$: for
instance, the least endo-relation $\omega \colon A \to A$ and the greatest endo-relation
$\Omega \colon A \to A$ give $\omega < \Omega$ (for $A \neq 0$), but ω cannot be obtained as Ωe with
$e \leqslant 1_A$, since Val $\Omega e = $ Val $\Omega = A$, while Val $\omega = 0$.

7.2.9 Solutions of 2.5.4

(a) The maps $u_i = pj_i$ form a commutative square (2.51), as $pj_1f = pj_2g$.
As to the universal property, giving a commutative square $v_1f = v_2g$ is
equivalent to giving a map $[v_1, v_2] \colon X_1 + X_2 \to B$ which coequalises the
maps $j_1f, j_2g \colon X_0 \to X_1 + X_2$.

(b) The topology of $\mathbb{S}^n$ is the finest that makes the embeddings $u_i \colon \mathbb{D}^n \to \mathbb{S}^n$
(as lower and upper hemisphere) continuous.

(c) In the new definition, X/A is the quotient of the sum $X + \{*\}$ which
identifies every $x \in A$ with the added point $*$. If $A = \emptyset$ we simply get
$X + \{*\}$. Otherwise, we come back to the quotient of X defined in 2.5.2(a).

7.2.10 Solutions of 2.5.5

(a) The components of sum and additive inverse in $\mathbb{R}^n$ are linear maps (in
$2n$ or n real variables). The components of multiplication and inverse in
$\mathbb{C}^*$ are rational functions (in 4 or 2 real variables).

(b) The continuity of φ is obvious. The property $\varphi(t + t') = \varphi(t).\varphi(t')$ follows from the properties of the complex exponential; more elementarily, it can be derived from the sum-formulas of sine and cosine.

(c) As to the restriction $\varphi' \colon \mathbb{R} \to \mathbb{S}^1$, we note that the open intervals of $\mathbb{R}$ form a basis of open sets, whose φ-images in $\mathbb{S}^1$ are open arcs (possibly total). Then the continuous bijection ψ is also open: an open subset V of $\mathbb{R}/\mathbb{Z}$ has an open preimage U in $\mathbb{R}$, and $\psi(V) = \varphi'(U)$.

The isomorphism ψ acts as the homeomorphism $\mathbb{I}/\partial\mathbb{I} \to \mathbb{S}^1$ of 2.5.2(a).

(d) The consistency follows from the fact that the projection $p \colon G \to G/H$ is an open mapping: the saturated $p^{-1}(p(U))$ of an open set U of G is $\bigcup_{h \in H} hU$, an open subset. Therefore, letting $G' = G/H$, we have a commutative diagram (in Set)

$$
\begin{array}{ccc}
G \times G & \xrightarrow{\ m\ } & G \\
{\scriptstyle p \times p}\big\downarrow & & \big\downarrow{\scriptstyle p} \\
G' \times G' & \xrightarrow[\ m'\ ]{} & G'
\end{array}
\qquad (7.16)
$$

where $m' \colon G' \times G' \to G'$ is the induced multiplication. Now $m'(p \times p) = pm$ is continuous and $p \times p$ is open, which implies that m' is continuous. The continuity of the inverse is even simpler.

(e) The division mapping $d \colon G \times G \to G$, $d(x, y) = xy^{-1}$ is continuous. If 1_G is a closed point, the diagonal $\Delta = d^{-1}\{1_G\}$ is closed in $G \times G$ and G is a Hausdorff space. The converse is obvious.

(f) Follows from (e), as the projection $p \colon G \to G/H$ gives $H = p^{-1}\{1\}$.

7.2.11 Solutions of 2.7.5

(b) The functor $\mathcal{P} \colon \mathcal{I} \to \mathsf{Slt}$ is faithful, because the partial bijection $f = (X_0, Y_0, f_0) \colon X \dashrightarrow Y$ is determined by $\mathcal{P}f$

$$
X_0 = (\mathcal{P}f)^{\bullet}(Y), \qquad Y_0 = (\mathcal{P}f)_{\bullet}(X),
$$

$$
\{f_0(x)\} = (\mathcal{P}f)_{\bullet}(\{x\}) \qquad (\text{for } x \in X_0).
$$

The transfer functor $\mathsf{Prj} \colon \mathcal{I} \to \mathsf{Slt}$ of $\mathcal{I}$ is also faithful, being related to $\mathcal{P}$ by an obvious functorial isomorphism

$$
\pi \colon \mathcal{P} \to \mathsf{Prj} \colon \mathcal{I} \to \mathsf{Slt}, \qquad \pi X \colon \mathcal{P}X \dashrightarrow \mathsf{Prj}_{\mathcal{I}}(X),
$$

$$
(\pi X)_{\bullet}(A) = (A, A, \mathrm{id}\,A) \colon X \dashrightarrow X, \qquad (\pi X)^{\bullet}(e) = X_0,
$$
(7.17)

for $A \subset X$ and $e = (X_0, X_0, \mathrm{id}\,X_0) \in \mathsf{Prj}\,X$.

(c) The categories I$\mathcal{T}$, I$\mathcal{C}$ and I$\mathcal{D}$ are concrete over $\mathcal{I}$. Letting C be any of them, its transfer functor is a composite of faithful functors C $\to$ $\mathcal{I}$ $\to$ Slt.

7.3 Exercises of Chapter 3

7.3.1 Solutions of 3.1.5

(a) The join of a family of relations $a_i \colon X \nrightarrow Y$ is their union $\bigcup_i a_i \subset X \times Y$. Distributivity of composition over these unions is easily checked. Note that composition does not distribute over (even binary) meets, generally.

(b) The join $\vee\, a_i$ of a family of subgroups of $A \oplus B$ always exists, but composition need not distribute over a join, even a binary one.

For instance, every subgroup H of the abelian group A has an associated projector $e_H \colon A \nrightarrow A$, the restriction of 1_A on H. Joins and composites of these projectors amount to joins and meets in the lattice Sub(A)

$$e_H \vee e_K = e_{(H \vee K)}, \qquad e_H e_K = e_H \wedge e_K = e_{(H \wedge K)},$$

and we have seen that Sub(A) need not be distributive (in 1.1.4(j)).

(c) In all these categories, a set of partial maps $X \nrightarrow Y$, upper bounded by a partial map f, has a join in $\mathcal{S}$. In the topological cases, this partial mapping is a restriction of f, and therefore continuous. In all of them, two distinct total mappings $f, g \colon X \to Y$ have no upper bound.

(d) For $\mathcal{S}$ this fact has already been remarked in 3.1.1. For $\mathcal{C}$ and $\mathcal{C}^r$, the continuity of the join follows from the Open Cover Lemma 2.1.7(a). For $\mathcal{D}$, apply the Finite Closed Cover Lemma 2.1.7(b).

7.3.2 Solutions of 3.2.3

(a), (b) One can use the extension property 3.2.1(d) and the associative property 3.2.1(a).

7.3.3 Solutions of 3.2.6

(a) We already remarked, in 3.1.1, that linked joins and meets exist in $\mathcal{S}$ and are preserved by composition.

The join f of a linked family of partial mappings $f_i \colon X \nrightarrow Y$ is defined on Def $f = \bigcup_i$ Def f_i, as the common extension of all these partial mappings

$$f(x) = f_i(x), \qquad \text{if } x \in \text{Def } f_i. \tag{7.18}$$

If the family is non-empty, the meet is the common restriction of f_i on $\bigcap_i \mathrm{Def}\, f_i$.

(b) The structure of $\mathcal{C}$ is also totally cohesive, with binary linked meets and arbitrary linked joins lifted from $\mathcal{S}$. For linked meets, $\mathrm{Def}\, f_1 \cap \mathrm{Def}\, f_2$ is an open subspace, and $f_1 \wedge f_2$ is continuous, as a restriction of both f_i. For linked joins, $U = \bigcup \mathrm{Def}\, f_i$ is an open subspace of X, and the continuity of $\bigvee f_i$ on U follows from the Open Cover Lemma 2.1.7(a).

Again, the structure of $\mathcal{C}^r$ is also totally cohesive, with the same binary linked meets and arbitrary linked joins: being of class C^r on an open subspace of a euclidean space is a local property.

(c) As in (b): we apply now the Finite Closed Cover Lemma 2.1.7(b).

(d) A family $f_i\colon X \rightarrowtail Y$ $(i \in I)$ linked in $\mathcal{I}$ is also linked in $\mathcal{S}$, and their join in $\mathcal{S}$ is a partial bijection; the same holds for their meet, provided that $I \neq \emptyset$.

(e), (f) Follows from the previous points, or is obvious.

(g) By definition, a finite linked family $f_i\colon X \rightarrowtail Y$ in $\mathcal{T}$ has an upper bound $f\colon X \rightarrowtail Y$; their join in $\mathcal{S}$ is a restriction of f and continuous.

As to countable joins, we can take a sequence $f_n\colon \mathbb{R} \rightarrowtail \mathbb{R}$ of maps, each defined on the singleton $\{x_n\}$, with $x_n = 1/(n+1)$. The sequence is always pairwise linked, but has a join in $\mathcal{T}$ if and only if the sequence $f_n(x_n)$ has a limit in $\mathbb{R}$.

For the last claim one can use two partial maps $\mathbb{R} \rightarrowtail \mathbb{R}$ defined on the intervals $[0, 1]$ and $]1, 2]$, respectively, whose join in $\mathcal{S}$ is not continuous at the point 1.

7.3.4 Solutions of 3.2.8

(b) The join of a countable linked set $\{a_n \mid n \in \mathbb{N}\}$ in $\mathsf{C}(X, Y)$ can be obtained as the join of an increasing sequence of finite linked joins $b_n = \bigvee \{a_k \mid k \leqslant n\}$.

(c) A consequence of the fact that every ρ-cohesive category is link-filtered (see 3.2.2).

7.3.5 Solutions of 3.3.6

(a) By 3.3.2(c).

(c) Apply 3.3.3 and (3.21).

(d) The projector $e = \bigvee_{a \in \alpha} \underline{e}(a)$ is well defined, as a ρ-join of projectors.

For every $a \in \alpha$, $a \leqslant \vee \alpha$; applying (b), we get $\underline{e}(a) \leqslant \underline{e}(\vee \alpha)$, and $e \leqslant \underline{e}(\vee \alpha)$. These projectors coincide, because $(\vee \alpha).e = \vee \alpha$:

$$(\vee \alpha).e = \vee_{a,b \in \alpha}\, ae(b) \geqslant \vee_{a \in \alpha}\, a\underline{e}(a) = \vee \alpha \geqslant (\vee \alpha).e.$$

(e) The axiom (ECH.0) is obviously inherited from **D**. (ECH.1, 2) follow from (i). The equivalence of (i) and (i′) is obvious. Let us note that **C** has binary linked meets, preserved (and lifted) by U.

(f) The ρ-cohesive case is a consequence of 3.2.5(b), as we have already taken care of binary linked meets.

7.3.6 Solutions of 3.3.7

(a) For a projector $g \in \mathsf{Prj}\,(Z)$, we have:

$$a^P b^P (g) = a^P (\underline{e}(gb)) = \underline{e}(\underline{e}(gb).a) = \underline{e}(a.\underline{e}(gba)) = \underline{e}(a).\underline{e}(gba)$$
$$= \underline{e}(g.ba) = (ba)^P (g).$$

$$a^P (ff') = a^P (f^P (f')) = (fa)^P (f') = (a.a^P (f))^P (f')$$
$$= (a^P (f))^P (a^P (f')) = a^P (f).a^P (f').$$

(b) Property (3.32) follows from (3.20): $fa = a.\underline{e}(fa)$. Then we have (3.33)

$$ee' = e.e^P (e') = e^P (1).e^P (e') = e^P (e').$$

*(c) We have seen that property (3.33) follows from (3.30)–(3.32). Now, $a^P (1)$ is the support of the morphism a, because $a.a^P (1) = a$, and $ae = a$ implies $a^P (1) \leqslant e$:

$$a^P (1) = (ae)^P (1) = e^P a^P (1) = e.a^P (1).$$

Finally, to verify (ECH.2), we let $f = b^P (1) = \underline{e}(b)$

$$\underline{e}(b).a = fa = a.a^P (f) = a.a^P b^P (1) = a.(ba)^P (1) = a.\underline{e}(ba).$$

7.3.7 Solutions of 3.4.2

(a) We know that the projectors of any object form a 1-semilattice, and $a\underline{e}(a) = aa^\sharp a = a$; if $a = ae$, for a projector e, then $a^\sharp a = a^\sharp ae$ and $\underline{e}(a) \leqslant e$. Finally, if a and b are consecutive morphisms

$$a\underline{e}(ba) = a(a^\sharp b^\sharp ba) = (b^\sharp b)(aa^\sharp)a = \underline{e}(b)a.$$

(b) If $a(b^\sharp b) = b(a^\sharp a)$, then

$$(ba^\sharp)(ba^\sharp) = b(b^\sharp ba^\sharp)(ba^\sharp) = b(a^\sharp ab^\sharp)(ba^\sharp) = (bb^\sharp b)(a^\sharp aa^\sharp) = ba^\sharp.$$

Conversely, if $ba^\sharp$ is a projector, then

$$b(a^\sharp a) = (ba^\sharp)a = (ab^\sharp ba^\sharp)a = a(b^\sharp b)(a^\sharp a) = a(a^\sharp a)(b^\sharp b) = a(b^\sharp b).$$

(c) The projectors of $\mathcal{I}$ are those of $\mathcal{S}$, namely the partial identities of sets.

The relation $a\,!'\,b$ means that the partial bijections a and b are linked in $\mathcal{S}$, as partial mappings, i.e. coincide where they are both defined, or have an upper bound in $\mathcal{S}$. The relation $a\,!''\,b$ means that the same holds for the partial mappings $a^\sharp$ and $b^\sharp$.

In other words, $a\,!'\,b$ means that a and b have a join in $\mathcal{S}$, which is a partial mapping; it belongs to $\mathcal{I}$ if and only if $a\,!''\,b$ also holds.

For any pair of partial bijections $a, b\colon X \to Y$ with

$$\operatorname{Def} a \cap \operatorname{Def} b = \emptyset, \qquad \operatorname{Val} a \cap \operatorname{Val} b \neq \emptyset, \tag{7.19}$$

$a\,!'\,b$ is (trivially) true, but $a\,!''\,b$ cannot be.

Therefore, the left cohesive structure $\mathcal{I}_L$ lacks binary linked joins.

7.3.8 Solutions of 3.5.3

(a) Condition (ii') gives

$$a_k^i\,\underline{e}(a_h^i) = a_k^i a_i^h a_h^i \leqslant v_k^h a_h^i = v_k^h a_h^i\,\underline{e}(a_h^i) \leqslant a_k^i\,\underline{e}(a_h^i).$$

Conversely, from (iii') we have:

$$a_k^i a_i^h = a_k^i\underline{e}(a_h^i)a_i^h = v_k^h a_h^i a_i^h \leqslant v_k^h.$$

(b) For $i \in I$ and $m \in M$, the family $(b_m^h\,a_h^i)_{h\in H}$ is linked

$$(b_m^k a_k^i)^\sharp(b_m^h a_h^i) = (a_i^k b_k^m)(b_m^h a_h^i) \leqslant a_i^k v_k^h a_h^i \leqslant a_i^h a_h^i \leqslant \operatorname{id} U_i,$$

$$(b_m^k a_k^i)(b_m^h a_h^i)^\sharp \leqslant \ldots \leqslant \operatorname{id} W_m,$$

and has a linked join c_m^i. To verify that the family of the latter is a symmetric profunctor, we write down half of the computations (for $i, j \in I$, $h, k \in H$, $m \in M$)

$$c_m^j u_j^i = \bigvee{}_h (b_m^h a_h^j u_j^i) \leqslant \bigvee{}_h (b_m^h a_h^i) = c_m^i,$$

$$c_j^m c_m^i = \bigvee{}_k (a_j^k b_k^m) \bigvee{}_h (b_m^h a_h^i) = \bigvee{}_{h,k} (a_j^k b_k^m b_m^h a_h^i)$$
$$\leqslant \bigvee{}_{h,k} (a_j^k v_k^h a_h^i) \leqslant \bigvee{}_k (a_j^k a_k^i) \leqslant u_j^i.$$

(c) We have already seen, in 3.4.2(c), that $\mathcal{I}_L$ lacks binary linked joins.

7.3.9 Solutions of 3.8.3

(b) Letting $a = c.e_a$ (for $a \in \alpha$), we have

$$a.e_a = c.e_a.e_a = a, \qquad a.e_b = c.e_a.e_b = c.e_b.e_a = b.e_a. \qquad (7.20)$$

(c) With the notation of (3.108), the family $e_a = \bigwedge_{b \in \alpha} e_{ab}$ (for $a \in \alpha$) is a resolution of α

$$a.e_a = a.\bigwedge_b e_{ab} = \bigwedge_b a.e_{ab} = \bigwedge_b a = a, \qquad (7.21)$$

$$a.e_b = ae_a.e_b = (a.e_{ba}e_b)e_a = (b.e_{ab}e_b)e_a = b.e_b.e_{ab}e_a = b.e_a. \qquad (7.22)$$

As to property (3.115), we fix some $b \in \alpha \neq \emptyset$. Then $b.\bigwedge e_a \leqslant b.e_a = a.e_b \leqslant a$ for all $a \in \alpha$; if $x \leqslant a$ for all $a \in \alpha$, then $x \leqslant a \wedge b = be_a$, hence $x \leqslant b.\bigwedge e_a$ (since, by hypothesis, $b.\bigwedge e_a = \bigwedge be_a$).

The compositive property of $\bigwedge \alpha$ is a straightforward consequence of the transfer of resolutions, in 3.8.2.

(d) Follows from (c), as finite meets of projectors in the 1-semilattice $\mathsf{Prj}\,(X)$ do exist.

7.4 Exercises of Chapter 4

7.4.1 Solutions of 4.1.3

(a) One can take the half-disc U, open in H^n

$$U = \{(t_1, t_2, ...t_n) \in \mathbb{R}^n \mid t_1 \geqslant 0, \|t\| < 1\}, \qquad (7.23)$$

and represented below for $n = 2$, with $\underline{\mathrm{Int}}\,U$, $\underline{\partial}\,U$ and $\partial_{H^n}U$ (as usual, a dotted line is not a part of the zone of the plane we are considering)

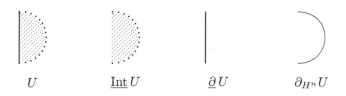

$$U \qquad\qquad \underline{\mathrm{Int}}\,U \qquad\qquad \underline{\partial}\,U \qquad\qquad \partial_{H^n}U$$

$$\underline{\mathrm{Int}}\,U = \{(t_1, t_2, ...t_n) \in U \mid t_1 > 0\},$$

$$\underline{\partial}\,U = \{(t_1, t_2, ...t_n) \in U \mid t_1 = 0\}, \qquad (7.24)$$

$$\partial_{H^n}U = \{(t_1, t_2, ...t_n) \in \mathbb{R}^n \mid t_1 \geqslant 0, \|t\| = 1\}.$$

Finally, $\partial_{\mathbb{R}^n}U = \underline{\partial}\,U \cup \partial_{H^n}U$.

7.4.2 Solutions of 4.1.4

(a) If condition (i) is satisfied, the partial mapping $\overline{f}p$ is constant on each leaf and $p'f$ is also, which means that f preserves leaves.

Conversely, we suppose that f belongs to $C^r\mathcal{F}$. If $x \in p(\text{Def } f)$, there is some $(x, y) \in \text{Def } f$, and we define $\overline{f}(x) = p'f(x, y)$, which we know to be independent of y.

We have thus a partial mapping $\overline{f}\colon U \twoheadrightarrow U'$ defined on $p(\text{Def } f)$, an open subset of U (because each cartesian projection is open). The inequality $p'f \leqslant \overline{f}p$ holds: if $(x, y) \in \text{Def } f$, we have

$$p'f(x, y) = \overline{f}(x) = \overline{f}p(x, y).$$

On the other hand, the inequality requires $\overline{f}$ to be defined as above. (Note that $\overline{f}p$ is defined on the union of leaves which meet $\text{Def } f$, but f can have a smaller definition set.)

To prove that $\overline{f}\colon U \twoheadrightarrow U'$ is a partial C^r-mapping, take $x_0 \in \text{Def } \overline{f}$ and $(x_0, y_0) \in \text{Def } f$. The open set $\text{Def } f$ contains a basic open nbd $U_0 \times V_0$ of (x_0, y_0) (a product of open sets), and $\overline{f}(x) = p'f(x, y)$ (for $x \in U_0$) is of class C^r on U_0.

(b) If $e\colon U \times V \twoheadrightarrow U \times V$ is idempotent in $C^r\mathcal{F}$, then it is a partial identity on an open set $\text{Def } e$ (and preserves leaves). The induced $\overline{e}\colon U \to U$ is the partial identity on the open set $p(\text{Def } e)$. The functor P preserves supports.

As to joins of parallel morphisms, it is sufficient to prove that P preserves all joins of projectors (by Proposition 3.8.5(b)). In fact, if $e_i\colon U \times V \twoheadrightarrow U \times V$ is the partial identity on some open subset W_i of $U \times V$ (for $i \in I$), then $\overline{e}_i\colon U \to U$ is the partial identity on $p(W_i)$, and it is sufficient to remark that

$$p(\bigcup_i W_i) = \bigcup_i p(W_i).$$

7.4.3 Solutions of 4.1.5

(a) We fix a countable basis of open sets of X. Let $\mathcal{U}$ be an open cover of X, and let (V_n) be a countable family of open sets of the basis, such that any $U \in \mathcal{U}$ is a union of some of them. Then (V_n) is a cover of X, and each V_n is contained in some $U_n \in \mathcal{U}$; the latter form a countable subcover of $\mathcal{U}$.

(b) A countable basis of $\mathbb{R}^n$ can be obtained by fixing a countable dense subset D (for instance $\mathbb{Q}^n$), and the family of all open discs with centre in D and radius $1/n$ ($n > 0$). Any subspace of a second countable space is also second countable.

(c) More generally, a paracompact space X where every point has a second countable open nbd $V(x)$ is always second countable.

In fact, the family $V(x)$ is an open cover of X, from which we can extract a countable subcover $(V(x_n))_{n \in \mathbb{N}}$. Taking a countable basis of open sets of each $V(x_n)$, and putting them together, we get a countable basis of open sets of X.

7.4.4 Solutions of 4.2.2

(a) Condition (i) is obviously necessary. If it holds, $f: X \rightarrowtail X'$ is defined on an open subset $\operatorname{Def} f = W$ of X which is saturated for p. We define the partial map $\overline{f}: B \rightarrowtail B'$ on the open subset $p(W) \subset B$, letting $\overline{f}(p(x)) = p'(f(x))$.

(b) If e is idempotent, $(\overline{e}\overline{e})p = p(ee) = pe = \overline{e}p$, and $\overline{e}\overline{e} = \overline{e}$.

All the rest follows from (a).

(c)–(e). Obvious.

(f) We have to verify two points.

First, the family $p^i = \overline{u}^i p_i: X_i \to B$ $(i \in I)$ forms a lax cocone of $(X_i, (u^i_j))_I$ (see 3.6.1(b)):

$$p^j . u^i_j = \overline{u}^j p_j u^i_j = \overline{u}^j \overline{u}^i_j p_i \leqslant \overline{u}^i p_i = p^i.$$

By the universal property of $(X, (u^i))$ as the gluing of $(X_i, (u^i_j))_I$ there is a unique $p: X \rightarrowtail B$ in $\mathcal{C}$ such that $pu^i = p^i$ (for all i). We have thus the commutative squares (4.26).

Second, this partial map is everywhere defined and surjective, because

- each $p_i: X_i \to B_i$ is everywhere defined and surjective,

- the commutative squares (4.26) give: $X = \bigcup_i \operatorname{Val} u^i$ and $B = \bigcup_i \operatorname{Val} \overline{u}^i$.

Finally, p is open because each $p_i: X_i \to B_i$ is, and $X = \bigcup_i \operatorname{Val} u^i$ is an open cover.

7.4.5 Solutions of 4.2.7

(a) If the mapping f is a local homeomorphism, then it is continuous and open.

Conversely, assuming (ii), every $x \in X$ has an open nbd U such that the restriction $f': U \to f(U)$ is a bijective continuous mapping onto an open subspace of Y. If U' is open in U (and in X), then $f'(U')$ is open in Y (and in $f(U)$), which proves that f' is a homeomorphism.

(b) If $p\colon X \to B$ is a covering map, then it is surjective. Taking $x \in X$ and $b = p(X)$, there is an open nbd V of B evenly covered by p, and the (open) sheet $U \subset p^{-1}(V)$ that contains x is mapped homeomorphically onto V by the covering map p.

(c) For $x \in p^{-1}(V)$, the sheet U_x of x over V meets the fibre $F_b = p^{-1}\{b\}$ in one point $q(x)$. This gives a surjective mapping $q\colon p^{-1}(V) \to F_b$ and a bijection

$$h\colon p^{-1}(V) \to V \times F_b, \qquad h(x) = (p_V(x), q(x)).$$

This mapping is continuous and open, because each restriction

$$U_x \to V \times \{x\}$$

is a homeomorphism between open subsets of domain and codomain.

7.4.6 Solutions of 4.4.2

(a) This is obvious when the ordered space X is a topological sum $X_1 + X_2$ (of non-empty spaces): then the lm-structure of X only depends on the restriction of its preorder to the summands X_i.

The last claim follows from the characterisation of equivalences of categories, in 1.6.3.

(b) A finite product $X = \Pi X_i$ in pTop (with the product topology and preorder), equipped with the (monotone) projections $p_i\colon X \to X_i$, satisfies also the universal property in lmTop.

Indeed, for a family $f_i\colon Y \to X_i$ of locally monotone maps, defined on a preordered space, the mapping $f = (f_i)\colon Y \to X$ is continuous; moreover, for each $y \in Y$, the component f_i is monotone on a convenient nbd V_i of y, so that f is monotone on the (finite) intersection of all V_i, still a neighbourhood of y.

(d) By hypothesis there exists an open cover (U_i) of X such that f is monotone on each U_i. If $x \prec x'$ in X, a monotone path from x to x' can be decomposed (as in 4.3.6(b)) in a concatenation of paths $a = a_1 * a_2 * ... * a_n$, so that each path a_k has image contained in some U_i. It follows that $f(x) \prec f(x')$ in Y.

7.4.7 Solutions of 4.5.2

(a) If $y \prec_V y'$ and $f(V) \subset W$ (an open subset of Y'), let (W_i) be an open cover of W and $V_i = f^{-1}(W_i)$ the corresponding open cover of V.

Then there is a sequence $y = y_0 \prec y_1 \prec \ldots \prec y_n = y'$ consistent with the cover (V_i), and the sequence $f(y) \prec f(y_1) \prec \ldots \prec f(y')$ is consistent with (W_i).

(b) If $f \colon X \to GY$ is an lp-map, $x \prec_X x'$ implies $f(x) \prec_Y f(x')$, and this implies that $f(x) \prec f(x')$ in the original preorder of Y.

Conversely, we suppose that $f \colon FX \to Y$ is a map of preordered spaces: if $x \prec_X x'$ then $fx \prec fx'$ in Y; we have to prove that f is also an lp-map $X \to GY$.

Take $x \prec_U x'$, and $f(U) \subset W$, open in Y. If (W_i) is an open cover of W and $U_i = U \cap f^{-1}(W_i)$ the corresponding open cover of U, there is a sequence $x = x_0, x_1, \ldots, x_n = x'$ where each term precedes the consecutive one in some U_i (with respect to $\prec_{U_i}$, and also to $\prec_X$); therefore the sequence $f(x) \prec f(x_1) \prec \ldots \prec f(x')$ is consistent with the cover (W_i) of W, and we have proved that $f(x) \prec_W f(x')$.

7.4.8 Solutions of 4.5.4

(a) Suppose that $x \leqslant x'$ in a convex subspace X of $\uparrow\mathbb{R}^n$. Then there is a monotone path $a \colon \uparrow\mathbb{I} \to X$ from x to x'. For every open cover (U_i) of X we can decompose a in a finite concatenation $a_1 * a_2 * \ldots * a_n$ of paths, each with image contained in some U_i. Finally the sequence

$$x = a_1(0) \prec a_1(1) = a_2(0) \prec a_2(1) \prec \ldots \prec a_n(1) = x',$$

is consistent with the cover (U_i).

(b) As in (a).

(c) The preordered space FGX has the order relation of the categorical sum $[0,1] + [2,3]$, where the points of different connected components are not comparable.

*(d) A reader who does not already know this classical, important example should begin by drawing in the cartesian plane the subspace X, formed by the graph of the function $f(x) = \sin 1/x$, for $x > 0$. Its closure in $\mathbb{R}^2$ is the subspace $Y = X \cup A$, with $A = \{0\} \times [-1,1]$.

The space X is path-connected, but Y (although connected), has two path-connected components, X and A.

The drawing 'shows' that the space X is path-preordered (for the given preorder), but Y is not: any point $(0, b) \in A$ precedes any point $(x, y) \in X$, and these points are in different path-components.

However, Y is an lg-space: we only need to consider the previous relation $(0, b) \prec (x, y)$, because we already know that X is path-preordered (whence an lg-space, by (b)), and two points of A are always connected by a (trivially directed) path.

Now, if (V_i) is an open cover of Y, there is some index j such that $(0, b) \in V_j$; then there is some $(x', y') \in V_j \cap A$, and we may choose the latter so that $x' < x$. Now $(0, b) \prec (x', y')$ in V_j and $(x', y') \prec (x, y)$ in X. As X is path-preordered, we conclude that there is a finite sequence

$$(0, b) \prec (x', y') \prec (x_1, y_1) \prec \ldots \prec (x, y)$$

in Y, consistent with (V_i).

(e) There exists an open cover (U_i) of X such that f is monotone on each U_i. If $x \prec x'$ there is a sequence $x = x_0 \prec x_1 \prec \ldots \prec x_n = x'$ where each pair of consecutive terms belongs to some U_i. Then $f(x) \prec f(x')$ in Y.

*(f) $FG(C)$ is an ordered topological space, where

$$(x, y) \leqslant (x', y') \quad \Leftrightarrow \quad (x, x' \leqslant 0 \text{ and } y \leqslant y') \text{ or } (x, x' \geqslant 0 \text{ and } y \leqslant y').$$

In [G8] this is the d-space $\uparrow\mathbb{O}^1$, obtained as the unpointed suspension of $\uparrow\mathbb{S}^0$.

7.4.9 Solutions of 4.5.7

(a) For a d-space X, the lp-space $F'(X)$ has the family $(\preceq_U)$ of path preorders: $x \preceq_U x'$ means that there exists in X a distinguished path from x to x', with image in U. F' acts 'identically' on the morphisms, as in 4.3.7 and 4.5.2.

For an lp-space Y, the distinguished paths of the d-space $G'(Y)$ are the lp-maps $a \colon F'(\uparrow\mathbb{I}) \to Y$. This is equivalent to saying that, if U is an open subinterval of $\mathbb{I}$ and $a(U) \subset V$ (open in Y) then a is monotone for the natural order of U and $\prec_V$.

Again, G' acts 'identically' on the morphisms. The adjunction is expressed by a natural bijection, which is an identity at the level of topological spaces

$$\mathsf{lpTop}(F'X, Y) \to \mathsf{dTop}(X, G'Y),$$
$$(f \colon F'X \to Y) \mapsto (f \colon X \to G'Y). \tag{7.25}$$

(b) The d-space $\uparrow\mathbb{S}^1$ corresponds to the lp-space $\mathsf{lp}\mathbb{S}^1$ along the previous adjunction, in the sense that $F'(\uparrow\mathbb{S}^1) = \mathsf{lp}\mathbb{S}^1$ and $G'(\mathsf{lp}\mathbb{S}^1) = \uparrow\mathbb{S}^1$.

The functor $F' \colon \mathsf{dTop} \to \mathsf{lpTop}$ preserves coequalisers, as a left adjoint. It follows that the lp-space $\mathsf{lp}\mathbb{S}^1$ is the coequaliser in lpTop of the usual maps $\partial^0, \partial^1 \colon \{*\} \rightrightarrows \mathsf{lp}\mathbb{I}$ of (4.53); here $\mathsf{lp}\mathbb{I} = F'(\uparrow\mathbb{I}) = G(\uparrow\mathbb{I})$ is the lp-space associated to the directed interval.

7.4.10 Solutions of 4.6.2

(a) In every cohesive category, an upper bounded set of parallel morphisms is linked.

(b) In all these cases, binary meets automatically distribute over arbitrary joins, by Theorems 3.3.3(b) and 3.4.4(c).

(c) By 3.4.6, we know that $\mathsf{K} = I\mathcal{C}$ is an inverse category, whose canonical order is the restriction of the order of C. The linking relation of C is locally indiscrete, and the same is true of the linking relation $u \mathbin{!_\mathsf{K}} v$ of K (which means that $u\,!\,v$ and $u^\sharp\,!\,v^\sharp$ in C, by (3.46)).

Now it is sufficient to apply Theorem 3.4.7 and the previous point (b): C is totally cohesive, therefore K is also, and it is a quantaloid.

7.4.11 Solutions of 4.6.7

(a) We already remarked that the inclusion $x^i \colon X_i \subset X$ belongs to $I\mathcal{C}$ if and only if X_i is open in X. This granted, Proposition 3.5.2 says that $(X, (x^i))$ is the gluing of the manifold $(X_i, (w_j^i))_I$ if and only if

$$1_X = \bigvee{}_i x^i x^{i\sharp}, \qquad x^{j\sharp} x^i = w_j^i \qquad (i,j \in I). \qquad (7.26)$$

The first condition is met, because $X = \bigcup X_i$. The second means that w_j^i is the partial identity on $X_i \cap X_j$.

(b) We can take two homeomorphisms $u_i \colon U_i \to X_i$ $(i = 1, 2)$ from open intervals of $\mathbb{R}$, onto subspaces of $\mathbb{R}^2$; we let $X = X_1 \cup X_2$.

Then we define the partial identities $w_j^i \colon X_i \rightarrowtail X_j$ letting $W_i^i = X_i$ and $W_j^i = \emptyset$ for $i \neq j$. The conditions (4.97) are met, and we have a manifold $(U_i, u^i, w_j^i)_I$ on $\mathcal{K}^r(1)$. This is embedded in $\mathbb{R}^2$ if and only if $X_1 \cap X_2 = W_2^1 = \emptyset$, which of course need not be the case.

Concretely, the gluing of the underlying manifold $(X_i, (w_j^i))$ in $I\mathcal{C}$ is the topological sum $X_1 + X_2$. This corresponds to the inclusions $(x^i \colon X_i \subset X)_i$ if and only if $X_1 \cap X_2 = \emptyset$.

(c) In the previous example, let

$$U_i = \mathbb{R}, \qquad X_1 = \mathbb{R} \times \{0\}, \qquad X_2 = \{0\} \times \mathbb{R}.$$

Now the subspace $X = X_1 \cup X_2 \subset \mathbb{R}^2$ is not locally euclidean.

This evident fact can be proved remarking that, for every neighbourhood W of the origin (in X), the subspace $U \setminus \{(0,0)\}$ is contained in the four semi-axes of the euclidean plane, and has at least four connected components. Thus W cannot be homeomorphic to $\mathbb{R}$ (taking a point out of $\mathbb{R}$ we have two connected components), nor to any higher $\mathbb{R}^n$ (taking out a point, the space stays connected).

7.4.12 Solutions of 4.8.1

(b) Let G be the additive group $\mathbb{Z}^2$ and G_1, G_2 the subgroups generated by $(1,0)$ and $(0,1)$, respectively; the intersection $G_1 \cap G_2$ is the trivial subgroup.

Two partial isomorphisms $F_i \colon G \rightarrowtail \mathbb{Z}$ defined on G_i cannot have an upper bound in $\mathcal{G}pd$.

(c) The partial identity $f \colon \mathbb{R}^2 \rightarrowtail U$ defined on the pierced plane $U = \mathbb{R}^2 \setminus \{0\}$ gives the identity $\Pi_1(U) \to \Pi_1(U)$. This is not a partial isomorphism $\Pi_1(\mathbb{R}^2) \rightarrowtail \Pi_1(U)$, as $\Pi_1(\mathbb{R}^2)$ is an indiscrete groupoid, while $\Pi_1(U)$ is equivalent to the fundamental groupoid of the circle.

7.5 Exercises of Chapter 5

7.5.1 Solutions of 5.1.6 (Pullbacks and pushouts)

(a) Firstly, in the notation of (5.8), the pullback A can be constructed as the equaliser of the maps $fp_1, gp_2 \colon X_1 \times X_2 \rightrightarrows X_0$. The second point is obvious: the product $X_1 \times X_2$ is the pullback of the pair of morphisms $X_1 \to \top \leftarrow X_2$.

Dually, the pushout A in (5.10) can be obtained as the coequaliser of two maps $X_0 \rightrightarrows X_1 + X_2$, as written out in (2.52) for the category Top.

(e) In Set the property is easy to prove. In Top our property holds in two main situations: when the subspaces X_i are both open in X, or both closed, using the two Cover Lemmas of 2.1.7.

7.5.2 Solutions of 5.1.8

(a) A morphism $\varphi \colon X \rightarrowtail Y$ can be uniquely written as $\varphi = [m, f]$, where $m \colon A_m \rightarrowtail X$ is a subobject and belongs to M. We have supposed that these subobjects form a set, say M_X, so that each new morphism $X \rightarrowtail Y$ can be seen as an element of the union of the sets $\{m\} \times \mathsf{C}(A_m, X)$, for $m \in M_X$.

(b) The claim on identities is obvious. Associativity is proved by the following diagram, using the pasting property of pullbacks (in 5.1.6(d))

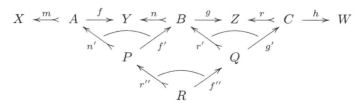

(c) The functorial properties are obvious. As to faithfulness: if, in diagram (5.15), $m = \operatorname{id} X = m'$, then $u = \operatorname{id} X$ and $f = f'$.

(d) The relation $m^\sharp m = 1$ is proved by the following pullback (m being mono)

$$A \xleftarrow{1} A \xrightarrow{m} X \xleftarrow{m} A \xrightarrow{1} A$$

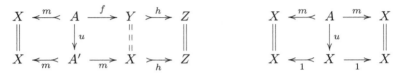

(e) If $h\colon Y \to Z$ is mono in C, the left diagram below shows that it is also mono in $\mathsf{P_M C}$: cancelling h, we get $f = f'u$

$$
\begin{array}{ccccccc}
X & \xleftarrow{\;m\;} & A & \xrightarrow{\;f\;} & Y & \xrightarrowtail{\;h\;} & Z \\
\| & & \downarrow{u} & & \| & & \| \\
X & \xleftarrow{\;m\;} & A' & \xrightarrow{\;m\;} & X & \xrightarrowtail{\;h\;} & Z
\end{array}
\qquad
\begin{array}{ccccc}
X & \xleftarrow{\;m\;} & A & \xrightarrow{\;m\;} & X \\
\| & & \downarrow{u} & & \| \\
X & \xleftarrow{\;1\;} & X & \xrightarrow{\;1\;} & X
\end{array}
$$

Conversely, if $fm^\sharp$ is mono, then

$$fm^\sharp(mm^\sharp) = fm^\sharp(\operatorname{id} X),$$

and $mm^\sharp = \operatorname{id} X$, which implies that m is invertible in C (by the right diagram above) and in $\mathsf{P_M C}$. Therefore $m^\sharp = m^{-1}$ and the monomorphism $fm^\sharp = fm^{-1}$ belongs to C.

(f) Every endomorphism $e = mm^\sharp\colon X \twoheadrightarrow X$ is idempotent.

These endomorphisms form a 1-semilattice $\mathsf{Prj}\,(X)$: if $e' = nn^\sharp\colon X \twoheadrightarrow X$, the composition pullback gives a diagonal morphism $h = mr = ns$ in M

$$X \xleftarrow{\;m\;} A \xrightarrowtail{\;m\;} X \xleftarrow{\;n\;} B \xrightarrowtail{\;n\;} X$$

so that $e'e = hh^\sharp = ee'$ is a projector of X. Note also that $n^\sharp m = sr^\sharp$. It is easy to see that $mm^\sharp \leqslant nn^\sharp$ if and only if $m \prec n$ within the monomorphisms in X.

As to (ECH.1), we let $\underline{e}(fm^\sharp) = mm^\sharp$. To verify that it is the smallest projector e such that $fm^\sharp.e = fm^\sharp$, let $fm^\sharp = fm^\sharp.nn^\sharp$; then, using the previous notation and the relation $n^\sharp m = sr^\sharp$

$$f = fm^\sharp m = (fm^\sharp nn^\sharp)m = frs^\sharp sr^\sharp = (fr)r^\sharp,$$

which implies (by definition of the morphisms of $\mathsf{P_M C}$) that r is invertible. Therefore $m \prec n$ and $mm^\sharp \leqslant nn^\sharp$.

Finally, for (ECH.2), consider the composition of $a = fm^\sharp\colon X \to Y$ and $b = gn^\sharp\colon Y \to Z$ (as in (5.16))

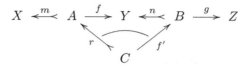

Here $\underline{e}(a) = mm^\sharp$, $\underline{e}(b) = nn^\sharp$ and $\underline{e}(ba) = mrr^\sharp m^\sharp$, so that

$$a.\underline{e}(ba) = fm^\sharp mrr^\sharp m^\sharp = frr^\sharp m^\sharp = nf'r^\sharp m^\sharp = nn^\sharp fm^\sharp = \underline{e}(b).a.$$

7.5.3 Solutions of 5.2.3

(a) We have an adjunction $F \dashv G$, with $F\colon \mathsf{C} \to \mathsf{D}$, and a functor $X\colon \mathsf{S} \to \mathsf{C}$. A universal cocone $(L, (u_i\colon X_i \to L))$ of the functor X is taken by F to a cocone $(FL, (Fu_i\colon FX_i \to FL))$ of the composed functor $FX\colon \mathsf{S} \to \mathsf{D}$.

To prove that this cocone is also universal, let $(Y, (g_i\colon FX_i \to Y))$ be a cocone of FX in D. The adjunction gives a cocone $(GY, (f_i\colon X_i \to GY))$ of X in C; then we have a unique map $f\colon L \to GY$ such that $fu_i = f_i$ (for all $i \in \mathrm{Ob}\,\mathsf{S}$), which corresponds to a unique map $g\colon FL \to Y$ such that $g.Fu_i = g_i$ (for all i).

(b) The universal property of the limit $(L, (u_i\colon L \to X_i))$ of the functor $X\colon \mathsf{S} \to \mathsf{C}$ can be rewritten in the following form:

- for every A in C, the set $\mathsf{C}(A, L)$ is the limit of the functor

$$\mathsf{C}(A, X(-))\colon \mathsf{S} \to \mathsf{Set},$$

with limit-cone $u_i.-\colon \mathsf{C}(A, L) \to \mathsf{C}(A, X_i)$.

The preservation of monomorphisms follows from their characterisation by a pullback, in 5.1.6(b).

7.5.4 Solutions of 5.2.5

(a) Properties (iii) and (iv) are trivially equivalent; (i) $\Rightarrow$ (ii) is obvious.

(ii) $\Rightarrow$ (iv) First, $em = mpm = m$. Second, any $f\colon Y \to X$ such that $ef = f$ factorises as $f = m(pf)$ through the monomorphism $m\colon A \to X$.

(iv) $\Rightarrow$ (i) If $m\colon A \to X$ is the equaliser of $1, e\colon X \to X$, the morphism $e\colon X \to X$ factorises as $e = mp$ for a unique $p\colon X \to A$, and $pm = 1_A$ because $mpm = em = m$, and m is mono.

The rest follows from duality, since (i) and (ii) are selfdual.

(e) We know that e splits as $e = mp$, with $pm = 1$. Then $mpm = m$ and $pmp = p$, so that $p = m^\sharp$. The rest follows from (a).

7.5.5 Solutions of 5.3.8

(a) Let $f \colon A \to B$ be an injective K-homomorphism. We take a basis $(a_i)_{i \in I}$ of A, and let $b_i = f(a_i)$. Then the family $(b_i)_{i \in I}$ is linearly independent in B, and can be extended to a basis $(b_i)_{i \in J}$ of B (with $I \subset J$). We define $g \colon B \to A$ as the unique linear mapping such that

$$g(b_i) = a_i, \text{ for } i \in I, \qquad g(b_i) = 0, \text{ for } i \in J \setminus I.$$

Then $gf \colon A \to A$ is the identity on each a_i, and therefore on A.

(We are splitting B in the internal direct sum of $f(A) \cong A$ and $\operatorname{Ker} g$, a vector subspace of B depending on our choices.)

(b) If $f \colon A \to B$ is a surjective K-homomorphism and (b_i) is a basis of B, we choose an element $a_i \in f^{-1}\{b_i\}$ for every $i \in I$, and let $g \colon B \to A$ be the unique linear mapping such that $g(b_i) = a_i$ (for all i).

Then $fg \colon B \to B$ is the identity on each b_i, and therefore on B.

(We are splitting A in the internal direct sum of $\operatorname{Ker} f$ and $g(B)$, a vector subspace of A isomorphic to B.)

(c) The mapping

$$\varphi \colon A \times B \to \operatorname{Hom}_K(A^*, B), \qquad \varphi(a,b)(\alpha) = \alpha(a).b \quad (\text{for } \alpha \colon A \to K),$$

is K-bilinear and induces our homomorphism $u \colon A \otimes_K B \to \operatorname{Hom}_K(A^*, B)$. (This also works for R-modules, but the sequel would not: note that, for $\mathbb{Z}$-modules, $\operatorname{Hom}(A, \mathbb{Z}) = 0$ for every finite abelian group A.)

Now, we choose a basis $(a_i)_{i \in I}$ of A. For every $i \in I$, we define a homomorphism α_i on this basis

$$\alpha_i \colon A \to K, \quad \alpha_i(a_i) = 1_K, \quad \alpha_i(a_j) = 0_K \quad (\text{for } j \neq i \text{ in } I),$$

which takes every vector $a = \sum_i \lambda_i a_i$ to the scalar λ_i. Note that $a = 0$ if and only if all homomorphisms α_i annihilate on it.

To prove that each u_{AB} is injective, let $u(a \otimes b) = 0$, so that $\alpha_i(a).b = 0$, for all $i \in I$. Now, either $a = 0$ or there is some index i such that $\alpha_i(a) \in K^*$, and then $b = 0$. In both cases $a \otimes b = 0$.

If A and B are finitely generated, the domain and codomain of u have the same finite dimension on K, namely the product $\dim(A).\dim(B)$, and therefore u is an isomorphism.

7.5.6 Solutions of 5.4.3

(a) In a pointed category any object A is isomorphic to $A \times \top = A \times \bot$. If A is exponentiable, $A \times -$ preserves the initial object and A is also initial.

(b) The first claim follows from (a): every object is initial and terminal, and there are some of them. Therefore any object X gives a functor $X \colon \mathbf{1} \to \mathsf{C}$ which is an equivalence of categories, by 1.6.3.

7.6 Exercises of Chapter 6

7.6.1 Solutions of 6.2.4

(a) In fact we have:

$$\bigvee_{y \in Y} (u_Y(y, f(x')).u_Y(f(x), y))$$
$$\geqslant u_Y(f(x), f(x')).u_Y(f(x), f(x))$$
$$\geqslant u_Y(f(x), f(x')) \geqslant u_X(x, x'),$$

$$f_*(x, y').f^*(y, x) = u_Y(f(x), y').u_Y(y, f(x)) \leqslant u_Y(y, y').$$

(b) An M-structure of the singleton $\{*\}$ amounts to an element $u(*, *) = e$ of M satisfying the conditions $e \geqslant 1$ and $ee \leqslant e$. Equivalently, $ee = e \geqslant 1$.

For $M = \operatorname{Rel}\mathsf{Set}(S)$, such an endo-relation $e \subset S \times S$ is the same as a preorder relation $\prec$ in the set S, writing $(x, x') \in e$ as $x \prec x'$.

In fact, the condition $e \geqslant 1$ amounts to reflexivity, while $ee \leqslant e$ amounts to transitivity.

7.6.2 Solutions of 6.2.9

(a) We apply a previous remark on the left inequality (6.41).

Given the pair $F \dashv G$, we define a mapping $f \colon X \to Y$ by *choosing*, for every $x \in X$, an element $f(x) = y(x, x)$ satisfying

$$F(x, f(x)) = G(f(x), x) = 1. \qquad (7.27)$$

This mapping is a functor, because

$$[x \prec x']_X = G(f(x), x) \wedge [x \prec x']_X \wedge F(x', f(x'))$$
$$\leqslant G(f(x), x) \wedge F(x, f(x')) \leqslant [f(x) \prec f(x')]_Y.$$

Finally, we have to prove that the associated adjoint profunctors $f_* \dashv f^*$ coincide with F and G.

In fact, $F(x,y) = [f(x) \prec y]_Y$ and $G(y,x) = [y \prec f(x)]_Y$, by the following inequalities:

$$[f(x) \prec y]_Y = F(x, f(x)) \wedge [f(x) \prec y]_Y \leqslant F(x,y)$$
$$= G(f(x), x) \wedge F(x,y) \leqslant [f(x) \prec y]_Y,$$
$$[y \prec f(x)]_Y = [y \prec f(x)]_Y \wedge G(f(x), x) \leqslant G(y,x)$$
$$= G(y,x) \wedge F(x, f(x)) \leqslant [y \prec f(x)]_Y.$$

7.6.3 Solutions of 6.4.2

(a) Plainly (i) implies (ii). Assuming (ii) and applying (6.43) we have

$$v_k^h \leqslant v_k^{k'} v_{k'}^h \leqslant v_k^k v_{k'}^h = v_{k'}^h \qquad \text{(for all } h \in H\text{)},$$

so that $v_k^h = v_{k'}^h$, and similarly $v_h^k = v_h^{k'}$.

*(b) We form a subset $I \subset H$ choosing one index in every equivalence class of the relation $k \sim k'$. Then the inclusion functor $V_I \to V_H$ gives inverse profunctors.

7.6.4 Solutions of 6.6.4

(a) As in 3.2.7.

(b), (c) As in Exercises 3.2.8(a), (b).

References

[AbM]   R. Abraham and J.E. Marsden, Foundations of mechanics, W.A. Benjamin Inc., 1967.

[Ad]   M. Adachi, Embeddings and immersions, Translations of Mathematical Monographs Vol. 124, Am. Math. Soc., 1993.

[AHS]   J. Adámek, H. Herrlich, G.E. Strecker, Abstract and concrete categories. The joy of cats, J. Wiley and Sons, 1990.

[AM]   L. Auslander and R.E. MacKenzie, Introduction to differentiable manifolds, Dover Publications, 2009.

[Ba]   M. Barr, Exact categories, in Lecture Notes in Mathematics Vol. 236, Springer, 1971, pp. 1–120.

[Be1]   J. Bénabou, Introduction to bicategories, in: Reports of the Midwest category seminar, Lecture Notes in Math., Springer, 1967, pp. 1–77.

[Be2]   J. Bénabou, Les distributeurs, Inst. de Math. Pure et Appliquée, Univ. Catholique de Louvain, Rapport n. 33, 1973.

[Bet]   R. Betti, Bicategorie di base, Ist. Mat. Univ. Milano, 2/S (1981).

[BetC]   R. Betti and A. Carboni, Cauchy completion and the associated sheaf, Cah. Topol. Géom. Différ. 23 (1982), 243–256.
Available at: www.numdam.org/

[BetW1]   R. Betti and R.F.C. Walters, The symmetry of the Cauchy-completion of a category, in: Category theory (Gummersbach, 1981), Lecture Notes in Math. Vol. 962, Springer 1982, pp. 8–12.

[BetW2]   R. Betti and R.F.C. Walters, Closed bicategories and variable category theory, Ist. Mat. Univ. Milano, 5 (1985).
Republished in: Reprints Theory Appl. Categ. 26 (2020), pp. 1–27.
http://www.tac.mta.ca/tac/reprints/articles/26/tr26.pdf

[Bi]   G. Birkhoff, Lattice theory, 3rd ed., Amer. Math. Soc. Coll. Publ. 25, 1973.

[Bir]   G.J. Bird, Morita theory for enriched categories, thesis, Univ. of Sydney, 1981.

[Bo]   F. Borceux, Handbook of categorical algebra 1–3, Cambridge Univ. Press, 1994.

[BoD]   F. Borceux and D. Dejean, Cauchy completion in category theory, Cah. Top. Géom. Diff. Catég. 27 (1986), 133–146.
Available at: www.numdam.org/

[BoT]   R. Bott and L.W. Tu, Differential forms in algebraic topology, Springer, 1982.

[Bou1]   N. Bourbaki, Algebra I, Chapters 1–3, Hermann and Addison–Wesley, 1974.

[Bou2]   N. Bourbaki, General Topology I, Chapters 1–4, Springer, 1989.

[Bou3]   N. Bourbaki, General Topology II, Chapters 5–10, Springer, 1989.

[Bra]   H. Brandt, Über eine Verallgemeinerung des Gruppenbegriffes, Math. Ann. 96 (1927), 360–366.

[Bw]   R. Brown, Topology and groupoids, Third edition of Elements of modern topology, 1968, BookSurge, LLC, Charleston, SC, 2006.

[BwH]   R. Brown and P.J. Higgins, Tensor products and homotopies for ω-groupoids and crossed complexes, J. Pure Appl. Algebra 47 (1987), 1–33.

[CaC]   A. Candel and L. Conlon, Foliations I, American Mathematical Society, 2000.

[CaG]   E. Carletti and M. Grandis, Generalised pushouts, connected colimits and codiscrete groupoids, Cah. Topol. Géom. Différ. Catég. 56 (2015), 232–240.
         Available at: www.numdam.org/

[CE]   H. Cartan and S. Eilenberg, Homological algebra, Princeton University Press, 1956.

[CP]   A.H. Clifford and G.B. Preston, The algebraic theory of semigroups, Vol. 1, Math. Surveys of the Amer. Math. Soc., 1961.

[CoG]   R. Cockett and R. Garner, Restriction categories as enriched categories, Theor. Comput. Sci. 523 (2014), 37–55.

[CoL]   J.R.B. Cockett and S. Lack, Restriction categories I: categories of partial maps, Theor. Comput. Sci. 270 (2002), 223–259.

[Di]   R.A. Di Paola, Creativity and effective inseparability in dominical categories, in: Atti degli incontri di Logica Matematica, 2 (1983–84), Dip. Mat. Univ. Siena, 477–478.

[DiH]   R.A. Di Paola and A. Heller, Dominical categories: recursion theory without elements, J. Symb. Log. 52 (1987), 594–635.

[DS]   N. Dunford and J.T. Schwartz, Linear operators, Part III, Wiley Interscience, 1971.

[E1]   C. Ehresmann, Espèces de structures locales, Sém. Ehresmann, tome 3 (1960–62), exp. n. 4, pp. 1–24.
         Available at:
         http://www.numdam.org/item?id=SE_1960-1962_3_A4_0

[E2]   C. Ehresmann, Catégories inductives et pseudogroupes, Ann. Inst. Fourier 10 (1960), 307–336.
         Available at: www.numdam.org/ Reprinted in [E6], pp. 155–180.

[E3]   C. Ehresmann, Elargissements de catégories, Cah. Top. Géom. Diff. 3 (1961), 25–73.
         Available at: www.numdam.org/ Reprinted in [E6], pp. 25–73.

[E4]   C. Ehresmann, Catégories structurées, Ann. Sci. Ecole Norm. Sup. 80 (1963), 349–425.

[E5]   C. Ehresmann, Catégories et structures, Dunod, 1965.

[E6]   C. Ehresmann, Œuvres complètes et commentées. II-1. Structures locales. (French), Edited and with commentary in English by Andrée Charles Ehresmann. Cah. Top. Géom. Diff. 22 (1981), suppl. 2, xxiv + 431 pp. (1982).
         Available at:
         https://ehres.pagesperso-orange.fr/C.E.WORKS_fichiers/C.E_Works.htm

[Ei]   S. Eilenberg, La suite spectrale, I: Construction générale, Sém. Cartan 1950–51, Exp. 8.

[EiK]   S. Eilenberg and G.M. Kelly, Closed categories, in Proc. Conf. Categorical Algebra, La Jolla 1965, Springer, 1966, pp. 421–562.

[EiM]   S. Eilenberg and S. Mac Lane, General theory of natural equivalences, Trans. Amer. Math. Soc. 58 (1945), 231–294.

[FGR]   L. Fajstrup, E. Goubault and M. Raussen, Algebraic topology and concurrency, Theor. Comput. Sci. 357 (2006), 241–178. (Revised version of a preprint at Aalborg, 1999.)

[Fr1]   P. Freyd, Abelian categories. An introduction to the theory of functors, Harper & Row, 1964.
          Republished in: Reprints Theory Appl. Categ. 3 (2003).

[Fr2]   P. Freyd, On the concreteness of certain categories, in: Symposia Mathematica, Vol. IV (INDAM, Rome, 1968/69), Academic Press, 1970, pp. 431–456.

[G1]   M. Grandis, Transfer functors and projective spaces, Math. Nachr. 118 (1984), 147–165.

[G2]   M. Grandis, On distributive homological algebra, I. RE-categories, Cah. Top. Géom. Diff. 25 (1984), 259–301.
          Available at: www.numdam.org/

[G3]   M. Grandis, Manifolds as enriched categories, in: Categorical Topology, Prague 1988, pp. 358–368, World Scientific 1989.

[G4]   M. Grandis, Cohesive categories and manifolds, Ann. Mat. Pura Appl. 157 (1990), 199–244.
          Available at: https://link.springer.com/journal/10231/157/1

[G5]   M. Grandis, Cohesive categories and measurable operators, Rend. Accad. Naz. Sci. XL Mem. Mat. 14 (1990), 195–234.
          Available at: https://www.accademiaxl.it/pubblicazioni-2/rendiconti-on-line/

[G6]   M. Grandis, Finite sets and symmetric simplicial sets, Theory Appl. Categ. 8 (2001), No. 8, 244–252.
          Available at: http://www.tac.mta.ca/tac/

[G7]   M. Grandis, Directed homotopy theory, I. The fundamental category, Cah. Topol. Géom. Différ. Catég. 44 (2003), 281–316.
          Available at: www.numdam.org/

[G8]   M. Grandis, Directed Algebraic Topology: Models of non-reversible worlds, Cambridge Univ. Press, 2009.
          Available at: http://www.dima.unige.it/~grandis/BkDAT_page.html

[G9]   M. Grandis, Homological Algebra: The interplay of homology with distributive lattices and orthodox semigroups, World Scientific Publishing Co., 2012.

[G10]   M. Grandis, Category Theory and Applications: A textbook for beginners, World Scientific Publishing Co., 2018.

[G11]   M. Grandis, An Elementary Overview of Mathematical Structures: Algebra, Topology and Categories, World Scientific Publishing Co., 2021.

[GP1]   M. Grandis and R. Paré, Limits in double categories, Cah. Topol. Géom. Différ. Catég. 40 (1999), 162–220.
          Available at: www.numdam.org/

[GP2]   M. Grandis and R. Paré, Adjoint for double categories, Cah. Topol. Géom. Différ. Catég. 45 (2004), 193–240.
          Available at: www.numdam.org/

[Gr]   G. Grätzer, General lattice theory, Academic Press, 1978.

[Ha]   T.E. Hall, On regular semigroups whose idempotents form a subsemigroup, Bull. Austral. Math. Soc. 1 (1969), 195–208.

[Hat]   A. Hatcher, Algebraic Topology, Cambridge Univ. Press, 2002.
          Available at: www.math.cornell.edu/~hatcher/

[He]   A. Heller, Dominical categories and recursion theory, in: Atti degli

354 *References*

incontri di Logica Matematica 2 (1983–84), Dip. Mat. Univ. Siena, 339–344.

[HS]   H. Herrlich and G.E. Strecker, Category theory, an introduction, Allyn and Bacon, 1973.

[Hi]   P. Hilton, Correspondences and exact squares, in: Proc. of the Conf. on Categorical Algebra, La Jolla 1965, Springer, 1966, 255–271.

[HiW]   P.J. Hilton and S. Wylie, Homology theory, Cambridge Univ. Press, 1962.

[Ho]   J.M. Howie, An introduction to semigroup theory, Academic Press, 1976.

[Hus]   D. Husemoller, Fibre Bundles, 3rd edition, Springer, 1994.

[Jo1]   P.T. Johnstone, Stone spaces, Cambridge University Press, 1982.

[Jo2]   P.T. Johnstone, The point of pointless topology, Bull. Amer. Math. Soc. 8 (1983), 41–53.

[Ka]   D.M. Kan, Adjoint functors, Trans. Amer. Math. Soc. 87 (1958), 294–329.

[KaW]   S. Kasangian and R.F.C. Walters, An abstract notion of glueing, unpublished manuscript, 1982.

[Kas]   J. Kastl, Inverse categories, in: Algebraische Modelle, Kategorien und Gruppoide, Akademie Verlag, 1979, pp. 51–60.

[Ke]   J.L. Kelley, General topology, Van Nostrand, 1955.

[Kl1]   G.M. Kelly, On Mac Lane's conditions for coherence of natural associativities, commutativities, etc., J. Algebra 1 (1964), 397–402.

[Kl2]   G.M. Kelly, Basic concepts of enriched category theory, Cambridge Univ. Press, 1982.

[KlS]   G.M. Kelly and R. Street, Review of the elements of 2-categories, in: Category Seminar, Sydney 1972/73, Lecture Notes in Math. Vol. 420, Springer, 1974, pp. 75–103.

[KN]   S. Kobayashi and K. Nomizu, Foundations of Differential Geometry, Vol. I, J. Wiley & Sons, 1963.

[Kr]   S. Krishnan, A convenient category of locally preordered spaces, Appl. Categ. Structures, 17 (2009), 445–466.

[La]   S. Lang, Introduction to differentiable manifolds, Interscience Publishers, 1962.

[Ls]   M.V. Lawson, Inverse semigroups, The theory of partial symmetries, World Scientific Publishing Co., 1998.

[Lw]   F.W. Lawvere, Metric spaces, generalized logic and closed categories, Rend. Sem. Mat. Fis. Univ. Milano 43 (1974), 135–166.
        Republished in: Reprints Theory Appl. Categ. 1 (2002), pp. 1–37.
        http://www.tac.mta.ca/tac/reprints/articles/1/tr1.pdf

[Li]   A.E. Liber, On the theory of generalised groups, Doklady Akad. Nauk SSSR 97 (1954), 25–28.

[LoR]   E. Lowen-Colebunders and G. Richter, An elementary approach to exponential spaces, Appl. Categ. Structures 9 (2001), 303–310.

[M1]   S. Mac Lane, Homology, Springer, 1963.

[M2]   S. Mac Lane, Natural associativity and commutativity, Rice Univ. Studies 49 (1963), 28–46.

[M3]   S. Mac Lane, Categories for the working mathematician, Springer, 1971.

[Mas]   W.S. Massey, Singular homology theory, Springer, 1980.

[May]   J.P. May, Simplicial objects in algebraic topology, Van Nostrand, 1967.

[Mi]   B. Mitchell, Theory of categories, Academic Press, 1971.

[Mo]   K. Morita, Duality for modules and its applications to the theory of rings with minimum condition, Sci. Rep. Tokyo Kyoiku Daigaku, Sect. A 6 (1958), 83–142.

[Mu]   J.R. Munkres, Topology: a first course, Prentice-Hall, 1975.

[ReS]   N.R. Reilly and M.E. Scheiblich, Congruences on regular semigroups, Pacific J. Math. 23 (1967), 349–360.

[RoR]   E. Robinson and G. Rosolini, Categories of partial maps, Inform. and Comput. 79 (1988), 95–130.

[Ro]   K. Rosenthal, The theory of quantaloids, Longman, 1996.

[Rs]   G. Rosolini, Continuity and effectiveness in topoi, D.Phil. thesis, University of Oxford, 1986.

[Sc]   B.M. Schein, On the theory of generalised groups and generalised heaps (Russian), Theory of semigroups and its applications (Russian) 1 (1965), 286–324.

[Se]   Z. Semadeni, Banach spaces of continuous functions, Polish Sci. Publ., 1971.

[Sp]   E.H. Spanier, Algebraic topology, Mc Graw–Hill, 1966.

[St]   N. Steenrod, The Topology of Fibre Bundles, Princeton University Press, 1951.

[Str]   R. Street, Limits indexed by category-valued 2-functors, J. Pure Appl. Alg. 8 (1976), 149–181.

[Ta]   I. Tamura, Topology of foliations: an introduction, Translations of Mathematical Monographs 97, Amer. Math. Soc., 1992.

[Va1]   V.V. Vagner, On the theory of partial transformations, Doklady Akad. Nauk SSSR 84 (1952), 653–656.

[Va2]   V.V. Vagner, Generalised groups, Doklady Akad. Nauk SSSR 84 (1952), 1119–1122.

[Vi]   J.W. Vick, Homology theory. An introduction to algebraic topology, Second edition, Springer, 1994.

[vN]   J. von Neumann, On regular rings, Proc. Nat. Acad. Sci. USA 22 (1936), 296–300.

[Wa1]   R.F.C. Walters, Sheaves and Cauchy-complete categories, Cah. Top. Géom. Diff. 22 (1981), 282–286.
Available at: www.numdam.org/

[Wa2]   R.F.C. Walters, Sheaves on sites as Cauchy-complete categories, J. Pure Appl. Algebra 24 (1982), 95–102.

[Ya]   M. Yamada, Note on a certain class of idempotent semigroups, Semigroup Forum 6 (1973), 180–188.

Index

CPSIA information can be obtained
at www.ICGtesting.com
Printed in the USA
JSHW041909130221
11914JS00002B/11